Selected Papers from the 16th International Conference on Squeezed States and Uncertainty Relations (ICSSUR 2019)

Selected Papers from the 16th International Conference on Squeezed States and Uncertainty Relations (ICSSUR 2019)

Editors

Margarita A. Man'ko
Luis L. Sánchez-Soto

MDPI • Basel • Beijing • Wuhan • Barcelona • Belgrade • Manchester • Tokyo • Cluj • Tianjin

Editors
Margarita A. Man'ko
Lebedev Physical Institute
Russia

Luis L. Sánchez-Soto
Universidad Complutense
Spain

Editorial Office
MDPI
St. Alban-Anlage 66
4052 Basel, Switzerland

This is a reprint of articles from the Special Issue published online in the open access journal *Quantum Reports* (ISSN 2624-960X) (available at: https://www.mdpi.com/journal/quantumrep/special_issues/quantumrep_ICSSUR2019).

For citation purposes, cite each article independently as indicated on the article page online and as indicated below:

LastName, A.A.; LastName, B.B.; LastName, C.C. Article Title. *Journal Name* **Year**, *Article Number*, Page Range.

ISBN 978-3-03943-424-4 (Hbk)
ISBN 978-3-03943-425-1 (PDF)

Contents

About the Editors

Margarita A. Man'ko was born in Moscow on December 4, 1939. She graduated from the high school in Moscow in 1957 being awarded by Golden medal. She graduated from the Lomonosov Moscow State University, the Faculty of Physics in 1963. Starting 1963 and till now, she has the permanent research position with the Lebedev Physical Institute in Moscow, now as the Leading Senior Researcher. She defended Ph.D. Thesis at the Lebedev Institute in 1970 in the group of Professor N.G. Basov, the Nobel Prize Winner, in the Laboratory of Professor A. M. Prokhorov, the Nobel Prize Winner. She has more than 400 scientific publications in Russian and International scientific journals. She organized and participated in many international conferences with invited talks. At present time, Margarita Man'ko is Scientific and Managing Editor of the Journal of Russian Laser Research published by Springer in New York, the Editorial Board member of Physica Scripta of the Royal Swedish Academy in Stockholm published by Institute of Physics in Bristol and the Editorial Board member of Quantum Reports, MDPI, Basel.

Luis L. Sánchez-Soto received his MSc (1984) and the Ph.D. (1988) in Physics from the Complutense University of Madrid. He has been visiting researcher at Boston, Paris, and Olomouc. He is a full professor of Quantum Optics in Madrid since 2002. In 2009, he joined the Max Planck Institute for the Science of Light, in Erlangen, where he was leading the theoretical group in the Division of Optics and Information, headed by Gerd Leuchs. His main research interests are quantum optics and quantum information.

Editorial

Editorial for the Special Issue "Selected Papers from the 16th International Conference on Squeezed States and Uncertainty Relations (ICSSUR 2019)"

Luis L. Sánchez-Soto [1,*] and Margarita A. Man'ko [2,*]

[1] Departamento de Óptica, Facultad de Fisica, Universidad Complutense, 28040 Madrid, Spain
[2] Lebedev Physical Institute, Leninskii Prospect 53, 119991 Moscow, Russia
* Correspondence: lsanchez@ucm.es (L.L.S.-S.); mmanko@sci.lebedev.ru (M.A.M.)

Received: 28 August 2020; Accepted: 1 September 2020; Published: 2 September 2020

The first quantum revolution started in the early 20th century and gave us new rules that govern physical reality. Accordingly, many devices that changed dramatically our lifestyle, such as transistors, medical scanners and lasers, appeared in the market. This was the origin of quantum technology, which allows us to organize and control the components of a complex system governed by the laws of quantum physics. This is in sharp contrast to conventional technology, which can only be understood within the framework of classical mechanics.

We are now in the middle of a second quantum revolution. Although quantum mechanics is nowadays a mature discipline, quantum engineering as a technology is now emerging in its own right. We are about to manipulate and sense individual particles, measuring and exploiting their quantum properties. This is bringing major technical advances in many different areas, including computing, sensors, simulations, cryptography and telecommunications.

The present collection of selected papers, all of them being invited talks at the16th International Conference on Squeezed States and Uncertainty Relations (ICSSUR 2019), held in Madrid from 17 to 21 June 2019, is a clear demonstration of the tremendous vitality of the field. The issue is composed of contributions from world leading researchers in quantum optics and quantum information, and presents viewpoints, both theoretical and experimental, on a variety of modern problems. We are confident that the reader will enjoy the present collection; it will be useful to experts working in all branches of quantum science.

The contributions to this collection deal with the modern results obtained in connection with the development of quantum technologies based on quantum mechanics, quantum information and quantum optics, as well as a better understanding of the foundations of quantum physics. It contains contemporarytheory of quantum phenomena, including practical applications and discussions of measurement procedures; see the paper "On the Prospects of Multiport Devices for Photon-Number-Resolving Detection" by Teo, Y.S.; Jeong, H.; Řeháček, J.; Hradil, Z.; Sanchez-Soto, L.L.; Silberhorn, C. [1].

The important result is searching forand finding during the last decades the possibility of formulating the basic notion of the state of a quantum system, like an atom or spin system, conventionally presented in terms of the wave function or density matrix, by using the standard probability distribution of the classical-like random variables given in the paper "Probability Representation of Quantum Mechanics Where System States Are Identified with Probability Distributions" by Chernega, V.N.; Man'ko, O.V.; Man'ko, V.I. [2]. The result is based on the existence of invertible maps of the vectors and density operators acting in Hilbert spaces onto the probability distributions.

The photon states, which in the previous period of quantum optics development were associated with quasidistributions like Wigner functions, Husimi-Kano functions and Glauber–Sudarshan functions, were found to be associated with the tomographicprobability distributions completely describing the states of quantum electromagnetic fields. An analogous association was also established for spin states and N-level atom states. In connection with discovering the possibility of describing quantum states by probability distributions, the superposition principle of quantum states responsible for the interference phenomenon was found to be formulated in terms of the nonlinear addition rule of probability distributions providing the probability distribution of the resulting state; see "Superposition Principle and Born's Rule in the Probability Representation of Quantum States" by Doskoch, I.Ya.; Man'ko, M.A. [3].

The other important development of quantum theory connected with the employment of coherent states and properties of oscillators, including their behavior in the presence of non-Hermitian Hamiltonians, is discussed in papers "Nonclassical States for Non-Hermitian Hamiltonians with the Oscillator Spectrum" by Zelaya, K.; Dey, S.; Hussin, V.; Rosas-Ortiz, O. [4], and "Coherent States for the Isotropic and Anisotropic 2D Harmonic Oscillators" by Moran, J.; Hussin, V. [5].

A recently developed group and graph theoretical approach aiming at the construction of large sets of mutually unbiased bases in finite-dimensional Hilbert spaces is discussed in the paper "Mutually Unbiased Bases and Their Symmetries" by Alber, G.; Charnes, C. [6]. In this approach, the construction of mutually unbiased bases in a Hilbert space of given dimension is reformulated as a clique-finding problem of a Cayley graph associated with a finite basis group. This approach offers the possibility to enlarge and possibly also to complete already known systems of mutually unbiased basis systems.

This collection of papers contains also the development of the theory of classical mechanics and relativistic dynamics, and its relation with the Poincare group; see the paper "Descriptions of Relativistic Dynamics with World Line Condition" by Ciaglia, F.M.; Di Cosmo, F.; Ibort, A.; Marmo, G. [7].

Furthermore, group theory, and oscillator creation and annihilation operator algebra, discussed in connection with Einstein's formula relating the particle energy with its mass and light velocity, are given in the contribution "Einstein's $E = mc^2$ Derivable from Heisenberg's Uncertainty Relations" by Başkal, S.; Kim, Y.S.; Noz, M.E. [8]. In this connection of classical mechanics with the quantum approach, one can point out that the probability distributions identified with quantum states obey kinetic equations of classical-like type, which presentother forms of the Schrödinger and von Neumann equations of the conventional formulation of quantum mechanics.

The phenomena of quantum correlations, including entanglement and its characteristics, are reviewed in detail in the following papers: "Distance between Bound Entangled States from Unextendible Product Bases and Separable States" by Wieaniak, M.; Pandya, P.; Sakarya, O.; Woloncewicz, B. [9] and "Selective Engineering for Preparing Entangled Steady States in Cavity QED Setup" by Sousa, E.H.S.; Roversi, J.A. [10]. Entanglement is considered for different kinds of quantum systems, both for systems with continuous variables and systems with discrete variables. The properties of nonclassical states of light are discussed in paper "Resource Theories of Nonclassical Light" by Tan, K.C.; Jeong, H. [11].

New important results describing the behavior of charged particles moving in the time-dependent magnetic field in solenoid are presented for new kinds of solenoid forms and different time-dependencies of magnetic fields; see the paper "A Quantum Charged Particle under Sudden Jumps of the Magnetic Field and Shape of Non-Circular Solenoids" by Dodonov, V.V.; Horovits, M.B. [12].

It should be pointed out that here one can find biographic memorial details of Roy Glauber and George Sudarshan, pioneers of quantum optics, and a kaleidoscope of the ICSSUR participants during the last three decades, presented in the above-mentioned contribution [3]. The scientific collaboration of George Sudarshan with friends and colleagues from European Universities and theirfundamental results in quantum optics, quantum mechanics and quantum field theory are reviewed in the paper "Remembering George Sudarshan" by Ibort, A.; Marmo, G. [13].

We hope that this collection of papers will be useful to get new results to be discussed at the next meeting of the ICSSUR series. This event started in 1991 as the Joint Seminar on Squeezed States and Uncertainty Relations of the University of Maryland at College Park and the Lebedev Physical

Institute in Moscow, with Young Suh Kim and Vladimir I. Man'ko as its Founders; later on, it was converted into the International Conference on Squeezed States and Uncertainty Relations, which takes place over the world every two years; see https://www.mdpi.com/journal/quantumrep/special_issues/quantumrep_ICSSUR2019.

Funding: This research received no external funding.

Conflicts of Interest: The authors declare no conflict of interest.

References

1. Teo, Y.S.; Jeong, H.; Řeháček, J.; Hradil, Z.; Sánchez-Soto, L.L.; Silberhorn, C. On the Prospects of Multiport Devices for Photon-Number-Resolving Detection. *Quantum Rep.* **2019**, *1*, 15. [CrossRef]
2. Chernega, V.N.; Man'ko, O.V.; Man'ko, V.I. Probability Representation of Quantum Mechanics Where System States Are Identified with Probability Distributions. *Quantum Rep.* **2020**, *2*, 6. [CrossRef]
3. Doskoch, I.Y.; Man'ko, M.A. Superposition Principle and Born's Rule in the Probability Representation of Quantum States. *Quantum Rep.* **2019**, *1*, 13. [CrossRef]
4. Zelaya, K.; Dey, S.; Hussin, V.; Rosas-Ortiz, O. Nonclassical States for Non-Hermitian Hamiltonians with the Oscillator Spectrum. *Quantum Rep.* **2020**, *2*, 2. [CrossRef]
5. Moran, J.; Hussin, V. Coherent States for the Isotropic and Anisotropic 2D Harmonic Oscillators. *Quantum Rep.* **2019**, *1*, 23. [CrossRef]
6. Alber, G.; Charnes, C. Mutually Unbiased Bases and Their Symmetries. *Quantum Rep.* **2019**, *1*, 20. [CrossRef]
7. Ciaglia, F.M.; di Cosmo, F.; Ibort, A.; Marmo, G. Descriptions of Relativistic Dynamics with World Line Condition. *Quantum Rep.* **2019**, *1*, 16. [CrossRef]
8. Başkal, S.; Kim, Y.S.; Noz, M.E. Einstein's $E = mc^2$ Derivable from Heisenberg's Uncertainty Relations. *Quantum Rep.* **2019**, *1*, 21. [CrossRef]
9. Wieśniak, M.; Pandya, P.; Sakarya, O.; Woloncewicz, B. Distance between Bound Entangled States from Unextendible Product Bases and Separable States. *Quantum Rep.* **2020**, *2*, 4. [CrossRef]
10. Sousa, E.H.S.; Roversi, J.A. Selective Engineering for Preparing Entangled Steady States in Cavity QED Setup. *Quantum Rep.* **2019**, *1*, 7. [CrossRef]
11. Tan, K.C.; Jeong, H. Resource Theories of Nonclassical Light. *Quantum Rep.* **2019**, *1*, 14. [CrossRef]
12. Dodonov, V.V.; Horovits, M.B. A Quantum Charged Particle under Sudden Jumps of the Magnetic Field and Shape of Non-Circular Solenoids. *Quantum Rep.* **2019**, *1*, 17. [CrossRef]
13. Ibort, A.; Marmo, G. Remembering George Sudarshan. *Quantum Rep.* **2019**, *1*, 24. [CrossRef]

quantum reports

MDPI

Article

Probability Representation of Quantum Mechanics Where System States Are Identified with Probability Distributions †

Vladimir N. Chernega [1], Olga V. Man'ko [1,2] and Vladimir I. Man'ko [1,3,*]

1. Lebedev Physical Institute, Leninskii Prospect 53, Moscow 119991, Russia; vchernega@gmail.com (V.N.C.); mankoov@lebedev.ru (O.V.M.)
2. Bauman Moscow State Technical University, The 2nd Baumanskaya Str. 5, Moscow 105005, Russia
3. Moscow Institute of Physics and Technology (State University), Institutskii per. 9, Dolgoprudnyi, Moscow Region 141700, Russia

* Correspondence: mankovi@lebedev.ru

† Based on the invited talk presented by Vladimir I. Man'ko at the 16th International Conference on Squeezed States and Uncertainty Relations {ICSSUR} (Universidad Complutense de Madrid, Spain, 17–21 June 2019).

Academic Editor: Ángel Santiago Sanz Ortiz
Received: 15 December 2019; Accepted: 14 January 2020; Published: 21 January 2020

Abstract: The probability representation of quantum mechanics where the system states are identified with fair probability distributions is reviewed for systems with continuous variables (the example of the oscillator) and discrete variables (the example of the qubit). The relation for the evolution of the probability distributions which determine quantum states with the Feynman path integral is found. The time-dependent phase of the wave function is related to the time-dependent probability distribution which determines the density matrix. The formal classical-like random variables associated with quantum observables for qubit systems are considered, and the connection of the statistics of the quantum observables with the classical statistics of the random variables is discussed.

Keywords: quantum tomography; probability representation; quantizer–dequantizer; qubit; quantum suprematism

---

1. Introduction

The goal of this work is to discuss some aspects of the new formulation of quantum mechanics where system states are described by the probability distributions. The Schrödinger and von Neumann equations take the form of kinetic equations for the probability distributions. In the conventional formulation of quantum mechanics, a system state (e.g., the oscillator state) is associated with the wave function [1]. In the presence of interaction of the system with the heat bath, the system state is identified with the density matrix [2,3]. The wave functions and density matrices are associated with state vectors in the Hilbert spaces and density operators acting on the vectors [4]. The physical observables—e.g., the energy of the oscillator—in the conventional formulation of quantum mechanics are described by the Hermitian matrices or Hermitian operators—e.g., the Hamiltonian of the oscillator—acting in the Hilbert space of the system state vectors. The aim of this work is to discuss the probability representation of quantum states—e.g., of the oscillator states [5–8].

In this representation, the quantum states are identified with fair probability distributions containing the same information on the states, which is contained in the state density operator. In this probability representation of quantum mechanics, physical observables are associated with the sets of formal classical-like random variables. The evolution of the quantum states described in the conventional formulation of quantum mechanics by the Schrödinger equation for the wave function or by the von Neumann equation for the density operator, as well as by the Gorini–Kossakowskii–Sudarshan–Lindblad (GKSL) equation [9,10], is described by the kinetic equations for the probability distributions identified with the quantum states.

The suggested probability representation is constructed using the invertible map of the probability distributions on the density operators acting in the Hilbert space. For example, in the case of systems with continuous variables like the harmonic oscillator, the invertible map is given either by the Radon transform [11] of the state Wigner function [12] or by the fractional Fourier transform of the wave function [13]. The tomographic probability distributions of spin states were discussed in [14–16]. We point out that different representations—like the Wigner function representation for the system states with discrete variables based on using the formalism of the quantizer operators—were studied in [17–20]. Gauge invariance of quantum mechanics in the probability representation was studied in [21].

This paper is organized as follows.

In Section 2, the case of the tomographic probability distribution of a system with continuous variables like the harmonic oscillator is considered. In Section 3, the generic method of quantization, based on using the quantizer–dequantizer operators [22,23] to introduce the associative product (star product) of the symbols of the operators, is discussed. In Section 4, the gauge invariance of the von Neumann equation and the gauge transform of the wave functions (state vectors) are studied. The gauge invariance is used to investigate the phase factor of the state vector depending on time in the probability representation of quantum mechanics. The qubit (spin-1/2) states are described in the probability representation of quantum mechanics in Section 5. Quantum observables are mapped onto the set of formal classical-like random variables in Section 6. The notion of distance between different quantum states is characterized using the standard notion of the difference between the probability distributions in Section 7. In Section 8, the evolution of quantum states is considered in the probability representation and in the other representations using the quantizer–dequantizer formalism. Conclusions are presented in Section 9.

2. Quantum State Description by Probability Distribution for the Case of Continuous Variables

The possibility of describing the states of systems with continuous variables—like the oscillator system—by probability distributions can be demonstrated on an example of the use of tomographic probability distribution [5–7]. For the harmonic oscillator state, the symplectic tomographic probability distribution can be introduced using the fractional Fourier transform of the wave function $\langle y|\psi\rangle = \psi(y)$. The function [13]

$$w_\psi(X|\mu,\nu) = \frac{1}{2\pi|\nu|}\left|\int \psi(y)\exp\left(\frac{i\mu}{2\nu}y^2 - \frac{iX}{\nu}y\right)dy\right|^2, \qquad -\infty \le X,\mu,\nu < \infty \tag{1}$$

has the following properties. It is nonnegative and normalized for arbitrary parameters μ and ν, i.e.,

$$\int w_\psi(X|\mu,\nu)dX = 1, \tag{2}$$

if the wave function is normalized, i.e., $\int |\psi(y)|^2 dy = 1$.

The density operator $\hat\rho_\psi = |\psi\rangle\langle\psi|$ of the pure state $|\psi\rangle$ can be reconstructed as follows [7]:

$$\hat\rho_\psi = \frac{1}{2\pi}\int dX\,d\mu\,d\nu\,w_\psi(X|\mu,\nu)\exp[i(X-\mu\hat q - \nu\hat p)], \tag{3}$$

where $\hat q$ and $\hat p$ are the position and momentum operators, respectively. For mixed states with the density operator $\hat\rho = \sum_k p_k\hat\rho_{\psi_k}$, $0 \le p_k \le 1$, $\sum_k p_k = 1$, the probability distribution $w_\rho(X|\mu.\nu)$ reconstructs the operator $\hat{\rho}$ in view of the same Formula (3). For a given operator $\hat{\rho}$, the tomographic probability distribution, called the symplectic tomogram of the state, is given by the relation [22,23]

$$w_\rho(X|\mu,\nu) = \mathrm{Tr}\,\hat{\rho}\,\delta(X - \nu\hat{q} - \nu\hat{p}). \tag{4}$$

If the real parameters μ and ν are described as $\mu = s\cos\theta$ and $\nu = s^{-1}\sin\theta$, where $s = 1$, the tomogram coincides with the optical tomogram $w^{\mathrm{opt}}(X|\theta)$ used to measure photon states [24], i.e.,

$$w^{\mathrm{opt}}(X|\theta) = \mathrm{Tr}\,\hat{\rho}\,\delta(X - \hat{q}\cos\theta - \hat{p}\sin\theta). \tag{5}$$

Here, θ is the local oscillator phase and X is the photon quadrature.

The optical tomogram of the pure state is related to the Wigner function [12] of the system via the Radon transform [25,26],

$$w_\psi^{\mathrm{opt}}(X|\theta) = \frac{1}{2\pi}\int W_\psi(q,p)\ delta(X - q\cos\theta - p\sin\theta)\,dq\,dp, \tag{6}$$

where

$$W_\psi(q,p) = \int \psi(q+u/2)\psi^*(q-u/2)\exp(-ipu)\,du, \quad \frac{1}{2\pi}\int W_\psi(q,p) = 1.$$

The optical tomogram is related to the symplectic tomogram $w^{\mathrm{opt}}(X|\theta) = w(X|\cos\theta,\sin\theta)$. If we know the optical tomogram, the symplectic tomogram is given by the formula

$$w(X|\mu,\nu) = \frac{1}{\sqrt{\mu^2+\nu^2}}w^{\mathrm{opt}}\left(\frac{X}{\sqrt{\mu^2+\nu^2}}\middle|\arctan\left(\frac{\nu}{\mu}\right)\right), \tag{7}$$

in view of the property of the Dirac delta-function $\delta(\lambda x) = \frac{1}{|\lambda|}\delta(x)$ used to define the tomogram (4).

The harmonic oscillator states, such as Fock states $|n\rangle$, $n = 0,1,2,\ldots$, are described in the probability representation of quantum mechanics by the distributions $(m = \hbar = \omega = 1)$

$$w_n(X|\mu,\nu) = \frac{\exp\left[-X^2/(\mu^2+\nu^2)\right]}{\sqrt{\pi(\mu^2+\nu^2)}}\frac{1}{2^n n!}H_n^2\left(\frac{X}{\sqrt{\mu^2+\nu^2}}\right). \tag{8}$$

Here, H_n is the Hermite polynomial.

The coherent states $|\alpha\rangle$ [27–29], the eigenfunctions of the oscillator annihilation operator $\hat{a}|\alpha\rangle = \alpha|\alpha\rangle$, $\hat{a} = (\hat{q}+i\hat{p})/\sqrt{2}$, are described by the normal distributions

$$w_\alpha(X|\mu,\nu) = \frac{1}{\sqrt{2\pi\sigma(\mu,\nu)}}\exp\left[-\frac{(X-\bar{X}(\mu,\nu))^2}{2\sigma(\mu,\nu)}\right], \tag{9}$$

$$\bar{X}(\mu,\nu) = \mu\sqrt{2}\,\mathrm{Re}\,\alpha + \nu\sqrt{2}\,\mathrm{Im}\,\alpha, \quad \sigma(\mu,\nu) = \frac{\mu^2+\nu^2}{2}. \tag{10}$$

The time evolution of the tomogram of the harmonic oscillator state is determined by the Green function [30], namely,

$$w(X|\mu,\nu,t) = \int w(X'|\mu',\nu',0)G(X,\mu,\nu,X',\mu',\nu',t)\,dX'\,d\mu'\,d\nu', \tag{11}$$

which can be expressed in terms of the Green function of the Schrödinger equation for the harmonic oscillator

$$g(y,y',t) = \frac{1}{\sqrt{2\pi i\sin t}}\exp\left(i\frac{y^2+y'^2}{2\tan t} - i\frac{yy'}{\sin t}\right). \tag{12}$$

The above function provides the formula for the evolution of the oscillator's wave function

$$\psi(y,t) = \int g(y,y',t)\psi(y',0)\,dy'. \tag{13}$$

Using this formula and relations (1), (3), and (4) for the tomogram expressed in terms of the wave function, we obtain the expression for the Green function in (11). Since the Green function in (13) can be given in the form of the path integral [31,32]

$$g(y,y',t) = \int D\left(u(t)\right)\exp\left\{i\,S_{\mathrm{cl};\,y,y'}\left(u(t)\right)\right\}, \tag{14}$$

where y and y' are the initial and final points of classical trajectories and $S_{\mathrm{cl};\,y,y'}\left(u(t)\right)$ is the classical action, the path integral formulation of quantum mechanics can be also applied in the probability representation of quantum states.

The generic evolution kernel for the tomogram in (11) reads in terms of the path integral as follows:

$$\begin{aligned} G(X,\mu,\nu,X',\mu',\nu',t) \;=\; & \tfrac{1}{2\pi}\int\left[\int D\left(u(t)\right)D\left(v(t)\right)\exp\left\{-iS_{\mathrm{cl};\,y,x}\left(u(t)\right)+iS_{\mathrm{cl};\,x',y'}\left(v(t)\right)\right\}\right] \\ & \times\langle x|\delta(X'-\mu'\hat{q}-\nu'\hat{p})|x'\rangle\langle y'|\exp\left(i(X-\mu\hat{q}-\nu\hat{p})\right)|y\rangle\,dx\,dy\,dx'\,dy'. \end{aligned} \tag{15}$$

Here, the matrix elements of the Weyl operator in the position representation are

$$\langle y'|\exp\left(i(X-\mu\hat{q}-\nu\hat{p})\right)|y\rangle = \exp\left(iX+\frac{i\mu\nu}{2}-i\mu y'\right)\delta(y'-y-\nu). \tag{16}$$

The matrix elements of the expression with the Dirac delta-function read

$$\langle x|\delta(X'-\mu'\hat{q}-\nu'\hat{p})|x'\rangle = \frac{1}{2\pi|\nu'|}\exp\left\{\frac{i(x-x')}{\nu'}\left[X'-\mu'\frac{x+x'}{2}\right]\right\}. \tag{17}$$

In view of the term in the matrix element of the Weyl operator (16) and the term with the Dirac delta-function in (17), after integrating in (15) over variable y', we arrive at the result

$$\begin{aligned} & G(X,\mu,\nu',X',\mu',\nu,t) \\ & = \frac{1}{4\pi^2|\nu'|}\int\left[\int D\left(u(t)\right)D\left(v(t)\right)\exp\left\{-i\,S_{\mathrm{cl};\,y,x}\left(u(t)\right)+i\,S_{\mathrm{cl};\,x',(y+\nu)}\left(v(t)\right)\right\}\right] \\ & \times\exp\left\{\frac{i\,(x-x')}{\nu'}\left[X'-\mu'\frac{x+x'}{2}\right]\right\}\exp\left[i\left(X-\frac{\mu\nu}{2}-\mu y\right)\right]\,dx\,dy\,dx'. \end{aligned} \tag{18}$$

For the harmonic oscillator, the Green function $g(y,y',t)$ (12) is expressed in terms of the path integral $g(y,y',t) = \int D\left(u(t)\right)\exp\left[iS_{\mathrm{cl};\,y,y'}\left(u(t)\right)\right]$. Thus, for the harmonic oscillator, we derive the integral expression for the Green function (propagator), which describes the evolution of symplectic tomographic probability distribution as follows:

$$\begin{aligned} & G(X,\mu,\nu,X',\mu',\nu',t) \\ & = \frac{1}{8\pi^3|\nu'|}\int\exp\left(-i\frac{y^2+x^2}{2\tan t}+i\frac{yx}{\sin t}+i\frac{x'^2+(y+\nu)^2}{2\tan t}-i\frac{x'(y+\nu)}{\sin t}\right) \\ & \times\exp\left\{\frac{i\,(x-x')}{\nu'}\left[X'-\mu'\frac{x+x'}{2}\right]\right\}\exp\left[i\left(X-\frac{\mu\nu}{2}-\mu y\right)\right)\,dx\,dy\,dx'. \end{aligned} \tag{19}$$

The propagator is expressed in the form of a Gaussian integral, and the result of the integration applied to the initial tomogram provides the same result, which one can obtain using the propagator,

$$G\left(X,\mu,\nu,\mu',\nu',t\right) = \delta\left(X-X'\right)\delta\left(\mu\cos t-\nu\sin t-\mu'\right)\delta\left(\nu\cos t+\mu\sin t-\nu'\right). \tag{20}$$

3. Quantizer–Dequantizer Formalism

The presented approach is the example of using the formalism of the quantizer–dequantizer method [22,23,33,34] applied for the quantization of classical system states in the tomographic probability representation of quantum mechanics. In this section, we formulate the relation of the path-integral method to the quantizer–dequantizer formalism by providing different representations of the quantum state evolution such as the evolution of the Wigner function [12] and the Glauber–Sudarshan [27,28] and Husimi–Kano [35,36] quasidistributions; for a review, see [37].

For a given Hilbert space $\mathcal{H}$, the sets of two operators $\hat{\mathcal{U}}(\vec{\gamma})$ and $\hat{\mathcal{D}}(\vec{\gamma})$—where $\hat{\gamma} = (\gamma_1, \gamma_2, \dots, \gamma_N)$—are continuous or discrete variables, and are called dequantizer and quantizer operators [22,38,39], respectively, if for any operator $\hat{A}$, one has the equalities

$$f_A(\vec{\gamma}) = \mathrm{Tr}\hat{A}\hat{\mathcal{U}}(\vec{\gamma}), \tag{21}$$

$$\hat{A} = \int f_A(\vec{\gamma})\hat{\mathcal{D}}(\vec{\gamma})\, d\vec{\gamma}. \tag{22}$$

The product of the operators $\hat{A}\hat{B}$ is mapped onto the associative product of functions $(f_A * f_B)(\vec{\gamma})$, called the star product. The function $f_A(\vec{\gamma})$ is called the symbol of the operator $\hat{A}$. In the considered representation of states, either by means of the Wigner function or by means of symplectic tomographic probability distribution, one uses known dequantizer–quantizer pairs of the operators.

In the case of the Wigner function $\vec{\gamma} = (q, p)$,

$$\hat{\mathcal{U}}(q,p) = \int |q + u/2\rangle\langle q - u/2|\, e^{-ipu}\, du, \tag{23}$$

and $\hat{\mathcal{D}}(q,p) = \frac{1}{2\pi}\hat{\mathcal{U}}(q,p)$.

In the case of the symplectic tomographic probability representation, $\vec{\gamma} = (X, \mu, \nu)$, and the dequantizer reads

$$\hat{\mathcal{U}}(X,\mu,\nu) = \delta\,(X - \mu\hat{q} - \nu\hat{p}). \tag{24}$$

The quantizer has the form of the Weyl operator

$$\hat{\mathcal{D}}(X,\mu,\nu) = \frac{1}{2\pi}\exp[i(X - \mu\hat{q} - \nu\hat{p})]. \tag{25}$$

Now we present the expression for the propagator of the symbol of the density operator $\hat{\rho}$ in terms of generic dequantizer–quantizer operators and the path integral which determines the wave function evolution; it is

$$\begin{aligned} G\,(\vec{\gamma},\vec{\gamma}',t) &= \int \left[\int D\,(u(t))\, D\,(v(t)) \exp\left\{-iS_{\mathrm{cl};\,y,x}\,(u(t)) + i\, S_{\mathrm{cl};\,x',y'}\,(v(t))\right\}\right] \\ &\quad \times \langle x|\hat{\mathcal{U}}(\vec{\gamma}')|x'\rangle\langle y'|\hat{\mathcal{D}}(\vec{\gamma})|y\rangle\, dx\, dy\, dx'\, dy'. \end{aligned} \tag{26}$$

This formula is a generalization of Formula (15) to the case of an arbitrary known representation of quantum states given in the form of expressions containing the dequantizer–quantizer operators. The formulated approach provides the relation of the path integral quantization of the classical mechanics with known star-product schemes of quantization.

4. Gauge Invariance and the Probability Representation of Quantum States

Using the dequantizer–quantizer formalism, gauge invariance in the probability representation of quantum mechanics was considered in [21].

The pure state vector $|\psi\rangle$ of a quantum system [4] satisfies the Schrödinger equation

$$i\frac{\partial|\psi(t)\rangle}{\partial t}=\hat{H}(t)|\psi(t)\rangle. \tag{27}$$

Here, we take Planck's constant $\hbar$ = 1; also, the state vector belongs to the Hilbert space; e.g., the N-dimensional Hilbert space of qudit states. The Hilbert space can be also infinite-dimensional, e.g., for the harmonic oscillator system states. In the generic case, the operator $\hat{H}(t)$ is the time-dependent operator (Hamiltonian of the system) acting in the Hilbert space of the system states. The density operator of the pure state $\hat{\rho}_\psi(t)$, which is determined by the state vector, i.e.,

$$\hat{\rho}_\psi(t)=|\psi(t)\rangle\langle\psi(t)|, \tag{28}$$

satisfies the von Neumann evolution equation [3]

$$i\frac{\partial\hat{\rho}_\psi(t)}{\partial t}+\left(\hat{H}(t)\hat{\rho}_\psi(t)-\hat{\rho}_\psi(t)\hat{H}(t)\right)=0. \tag{29}$$

The gauge transform of the state vector of the form

$$|\psi(t)\rangle=e^{i\chi(t)}|\psi_0(t)\rangle, \tag{30}$$

where $\chi(t)$ is the time-dependent phase, is the symmetry transform of the von Neumann equation, since the density operator of the pure state with the state vector $|\psi_0(t)\rangle$ is equal to the density operator of the pure state with the state vector $|\psi(t)\rangle$, i.e., $\hat{\rho}_\psi(t)=\hat{\rho}_{\psi_0}(t)$. In view of this fact, the density operator $\hat{\rho}_{\psi_0}(t)$ satisfies the same evolution Equation (29). In the case of a mixed state of the system with the density operator given as an arbitrary convex sum of the pure state density operators, i.e.,

$$\hat{\rho}(t)=\sum_k\lambda_k\hat{\rho}_{\psi_k}(t),\quad 0\leq\lambda_k\leq 1,\quad \sum_k\lambda_k=1, \tag{31}$$

the operator $\hat{\rho}(t)$ satisfies the evolution Equation (29).

Now we consider the influence of the gauge transform (30) and express the connection of the Schrödinger equations for state vectors $|\psi(t)\rangle$ and $|\psi_0(t)\rangle$, using the time dependence of the phase $\chi(t)$. In view of (30), we obtain the evolution equation for the state vector $|\psi_0(t)\rangle$ of the form

$$i\frac{\partial|\psi(t)_0\rangle}{\partial t}=\hat{H}_\chi(t)|\psi_0(t)\rangle, \tag{32}$$

where the Hamiltonian $\hat{H}_\chi(t)$ reads

$$\hat{H}_\chi(t)=[\hat{H}(t)+\dot{\chi}(t)\hat{1}]. \tag{33}$$

The unitary evolution operator $\hat{U}(t)$, which describes the evolution of the state vector $|\psi(t)\rangle$, namely,

$$|\psi(t)\rangle=\hat{U}(t)|\psi(t=0)\rangle, \tag{34}$$

satisfies the equation

$$i\frac{\partial\hat{U}(t)}{\partial t}=\hat{H}(t)\hat{U}(t) \tag{35}$$

with the initial condition $\hat{U}(t=0)=\hat{1}$.

In the case of the qubit, for the time-independent Hamiltonian

$$H=\begin{pmatrix}H_{11} & H_{12}\\ H_{12}^* & H_{22}\end{pmatrix}, \tag{36}$$

the unitary matrix $U(t)$ reads

$$U(t) = \exp\left[-it \begin{pmatrix} H_{11} & H_{12} \\ H_{12}^* & H_{22} \end{pmatrix}\right] = \exp\left(-\frac{it}{2}(H_{11}+H_{22})\right) M(t), \tag{37}$$

where the unitary matrix $M(t)$ is

$$M(t) = \begin{pmatrix} \cos\Omega t - \frac{i \sin\Omega t}{2\Omega}(H_{11}-H_{22}) & -\frac{i\sin\Omega t}{\Omega}H_{12} \\ -\frac{i\sin\Omega t}{\Omega}H_{12}^* & \cos\Omega t + \frac{i\sin\Omega t}{2\Omega}(H_{22}-H_{11}) \end{pmatrix}. \tag{38}$$

Here, the frequency Ω is given by formula

$$\Omega = \sqrt{H_{12}H_{12}^* + [(H_{11}-H_{22})/2]^2}. \tag{39}$$

Analogously, the evolution of the state vector $|\psi_0(t)\rangle$ is described by the unitary operator $U_\chi(t)$, which satisfies Equation (35), where the Hamiltonian $\hat{H}(t)$ is replaced by the operator $(\hat{H}(t) + \dot{\chi}(t)\hat{1})$. The unitary evolution operators $\hat{U}(t)$ and $\hat{U}_\chi(t)$ are different ones but, due to the gauge invariance of the von Neumann equation for the density operators satisfying Equation (29) with Hamiltonians $\hat{H}(t)$ and $\hat{H}_\chi(t)$, the density operators are equal. In the case of continuous variables, the wave function $\psi(x,t) = \langle x|\psi(t)\rangle$ and the function $\psi_0(x,t) = \langle x|\psi_0(t)\rangle$ are different. The Schrödinger equation for the system wave function $\psi(x,t)$, written in the form of an equation for the function $\psi_0(x,t)$ following from Equation (32) for the vector $|\psi_0(t)\rangle$, provides the relation

$$\dot{\chi}(t) = i\frac{\partial}{\partial t}\ln\psi_0(x,t) - \psi_0^{-1}(x,t)\hat{H}(t)\psi_0(x,t). \tag{40}$$

This relation contains information on the phase $\chi(t)$, determining the gauge transform of the wave function $\psi(x,t)$, which does not change the density matrix

$$\rho_\psi(x,x',t) = \langle x|\psi(t)\rangle\langle\psi(t)|x'\rangle = \rho_{\psi_0}(x,x',t) = \langle x|\psi_0(t)\rangle\langle\psi_0(t)|x'\rangle.$$

Analogous consideration can be presented in the case of N-dimensional Hilbert space, e.g., in the case of a qubit system with $N = 2$.

5. Qubit State in the Probability Representation

In the case of qubits (two-level atom, spin-1/2 system), the variable x takes two values $x \to m = \pm 1/2$, where m is the spin-1/2 projection in the z-direction. The pure state is described by the state vector $|\psi\rangle$ with two components $|\psi\rangle = \begin{pmatrix} \psi_{+1/2} \\ \psi_{-1/2} \end{pmatrix}$ satisfying the normalization condition $\langle\psi|\psi\rangle = |\psi_{+1/2}|^2 + |\psi_{-1/2}|^2 = 1$. The 2×2 density matrix of the pure state ρ_ψ reads

$$\rho_\psi = \begin{pmatrix} |\psi_{+1/2}|^2 & \psi_{+1/2}\psi^*_{-1/2} \\ \psi_{-1/2}\psi^*_{+1/2} & |\psi_{-1/2}|^2 \end{pmatrix}. \tag{41}$$

The 2×2 density matrix $\rho_{mm'}$ of the mixed state has the matrix elements expressed in terms of probabilities p_1, p_2, p_3 [40,41]:

$$\rho = \begin{pmatrix} p_3 & (p_1 - 1/2) - i(p_2 - 1/2) \\ (p_1 - 1/2) + i(p_2 - 1/2) & 1 - p_3 \end{pmatrix}. \tag{42}$$

Here, the numbers $0 \leq p_1, p_2, p_3 \leq 1$ determine three probability distributions— $(p_1, 1-p_1)$, $(p_2, 1-p_2)$, and $(p_3, 1-p_3)$—of dichotomic random variables. These numbers satisfy the Silvester criterion of nonnegativity of the matrix eigenvalues:

$$\sum_{j=1}^{N} (p_j - 1/2)^2 \leq 1/4. \tag{43}$$

For the pure state with the density matrix given by (41) satisfying the condition $\rho_\psi^2 = \rho_\psi$, we have equality $\sum_{j=1}^{N} (p_j - 1/2)^2 = 1/4$. Our aim is to consider the gauge invariance property of matrix (42). We introduce vector $|\psi_0(t)\rangle$ of the form

$$|\psi_0(t)\rangle = \begin{pmatrix} \sqrt{p_3(t)} \\ \frac{1}{\sqrt{p_3(t)}} [(p_1(t) - 1/2) + i\,(p_2(t) - 1/2)] \end{pmatrix}. \tag{44}$$

The vector $|\psi(t)\rangle = e^{i\chi(t)}|\psi_0(t)\rangle$ gives the density matrix of the form (42). Equation (40) provides the relation of the phase $\chi(t)$ to probabilities $p_1(t)$, $p_2(t)$, and $p_3(t)$. In fact, the probabilities satisfy the von Neumann evolution equation for the density matrix $|\psi_0(t)\rangle\langle\psi_0(t)|$ of the form

$$\begin{pmatrix} \frac{d}{dt}p_3(t) & \frac{d}{dt}p_1(t) - i\frac{d}{dt}p_2(t) \\ \frac{d}{dt}p_1(t) + i\frac{d}{dt}p_2(t) & -\frac{d}{dt}p_3(t) \end{pmatrix}$$
$$= \begin{pmatrix} H_{11}(t) & H_{12}(t) \\ H_{21}(t) & H_{22}(t) \end{pmatrix} \begin{pmatrix} p_3(t) & (p_1(t) - 1/2) - i\,(p_2(t) - 1/2) \\ (p_1(t) - 1/2) + i\,(p_2(t) - 1/2) & 1 - p_3(t) \end{pmatrix} \tag{45}$$
$$- \begin{pmatrix} p_3(t) & (p_1(t) - 1/2) - i\,(p_2(t) - 1/2) \\ (p_1(t) - 1/2) + i\,(p_2(t) - 1/2) & 1 - p_3(t) \end{pmatrix} \begin{pmatrix} H_{11}(t) & H_{12}(t) \\ H_{21}(t) & H_{22}(t) \end{pmatrix}.$$

This matrix equation provides the connection of the time derivatives $\frac{dp_j(t)}{dt}$; $j = 1,2,3$ with probabilities and matrix elements of the Hamiltonian $H_{jk}(t)$; $j,k = 1,2$. In addition, Equation (40) written for probabilities $p_j(t)$; $j = 1,2,3$ takes the form

$$i\frac{d}{dt}\sqrt{p_3(t)} = [H_{11}(t) + \dot{\chi}(t)]\sqrt{p_3(t)} + H_{12}(t)\frac{(p_1(t) - 1/2) + i\,(p_2(t) - 1/2)}{\sqrt{p_3(t)}}. \tag{46}$$

Thus, using (40), we can obtain the connection of the phase $\chi(t)$ with probabilities $p_1(t)$, $p_2(t)$, and $p_3(t)$. For the case of the spin-1/2 pure state with matrix elements of the Hamiltonian H_{jk}; $j = 1,2$, one obtains the evolution equation for the phase $\chi(t)$ of the Pauli spinor of the form

$$\frac{d\chi(t)}{dt} = p_3^{-1}(t)\left[(p_2(t) - 1/2)\,\mathrm{Im}\,H_{12}(t) - (p_1(t) - 1/2)\,\mathrm{Re}\,H_{12}(t)\right] - H_{11}(t). \tag{47}$$

The probabilities $p_1(t)$, $p_2(t)$, and $p_3(t)$ are given as solutions of the kinetic Equation (45), which follows from the von Neumann equation for the density matrix $\rho(t)$ of spin-1/2 states, and it does not contain the phase $\chi(t)$. The evolution Equation (47) can be rewritten using Bloch parameters $B_j(t) = 2p_j(t) - 1$. These parameters have the physical meaning connected with mean values of the spin-1/2 projections onto three perpendicular directions—x, y, and z, respectively. The equation has the form

$$\frac{d\chi(t)}{dt} = [B_3(t) + 1]^{-1}\,[B_2(t)\,\mathrm{Im}\,H_{12}(t) - B_1(t)\,\mathrm{Re}\,H_{12}(t)] - H_{11}(t). \tag{48}$$

The values $B_j(t)$ and probabilities $p_j(t)$ can be measured. This means that one can find the evolution of the phase $\chi(t)$ of the spin-1/2 wave function given by the formula

$$\chi(t)=\int_0^t d\tau\left\{p_3^{-1}(\tau)\left[(p_2(\tau)-1/2)\,\mathrm{Im}\,H_{12}(\tau)-(p_1(\tau)-1/2)\,\mathrm{Re}\,H_{12}(\tau)\right]-H_{11}(\tau)\right\},\quad \chi(0)=0. \tag{49}$$

For the Hamiltonian describing the evolution of the spin-1/2 particle with magnetic moment $\vec{\mathcal{M}}$ in a constant magnetic field $\vec{\mathcal{H}}=(0,0,\mathcal{H})$, i.e., $H(t)=-\mathcal{M}\mathcal{H}\sigma_z$, the probabilities $p_1(t)$, $p_2(t)$, and $p_3(t)$ depend on the initial values of the state, $p_1(0)$, $p_2(0)$, and $p_3(0)$. For example, if $p_1(0)=p_2(0)=1/2$ and $p_3(0)=1$, one has Equation (47) of the form

$$\frac{d\chi(t)}{dt}=\mathcal{M}\,\mathcal{H}. \tag{50}$$

The solution (49) of this equation gives the phase

$$\chi(t)=\mathcal{M}\,\mathcal{H}t, \tag{51}$$

and this phase corresponds to the solution of the Schrödinger equation for the Pauli spinor $|\psi(t)\rangle$,

$$|\psi(t)\rangle=\begin{pmatrix} e^{-i\mu\mathcal{H}t} & 0\\ 0 & e^{i\mu\mathcal{H}t}\end{pmatrix}\begin{pmatrix}1\\0\end{pmatrix}, \tag{52}$$

where the initial value $|\psi(0)\rangle$ corresponds to the pure state density matrix $\rho(0)=\begin{pmatrix}1&0\\0&0\end{pmatrix}$, and the initial phase $\chi(0)=0$.

The density matrix $\rho(t)$ corresponding to the pure state (52) does not depend on the magnetic field $\mathcal{H}$ due to the stationarity of probabilities $p_1(t)=p_2(t)=1/2$ and $p_3(t)=1$ in this particular case. However, in the case of the initial state with density matrix $\rho(0)=\frac{1}{2}\begin{pmatrix}1&1\\1&1\end{pmatrix}$ and $|\psi(0)\rangle=\frac{1}{\sqrt{2}}\begin{pmatrix}1\\1\end{pmatrix}$, one has the evolution of probabilities

$$p_1(t)=\frac{1+\cos(2\mathcal{M}\mathcal{H}t)}{2},\qquad p_2(t)=\frac{1+2\sin(2\mathcal{M}\mathcal{H}t)}{2},\qquad p_3(t)=\frac{1}{2}. \tag{53}$$

The phase $\chi(t)$ given by (49) takes into account the connection of the probabilities of the spin-1/2 projections $m=+1/2$ onto perpendicular directions x and y on the magnetic field. Thus, the value of the magnetic field can be found measuring the probability evolution.

6. The Quantum Observable as a Set of Dichotomic Random Variables

As we found, the density matrix of a spin-1/2 system (42) is expressed in terms of three probability distributions, $(p_1,\,1-p_1)$, $(p_2,\,1-p_2)$, and $(p_3,\,1-p_3)$, where the probabilities p_j, $j=1,2,3$ satisfy inequality (43). For an observable

$$A=\begin{pmatrix}A_{11}&A_{12}\\A_{21}&A_{22}\end{pmatrix}, \tag{54}$$

one can determine the set of formal classical-like random variables using the following rule.

Let us determine a random dichotomic variable $\vec{x}$ taking two values, $(x_1=x,\,x_2=-x)$, and another random variable $\vec{y}$ taking two values, $(y_1=y,y_2=-y)$, where $x+iy=A_{12}$. The third random variable $\vec{z}$ takes two real values, $(z_1=A_{11},\,z_2=A_{22})$. The introduced random dichotomic variables can be interpreted as the rules in such a game as tossing three classical coins. The game with the first coin has a gain equal to x and a loss equal to $-x$. For the second coin, the gain is equal to y and the loss is equal to $-y$. For the third coin, the gain and loss are not equal; they are denoted as z_1 and z_2, respectively. We interpret the probability distributions $(p_1, 1-p_1)$, $(p_2, 1-p_2)$, and $(p_3, 1-p_3)$ as distributions describing the classical statistics of random variables $\vec{x}$, $\vec{y}$, and $\vec{z}$. This means that we define the moments for the dichotomic variables

$$M_n(\vec{x}) = x_1^n p_1 + (1-p_1)x_2^n, \quad M_n(\vec{y}) = y_1^n p_2 + (1-p_2)y_2^n, \quad M_n(\vec{z}) = z_1^n p_3 + (1-p_3)z_2^n, \qquad n = 1,2,\ldots \tag{55}$$

The quantum statistics of observable (54) are described by the matrix A and the density matrix ρ given by (42) as follows:

$$\langle A^n \rangle = \operatorname{Tr} \rho\, A^n, \quad n = 1,2,\ldots \tag{56}$$

One can check that, for the quantum observable A, the mean value $\langle A \rangle$ is expressed in terms of classical mean values $M_1(\vec{x})$, $M_1(\vec{y})$, and $M_1(\vec{z})$, namely,

$$\langle A \rangle = M_1(\vec{x}) + M_1(\vec{y}) + M_1(\vec{z}). \tag{57}$$

The highest moments (55) of classical-like random variables, which we used to interpret the matrix elements of the quantum observable A and the quantum moments, e.g., $\langle A^2 \rangle$, are not expressed in the form of a sum analogous to (57).

We consider now an example of the qubit thermal equilibrium state with the density matrix $\rho(T)$. For a given Hamiltonian H (36) with matrix elements H_{jk}; $j,k = 1,2$, the density matrix reads

$$\rho(\beta) = \frac{1}{Z(\beta)} \exp(-\beta H), \qquad \beta = \frac{1}{T}. \tag{58}$$

Using the properties of Pauli matrices $\sigma_1 = \begin{pmatrix} 0 & 1 \\ 1 & 0 \end{pmatrix}$, $\sigma_2 = \begin{pmatrix} 0 & -i \\ i & 0 \end{pmatrix}$, and $\sigma_3 = \begin{pmatrix} 1 & 0 \\ 0 & -1 \end{pmatrix}$, namely, $\sigma_j^2 = 1$, $\operatorname{Tr} \sigma_j = 0$; $j = 1,2,3$ and the relation $\exp \tau(\vec{\sigma}\vec{n}) = (\cosh \tau)1_2 + (\sinh \tau)\vec{\sigma}\vec{n}$, where $\vec{n}^2 = 1$, we arrive at

$$\exp \begin{pmatrix} a & b \\ c & d \end{pmatrix} = e^{(a+d)/2} \begin{pmatrix} \cosh \tau + \dfrac{a-d}{2\tau} \sinh \tau & b\dfrac{\sinh \tau}{\tau} \\ c\dfrac{\sinh \tau}{\tau} & \cosh \tau + \dfrac{d-a}{2\tau} \sinh \tau \end{pmatrix}, \tag{59}$$

where

$$\tau = \sqrt{[(a-d)/2]^2 + bc} \tag{60}$$

and

$$\operatorname{Tr} \left[\exp \begin{pmatrix} q & b \\ c & d \end{pmatrix} \right] = 2 \exp\left[(a+d)/2\right] \cosh \tau. \tag{61}$$

The thermal equilibrium state of the qubit at the temperature $T = 1/\beta$ is identified with three probabilities, $p_1(\beta)$, $p_2(\beta)$, and $p_3(\beta)$. For Hamiltonian (36), $-\beta H_{11} = a$, $-\beta H_{12} = b$, $-\beta H_{21} = c$, and $-\beta H_{22} = d$, the probabilities read

$$p_1(\beta) = \frac{1}{2}\left[1 + \frac{(\mathrm{Re}\, H_{12})\,\tanh\left(\beta\sqrt{[(H_{11}-H_{22})/2]^2 + |H_{12}|^2}\right)}{\sqrt{[(H_{11}-H_{22})/2]^2 + |H_{12}|^2}}\right],$$
$$p_2(\beta) = \frac{1}{2}\left[1 + \frac{(\mathrm{Im}\, H_{12})\,\tanh\left(\beta\sqrt{[(H_{11}-H_{22})/2]^2 + |H_{12}|^2}\right)}{\sqrt{[(H_{11}-H_{22})/2]^2 + |H_{12}|^2}}\right],$$
$$p_3(\beta) = \frac{1}{2}\left[1 + \frac{(H_{22}-H_{11})\,\tanh\left(\beta\sqrt{[(H_{11}-H_{22})/2]^2 + |H_{12}|^2}\right)}{\sqrt{(H_{11}-H_{22})^2 + 4|H_{12}|^2}}\right]. \tag{62}$$

We arrive at the function $Z(\beta) = \mathrm{Tr}\,[\exp(-\beta H)]$ of the form

$$Z(\beta) = 2\left[\exp\left(-\beta\left[(H_{11}+H_{22})/2\right]\right)\right]\cosh\left(\beta\sqrt{[(H_{11}-H_{22})/2]^2 + |H_{12}|^2}\right).$$

Thus, the three probabilities $p_j(\beta)$; $j = 1,2,3$ describe the statistical properties of thermal equilibrium states of qubits (two-level atoms).

The geometrical picture for all of the states of the qubit is given in terms of Bloch ball parameters. However, the states of the qubit could also be presented in the form of Triada of Malevich's squares (Figure 1) using the probabilities p_1, p_2, and p_3 [40,42,43].

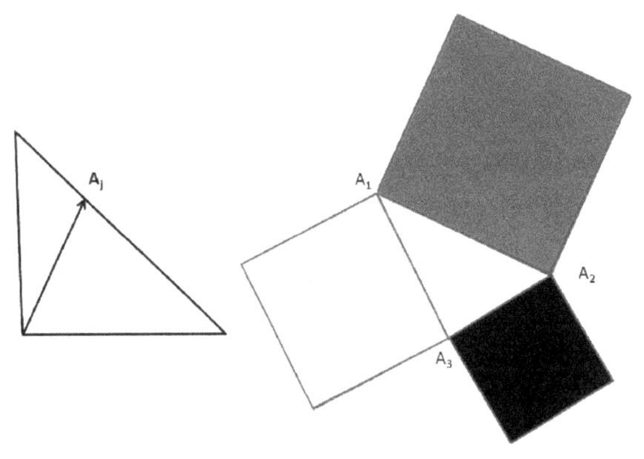

Figure 1. Triada of Malevich's squares as a geometrical interpretation of qubit states.

An equilateral triangle with a side length of $\sqrt{2}$ is given. This triangle is constructed by gluing vertices on hypotenuses of three rectangular triangles with equal legs and with the points A_j; $j = 1,2,3$. The points have coordinates p_j and $1 - p_j$. Inside this equilateral triangle, we have the triangle $A_1A_2A_3$ with three side lengths:

$$S_1 = \left(2 + 2p_1^2 - 4p_1 - 2p_2 + 2p_2^2 + 2p_1 p_2\right)^{1/2},$$
$$S_2 = \left(2 + 2p_2^2 - 4p_2 - 2p_3 + 2p_3^2 + 2p_2 p_3\right)^{1/2},$$
$$S_3 = \left(2 + 2p_3^2 - 4p_3 - 2p_1 + 2p_1^2 + 2p_3 p_1\right)^{1/2}. \tag{63}$$

We construct the squares on sides A_1A_2, A_2A_3, and A_3A_1 of triangle $A_1A_2A_3$. We paint the squares in different colors: The square with side A_1A_2 is red, the square with side A_2A_3 is black, and the square with side A_3A_1 is white. Thus, the state with density matrix ρ can be geometrically described either by the point in the Bloch ball or by the three squares on the plane, called the Triada of Malevich's squares. This approach was called quantum suprematism [40,42,43].

The quantum nature of the qubit state provides a difference with classical coins, which are also described by the probabilities p_1, p_2, and p_3. The maximum area of the Triada of Malevich's squares for three classical coins is equal to 6. However, for the qubit state (spin-1/2 state, two-level atom state), it is equal to 3. By measuring Bloch parameters or probabilities p_1, p_2, and p_3, one provides the possibility of checking this property. This measurement is analogous to checking the Bell inequality for a two spin-1/2 system, where the maximum value of specific correlations is equal to $2\sqrt{2}$. The classical-like maximum is equal to 2. However, in the case of the possible measurement of the Malevich's square areas (i.e., measuring probabilities p_1, p_2, and p_3), the system under investigation is a single spin-1/2 particle. In the case of the Bell inequality, the system under study is the two-qubit system. Thus, one can check the differences in the properties of classical and quantum correlations in the case of one spin-1/2 particle by measuring probabilities p_1, p_2, and p_3, which determine the area of Malevich's squares.

7. Distance between Quantum States

The probability representation of quantum states can be used to introduce classical-like characteristics of difference between the states in addition to the fidelity. We consider qubit states as an example. For density matrices ρ_1 and ρ_2, there exist six probability distributions

$$(p_1^{(1)}, 1 - p_1^{(1)}),\ (p_2^{(1)}, 1 - p_2^{(1)}),\ (p_3^{(1)}, 1 - p_3^{(1)}),\ (p_1^{(2)}, 1 - p_1^{(2)}),\ (p_2^{(2)}, 1 - p_2^{(2)}),\ (p_3^{(2)}, 1 - p_3^{(2)})$$

which determine the density matrices. In conventional probability theory, the difference between the two probability distributions $(P_1, P_2, \ldots, P_N)$ and $(Q_1, Q_1, \ldots, Q_N)$ is characterized by the distance parameter

$$S = [(P_1 - Q_1)^2 + (P_2 + Q_2)^2 + \cdots + (P_N - Q_N)^2].$$

This classical formula can be used to characterize the difference of quantum states with density matrices ρ_1 and ρ_2. We determine such characteristics for the two qubit states as follows:

$$S_{12} = 2 \sum_{j=1}^{3} (p_j^{(1)} - p_j^{(2)})^2. \tag{64}$$

In the case of continuous variables, the difference of the two states is characterized by an analogous relation; it reads

$$S_{12}(\mu, \nu) = \int [w_1(X|\mu, \nu) - w_2(X|\mu, \nu)]^2 \, dX, \tag{65}$$

where we use the tomographic probability distributions $w_{1,2}(X|\mu, \nu)$ corresponding to the states, e.g., of the oscillator with density operators ρ_1 and ρ_2, respectively. The difference of the two states depends on the parameters μ and ν, which describe the reference frames in the phase space where the position X is measured.

In the case of the classical oscillator, for which the tomographic probability distribution can be introduced [6] and expressed in terms of the Radon transform of the classical probability density function $f(q,p)$

$$w_{\rm cl}(X|\mu,\nu) = \int f(q,p)\,\delta(X-\mu q-\nu p)\,dq\,dp, \tag{66}$$

the parameters $\mu = s\cos\theta$ and $\nu = s^{-1}\sin\theta$ correspond to scale and rotation transforms of the position and momentum: $q \to q' = sq$, $p \to p' = s^{-1}p$, $q' \to q'' = q'\cos\theta + p'\sin\theta$, and $p' \to p'' = p'\cos\theta - q'\sin\theta$, respectively. Thus, we can use the notion of the difference of the states both in classical and quantum mechanics using the tomographic probability distributions.

In connection with the construction of the probability distributions identified with quantum states, one can apply other characteristics of the state difference used in classical probability theory to study the difference of two probability distributions. For example, in addition to (66), the Shannon relative entropy can be considered as a tool to characterize the difference of the two states of photons with two different tomographic probability distributions.

The relative entropy for quantum symplectic and optical tomograms must be nonnegative; i.e., for the symplectic tomogram, we have

$$H_{\rm relative} = \int w_1(X|\mu,\nu)\ln\left[\frac{w_1(X|\mu,\nu)}{w_2(X|\mu',\nu')}\right]dX \geq 0.$$

This relative entropy can be equal to zero only for equal probability distributions. Since optical tomograms of photon states are measured experimentally [24], tomograms $w_1(X|\theta_1)$ and $w_2(X|\theta_2)$ must satisfy the inequality for the relative entropy:

$$H(\theta_1,\theta_2) = \int w_1(X|\theta_1)\ln\left[\frac{w_1(X|\theta_1)}{w_2(X|\theta_2)}\right]dX \geq 0.$$

This inequality can be checked experimentally. An analogous entropic inequality can be checked for qubit states, namely,

$$\sum_{j=1}^{3}\left[p_j\ln\left(\frac{p_j}{\mathcal{P}_j}\right)+(1-p_j)\ln\left(\frac{1-p_j}{1-\mathcal{P}_j}\right)\right] \geq 0,$$

where $p_1, p_2, p_3, \mathcal{P}_1, \mathcal{P}_2$, and $\mathcal{P}_3$ are different probabilities of spin projections $m = +1/2$ on the x, y, z-directions in two different spin-1/2 states identified with the probability distributions.

For composite systems like that of the two qubits, quantum correlations between the subsystem degrees of freedom can be associated with the probability distributions and their entropies by describing the states of the system and its subsystems. This problem needs extra consideration.

8. The Evolution Equation in the Probability Representation

Given an arbitrary pair of dequantizer–quantizer operators $\hat{\mathcal{U}}(\gamma)$ and $\hat{\mathcal{D}}(\gamma)$, the von Neumann Equation (27) can be written for the symbol of operator $\hat{\rho}(t)$ determined by the dequantizer operator,

$$f_\rho(\vec{\gamma},t) = \mathrm{Tr}\left(\hat{\mathcal{U}}(\vec{\gamma})\hat{\rho}(t)\right); \tag{67}$$

the equation has the form

$$\frac{\partial}{\partial t}f_\rho(\vec{\gamma},t) = \int f_\rho(\vec{\gamma}',t)K(\vec{\gamma},\vec{\gamma}',t)\,d\vec{\gamma}'. \tag{68}$$

Here, the integral kernel is determined by the quantizer–dequantizer operators [21]

$$K(\vec{\gamma},\vec{\gamma}',t) = i\mathrm{Tr}\left[\hat{\mathcal{U}}(\vec{\gamma})\left(\hat{\mathcal{D}}(\vec{\gamma}')\hat{H}-\hat{H}\hat{\mathcal{D}}(\vec{\gamma}')\right)\right]. \tag{69}$$

In the case where the symbol of the density operator is a probability distribution, e.g., in the case of $\vec{\gamma} = (X, \mu, \nu)$, using dequantizer–quantizer operators (24) and (25), we arrive at the von Neumann equation of the form of a kinetic equation for tomographic probability distribution $w(X|\mu, \nu, t)$ [5]. Equations (68) and (69) provide the unitary evolution description for all known representations of quantum states in the literature, including the probability representation. For optical tomographic probability distribution, the evolution equation was found in [44].

9. Conclusions

To conclude, we point out the main results of our work.

We reviewed the probability representation of quantum states and considered the application of this approach on the examples of harmonic oscillator states and qubit states. In the case of systems with continuous variables, like the harmonic oscillator, we obtained the relation of the probability representation of quantum states with the path integral method. We expressed the propagator describing the evolution of the tomographic probability distribution identified with the quantum state of the system with continuous variables, like the oscillator, with the path integral determining the Green function of the Schrödinger evolution equation for the wave function of the system.

In the generic case of systems with continuous variables, we formulated the connection of the path integral representation of the quantum-state evolution with other representations of quantum states, using the quantizer–dequantizer approach to the quantization of classical systems studied in [22,23,38,39]. The notion of classical-like difference of quantum states was introduced, using the standard notion of the difference of two probability distributions for both the oscillator system and qubit system, where the state is identified with three probability distributions of dichotomic random variables.

We found the equation for the time-dependent phase determining the gauge transform of the wave function. For the qubit state, the equation is given in the form of a relation, where the time-dependent phase is expressed in terms of the probabilities of dichotomic random variables determining the evolution of the density matrix satisfying the von Neumann equation with a given Hamiltonian.

The statistical properties of qubit quantum observables were discussed in relation with the classical statistical properties of the classical-like dichotomic random variables. The different aspects of the relation of the probability theory with properties of quantum or quantum-like states were discussed in [45–49]. We will consider other examples of the connection of the path integral with star-product schemes and entropic inequalities for quantum systems based on the probability representation of quantum mechanics in future publications.

Author Contributions: This paper is partially based on the talk of V.I.M. at the 16th International Conference on Squeezed States and Uncertainty Relations (eventos.ucm.es/30364/detail/international-conference-on-squeezed-states-and-uncertainty-relations-2019.html) (Universidad Complutense de Madrid, Spain, June 17–21, 2019). Conceptualization, V.I.M.; writing-original draft preparation V.N.C. and O.V.M.; writing-review and editing V.N.C. and O.V.M.; supervision V.I.M. All authors have read and agreed to the published version of the manuscript.

Funding: This research received no external funding.

Acknowledgments: V.I.M. thanks the Organizers of the 16th International Conference on Squeezed States and Uncertainty Relations (Madrid, 17–21 June 2019) and especially Luis Sanchez-Soto and Alberto Ibort for their invitation and kind hospitality.

Conflicts of Interest: The authors declare no conflict of interest. The funders had no role in the design of the study; in the collection, analyses, or interpretation of data; in the writing of the manuscript, or in the decision to publish the results.

References

1. Schrödinger, E. Der stetige Übergang von der Mikro-zur Makromechanik. *Naturwissenchaften* **1926**, *14*, 664. [CrossRef] [CrossRef]
2. Landau, L. Das Dämpfungsproblem in der Wellenmechanik. *Z. Phys.* **1927**, *45*, 430–441. [CrossRef] [CrossRef]

3. von Neumann, J. Wahrscheinlichkeitstheoretischer Aufbau der Quantenmechanik. *Gött. Nach.* **1927**, *1927*, 245–272. [CrossRef]
4. Dirac, P.A.M. *The Principles of Quantum Mechanics*; Clarendon Press: Oxford, UK, 1981; ISBN 9780198520115.
5. Mancini, S.; Man'ko, V.I.; Tombesi, P. Symplectic Tomography as Classical Approach to Quantum Systems. *Phys. Lett. A* **1996**, *213*, 1–6. [CrossRef] [CrossRef]
6. Man'ko, O.V.; Man'ko, V.I. Quantum States in Probability Representation and Tomography. *J. Russ. Laser Res.* **1997**, *18*, 407–444. [CrossRef] [CrossRef]
7. Asorey, M.; Ibort, A.; Marmo, G.; Ventriglia, F. Quantum Tomography Twenty Years Later. *Phys. Scr.* **2015**, *90*, 074031. [CrossRef] [CrossRef]
8. Ibort, A.; Man'ko, V.I.; Marmo, G.; Simoni, A.; Ventriglia, F. An Introduction to the Tomographic Picture of Quantum Mechanics. *Phys. Scr.* **2009**, *79*, 065013. [CrossRef] [CrossRef]
9. Gorini, V.; Kossakowski, A.; Sudarshan, E.C.G. Completely Positive Dynamical Semigroups of N-Level Systems. *J. Math. Phys.* **1976**, *17*, 821–825. [CrossRef] [CrossRef]
10. Lindblad, G. On the Generators of Quantum Dynamical Semigroups. *Commun. Math. Phys.* **1976**, *48*, 119–130. [CrossRef] [CrossRef]
11. Radon, J. Über die Bestimmung von Funktionen durch ihre Integralwerte längs gewisser Mannigfaltigkeiten. In *Berichte Uber Die Verhandlungen der Koniglich-Sachsischen Akademie der Wissenschaften zu Leipzig, Mathematisch-Physische Klasse*; Weidmannsche Buchhandlung: Leipzig, Germany, 1917; Volume 69, pp. 262–277.
12. Wigner, E. On the Quantum Correction For Thermodynamic Equilibrium. *Phys. Rev.* **1932**, *40*, 749–759. [CrossRef] [CrossRef]
13. Man'ko, V.I.; Mendes, R.V. Non-Commutative Time-Frequency Tomography. *Phys. Lett. A* **1999**, *263*, 53–61. [CrossRef] [CrossRef]
14. Weigert, S. Quantum Time Evolution in Terms of Nonredundant Probabilities. *Phys. Rev. Lett.* **2000**, *84*, 802–806. [CrossRef] [CrossRef] [PubMed]
15. Amiet, J.P.; Weigert, S. Coherent States and the Reconstruction of Pure Spin States. *J. Opt. B Quantum Semiclass. Opt.* **1999**, *1*, L5. [CrossRef] [CrossRef]
16. D'Ariano, G.M.; Maccone, L.; Paini, M. Spin Tomography. *J. Opt. B Quantum Semiclass. Opt.* **2003**, *5*, 77–84. [CrossRef] [CrossRef]
17. Muñoz, C.; Klimov A.B.; Sánchez-Soto, L. Discrete Phase-Space Structures and Wigner functions for N Qubits. *Quantum Inf. Process* **2017**, *16*, 158. [CrossRef] [CrossRef]
18. Klimov, A.B.; Romero, J.L.; Björk, G.; Sánchez-Soto, L.L. Geometrical Approach to Mutually Unbiased Bases. *J. Phys. A Math. Theor.* **2007**, *40*, 3987–3998. [CrossRef] [CrossRef]
19. Stratonovich, R.L. On Distributions in Representation Space. *Sov. Phys. JETP* **1957**, *4*, 891–898. [CrossRef]
20. Ibort, A.; Man'ko, V.I.; Marmo, G.; Simoni, A.; Stornaiolo, C.; Ventriglia F. Realization of Associative Products in Terms of Moyal and Tomographic Symbols. *Phys. Scr.* **2013**, *87*, 038107. [CrossRef] [CrossRef]
21. Korennoy, Y.A.; Man'ko, V.I. Gauge Transformation of Quantum States in the Probability Representation. *J. Phys. A Math. Theor.* **2017**, *50*, 155302. [CrossRef] [CrossRef]
22. Man'ko, O.V.; Man'ko, V.I.; Marmo, G. Alternative Commutation Relations, Star Products and Tomography. *J. Phys. A Math. Gen.* **2002**, *35*, 699–719. [CrossRef] [CrossRef]
23. Man'ko, O.V.; Man'ko, V.I.; Marmo, G. Tomographic Map within the Framework of Star-Product Quantization. In *Second International Symposium "Quantum Theory and Symmetries" (Krakow, July 2001)*; Kapuscik, E., Horzela, A., Eds.; World Scientific: Singapore, 2002; pp. 126–133. [CrossRef]
24. Smithey, D.T.; Beck, M.; Raymer, M.G.; Faridani, A. Measurement of the Wigner Distribution and the Density Matrix of a Light Mode Using Optical Homodyne Tomography: Application to Squeezed States and the Vacuum. *Phys. Rev. Lett.* **1993**, *70*, 1244–1247. [CrossRef] [CrossRef]
25. Bertrand, J.; Bertrand, P. A Tomographic Approach to Wigner's Function. *Found. Phys.* **1989**, *17*, 397–405. [CrossRef] [CrossRef]
26. Vogel, K.; Risken, H. Determination of Quasiprobability Distributions in Terms of Probability Distributions for the Rotated Quadrature Phase. *Phys. Rev. A* **1989**, *40*, 2847–2849. [CrossRef] [CrossRef]
27. Glauber, R.J. Photon Correlations. *Phys. Rev. Lett.* **1963**, *10*, 84–86. [CrossRef] [CrossRef]
28. Sudarshan, E.C.G. Equivalence of Semiclassical and Quantum-Mechanical Descriptions of Statistical Light Beams. *Phys. Rev. Lett.* **1963**, *10*, 277–279. [CrossRef] [CrossRef]

29. Klauder, J.R.; Sudarshan, E.C.G. *Fundamentals of Quantum Optics*; Benjamin: New York, NY, USA, 1968; ISBN 978-0486450087.
30. Man'ko, O.V.; Man'ko, V.I. "Classical" Propagator and Path Integral in the Probability Representation of Quantum Mechanics. *J. Russ. Laser Res.* **1999**, *20*, 67–76. [CrossRef] [CrossRef]
31. Feynman, R.P. Space-Time Approach to Non-Relativistic Quantum Mechanics. *Rev. Mod. Phys.* **1948**, *20*, 367–387. [CrossRef] [CrossRef]
32. Feynman, R.P.; Hibbs, A. *Quantum Mechanics and Path Integrals*; McGraw Hill: New York, NY, USA, 1965; ISBN 0-07-020650-3.
33. Man'ko, O.V.; Man'ko, V.I.; Marmo, G.; Vitale, P. Star-Products, Duality and Double Lie Algebras. *Phys. Lett. A* **2007**, *360*, 522–532. [CrossRef] [CrossRef]
34. Belolipetskiy, S.N.; Chernega, V.N.; Man'ko, O.V.; Man'ko, V.I. Probability Representation of Quantum Mechanics and Star-Product Quantization. *J. Phys. Conf. Ser.* **2019**, *1348*, 012101. [CrossRef]
35. Husimi, K. Some Formal Properties of the Density Matrix. *Proc. Phys. Math. Soc. Jpn.* **1940**, *22*, 264–314. [CrossRef]
36. Kano, Y. A New Phase-Space Distribution Function in the Statistical Theory of the Electromagnetic Field. *J. Math. Phys.* **1965**, *6*, 1913–1915. [CrossRef] [CrossRef]
37. Schleich, W. *Quantum Optics in Phase Space*; Wiley-VCH: Hoboken, NJ, USA, 2001; ISBN-13 978-3527294350.
38. Lizzi, F.; Vitale, P. Matrix Bases for Star-Products: A Review. *Symmetryli Integr. Geom. Methods Appl.* **2014**, *10*, 086. [CrossRef] [CrossRef]
39. Ciaglia, F.M.; Di Cosmo, F.; Ibort, A.; Marmo, G. Dynamical Aspects in the Quantizer-Dequantizer Formalism. *Ann. Phys.* **2017**, **385**, 769–781. [CrossRef] [CrossRef]
40. Chernega, V.N.; Man'ko, O.V.; Man'ko, V.I. Triangle Geometry of the Qubit State in the Probability Representation Expressed in Terms of the Triada of Malevich's Squares. *J. Russ. Laser Res.* **2017**, *38*, 141–149. [CrossRef] [CrossRef]
41. Man'ko, V.I.; Marmo, G.; Ventriglia, F.; Vitale, P. Metric on the Space of Quantum States from Relative Entropy. Tomographic Reconstruction. *J. Phys. A Math. Gen.* **2017**, *50*, 335302. [CrossRef] [CrossRef]
42. Chernega, V.N.; Man'ko, O.V.; Man'ko, V.I. Probability Representation of Quantum Observables and Quantum States. *J. Russ. Laser Res.* **2017**, *38*, 324–333. [CrossRef] [CrossRef]
43. Chernega, V.N.; Man'ko, O.V.; Man'ko, V.I. Triangle Geometry for Qutrit States in the Probability Representation. *J. Russ. Laser Res.* **2017**, *38*, 416–425. [CrossRef] [CrossRef]
44. Amosov, A.A.; Korennoy, Y.A.; Man'ko, V.I. Description and Measurement of Observables in the Optical Tomographic Probability Representation of Quantum Mechanics. *Phys. Rev. A* **2012**, *85*, 052119. [CrossRef] [CrossRef]
45. Foukzon, J.; Potapov, A. A.; Menkova, E.; Podosenov, S.A. A New Quantum-Mechanical Formalism Based on the Probability Representation of Quantum States. *arXiv* **2016** arXiv:1612.0298(2016).
46. Khrennikov, A.; Alodjants, A. Classical (Local and Contextual) Probability Model for Bohm-Bell Type Experiments: No-Signaling as Independence of Random Variables. *Entropy* **2019**, *21*, 157. [CrossRef] [CrossRef]
47. Stornaiolo, C. Tomographic Represention of Quantum and Classical Cosmology. In *Accelerated Cosmic Expansion. Proceedings of the Fourth International Meeting on Gravitation and Cosmology*; Moreno Gonzälez, C., Madriz Aguilar, J., Reyes Barrera L., Eds.; Astrophysics and Space Science Proceedings; Springer: Berlin/Heidelberg, Germany, 2014; Volume 38. [CrossRef]
48. Facchi, P.; Ligabó, M. Classical and Quantum Aspects of Tomography. In *Proceedings of XVIII International Fall Workshop on Geometry and Physics, Benasque, Spain, 2009*; Asorey, M., Clemente-Gallardo, J., Martñez, E., Cariñena, J.F., Eds.; AIP: College Park, MD, USA, 2010; Volume 1260, p. 3. [CrossRef]
49. Elze, H.-T.; Gambarotta, G.; Vallone, F. General Linear Dynamics—Quantum, Classical or Hybrid. *J. Phys. Conf. Ser.* **2011**, *306*, 012010. [CrossRef] [CrossRef]

Article

Distance between Bound Entangled States from Unextendible Product Bases and Separable States

Marcin Wieśniak [1,2,*], Palash Pandya [1], Omer Sakarya [3] and Bianka Woloncewicz [1,2]

1. Institute of Theoretical Physics and Astrophysics, Faculty of Mathematics, Physics, and Informatics, University of Gdańsk, 80-308 Gdańsk, Poland; palashpandya.iiith@gmail.com (P.P.); bibivolo@gmail.com (B.W.)
2. International Centre for Theory of Quantum Technologies (ICTQT), University of Gdańsk, 80-308 Gdańsk, Poland
3. Institute of Informatics, Faculty of Mathematics, Physics, and Informatics, University of Gdańsk, 80-308 Gdańsk, Poland; osakarya@sigma.ug.edu.pl

* Correspondence: marcin.wiesniak@ug.edu.pl

Received: 30 November 2019; Accepted: 6 January 2020; Published: 13 January 2020

Abstract: We discuss the use of the Gilbert algorithm to tailor entanglement witnesses for unextendible product basis bound entangled states (UPB BE states). The method relies on the fact that an optimal entanglement witness is given by a plane perpendicular to a line between the reference state, entanglement of which is to be witnessed, and its closest separable state (CSS). The Gilbert algorithm finds an approximation of CSS. In this article, we investigate if this approximation can be good enough to yield a valid entanglement witness. We compare witnesses found with Gilbert algorithm and those given by Bandyopadhyay–Ghosh–Roychowdhury (BGR) construction. This comparison allows us to learn about the amount of entanglement and we find a relationship between it and a feature of the construction of UPBBE states, namely the size of their central tile. We show that in most studied cases, witnesses found with the Gilbert algorithm in this work are more optimal than ones obtained by Bandyopadhyay, Ghosh, and Roychowdhury. This result implies the increased tolerance to experimental imperfections in a realization of the state.

Keywords: bound entanglement; entanglement witness; Hilbert–Schmidt measure; optimization algorithms

---

1. Introduction

Entanglement is likely the most counter-intuitive feature of quantum mechanics. It allows quantum systems to exhibit correlations, which cannot be reconstructed by any set of prearranged local quantum states. As such, entanglement is seen as a resource responsible for advantage in various communication tasks (see, e.g., Ref. [1]). However, while it is trivial to describe entanglement for pure states [2], it is one of the most important open questions of quantum information theory to determine whether a given mixed state is entangled. This problem is additionally complicated by, e.g., existence of bound entanglement [3], which cannot be transformed in the local actions and classical communication (LOCC) regime into an ensemble of pure maximally entangled states. Hence, these states have limited applications in quantum communication.

For bipartite states with distillable entanglement, we again have a straightforward tool to detect their nonclassicality: they become non-positive under partial transposition [4]. Thus, the problem of certifying entanglement for the bipartite case is reduced to the case of bound entangled (BE) states [3]. In theory, any form of entanglement can be certified with a proper positive, but not completely positive (PNCP) map [5]. The Jamiołkowski–Choi isomorphism [6,7] links each such map to a linear operator called an entanglement witness [8]. A reference state, entanglement of which we want to confirm, should give the expectation value of the witness inaccessible with only product states.

Naturally, any entangled state has its own class of witnesses that detect its nonclassical correlations. This follows directly from the fact that it is represented by a point in some distance from a convex set of separable states—there are infinitely many states separating the two, but for an individual state they cannot define the whole boundary of the set. The witness that detects entanglement in the largest volume of the state set is optimal. Since neither the set of all the states nor of the separable ones is a polytope, an optimal entanglement witness is related to a hyperplane tangential to a line connecting the reference state and CSS. Contrary to a common belief, an optimal entanglement witness does not correspond to robustness against any particular form of noise.

The plethora of entanglement witnesses, or equivalently, PNCP maps, is responsible for the complexity of the problem of entanglement detection. Not only do we need to optimize the expectation value of the witness over all separable states, but also the witness itself needs to be a subject of optimization, as we have no a priori knowledge about its form. The *max-max* nature of the problem is the main difficulty in tackling the question, even just numerically.

In Ref. [9], Bandyopadhyay, Ghosh, and Roychowdhury proposed a construction of entanglement witnesses for UPB BE states. They have a strikingly simple structure comprised of the combination of a projection onto the support of the state and the unit operator with a weight equal to the maximal mean value of the UPB BE state in a separable state. Unfortunately, till date these witnesses have not been studied extensively, for example, in terms of optimality. From this point onwards, this family of entanglement witnesses will be referred to as BGR witnesses.

In our recent article [10] we have employed the Gilbert algorithm [11] to find approximation of a separable state closest (with respect to the Hilbert–Schmidt measure, $D(\rho,\sigma) = \sqrt{\mathrm{tr}(\rho-\sigma)^2}$) to a reference state. We argued that basing on decay of the distance (and other properties) we can classify a state as strongly entangled or practically separable. Here, we elaborate on this approach by asking if an implementation of the Gilbert algorithm can yield a separable state close enough to a reference state that it gives a state-tailored entanglement witness. As argued in Ref. [12], the optimal entanglement witness for state ρ_0 will be given by a hyperplane perpendicular to a vector in the state space $\rho_{CSS} - \rho_0$, where ρ_{CSS} is the closest separable state with respect to the Hilbert–Schmidt measure. The algorithm cannot reach ρ_{CSS}, but instead it gives an approximation ρ_1, which introduces a small tilt in the plane of the witness. Also, as subsequent corrections to the found state become exponentially small, it is not feasible to reach arbitrarily high precision. Thus, the question is: can this tilt be small enough, so that the hyperplane still defines a valid entanglement witness for ρ_0.

Computationally, our approach is substantially easier than solving a *max-max* problem. The latter requires that the two optimizations be conducted in turns. It may happen that a step taken in optimizing a witness actually degrades it which can be only verified by optimizing over separable states. On the other hand, the Gilbert algorithm guarantees convergence to the true solution, so it can be conducted independently.

To test this hypothesis, we chose BE states from unextendible product bases [13,14]. On one hand, they have a simple construction, which can be straight-forwardly generalized to a large number of cases. On the other—their bound entanglement is nontrivial to be certified. Their additional advantage is that their CSS can be found mixing only strictly real states, which improves the efficiency. Also, an entanglement witness can be constructed analytically for them [9].

In the next section we review the construction of BE states. In Section 3 we briefly discuss the concept of an entanglement witness. Section 4 describes the Gilbert algorithm. The numerical results are discussed in Section 5, followed by conclusions.

2. Bound Entangled States from Unextendible Product Bases

We continue with a brief introduction of bound entangled states from unextendible product bases (UPB BE states). It is one of most fundamental methods of generating bound entangled states, first presented in Refs. [13,14]. The idea is as follows. Given a two- (or more) party Hilbert space we choose a subspace, which can be represented by a factorizable set of projectors from the computational basis. Subsequently, we remove from this subspace the projector on the equal superposition of all computational basis states belonging to that region. The remaining is defined as a tile. For example, in Hilbert space $H = \text{span}(|1\rangle, |2\rangle, |3\rangle)^{\otimes 2}$, a possible tile would be $\sum_{i,j=1,2} |i,j\rangle\langle i,j| - \frac{1}{2}\sum_{i,j,k,l=1}^{2} |i,j\rangle\langle k,l|$. Figure 1 depicts such a construction in d-dimensions. The next step is to find a covering of the entire Hilbert space such that there is no subset of regions related to tiles, which can be merged into a region, which is, again, a factorizable region. Once such a covering is found, we sum subspaces of all the tiles, and add a projector on the uniform superposition of all states from the computational basis, $\Pi_{sym} = \frac{1}{d_1 d_2}\sum_{i,k=1}^{d_1}\sum_{j.,l=1}^{d_2} |i,j\rangle\langle k,l|$ (sometimes called a stopper state). As we have previously removed the symmetric states from each tile, they are orthogonal to this state. The bound entangled state is then the normalized projector onto the complement of the sum. The state is PPT, which can be argued from its construction, but it is entangled, since its kernel does not contain any product states. There is no product state orthogonal to the sum of all tiles and the symmetric state anymore.

Let us denote a projection on the subspace related to a k-th tile as $\Pi^{(k)}$, and onto the symmetrized state of that subspace as $\Pi^{(k)}_{sym}$. Then, any UPB BE state is given by:

$$\rho = \frac{1}{K}(\mathbb{1} - \Pi_{sym} - \sum_{k=1}^{K}(\Pi^{(k)} - \Pi^{(k)}_{sym})). \quad (1)$$

The rank of the state is hence $K-1$, where K is the number of tiles necessary to cover the whole Hilbert space. For two parties, the minimum number of tiles is 5, while for three parties, 9 tiles are necessary.

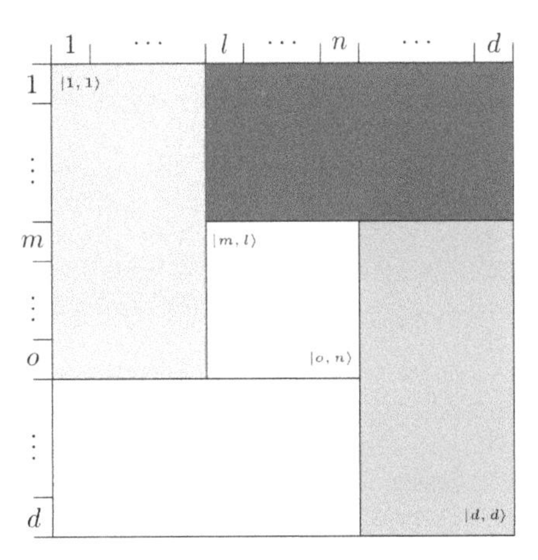

Figure 1. Visualization of the structure of bound entangled states from unextendible product bases considered in this article. The size of the central tile is $(n-l+1)(m-o+1)$.

Possible low rank of the UPB bound entangled states makes them attractive from the experimental point of view. Granted access to a source of an arbitrary pure state, a resource readily available with modern high-intensity sources of multidimensional entanglement, local filters and operations, one can relatively easily realize mixtures. One of the greatest challenges in such realizations is high sensitivity of the bound entanglement. For example, any deviation of weights of the eigenstates from the support of the state will transform it into an NPT (negative partial transpose) state. This aspect will play a role when the construction of the source cannot guarantee equal mixing. A possible solution is to add white noise, but PPT entanglement tends to be rather fragile.

Another approach is to *emulate* the state. Instead of looking at its eigenvectors, our aim is to reconstruct the probability distributions that it produces. This can be done by taking a proper combination of product states, possibly with negative coefficients. In case of UPB BE states we can sum up probabilities given by all the states $\Pi_{sym}^{(k)}$ and measure relevant probabilities Π_{sym} to be deducted. Naturally, to establish probabilities we need to measure the projections of complementary states. This can be done after the measurements, or in the following procedure. The observers are informed every time Π_{sym} is produced and a projection is successful, and they increase the denominator in the relative frequency, but decrease the numerator. In some cases, this might lead to "negative probabilities", which then should be taken to be zero, or the best fit to a physical density operator.

Naturally, such an emulated state cannot bring any advantage in communication tasks or distributed computation. In particular, the emulation cost is broadcasting a bit. However, emulation can be extremely useful for testing and perfecting techniques for certifying entanglement. As it is complicated to build a source of an entangled state, it is experimentally convenient to first master the accompanying techniques to be sure that the desired goal can be achieved.

3. Entanglement Witnesses

Let us start with a brief recall of the concept of an entanglement witness. Most generally, it would be a Hermitian operator with a certain range of eigenvalues accessible only in entangled states. In that sense, a Hamiltonian of an antiferromagnetic spin chain of any length is an entanglement witness [15]. By a common convention, the lowest mean value attainable of a witness with separable states is 0, and some entangled states may lead to negative mean values. We note, however, that this convention may not always be practical. For any entanglement witness W, we can define a linear functional $\Lambda_W(\cdot) = \mathrm{tr}(\cdot W)$, which divides the space of Hermitian operators with a hyperplane. On one side of this hyperplane ($\Lambda_W(\cdot) \geq 0$) lie all separable states, on the other ($\Lambda_W(\cdot) < 0$)—our reference state ρ_0 and its neighborhood. Since the set of separable states in a given Hilbert space is convex, there always exist such a hyperplane for any entangled state.

Then we can introduce a class of tight witnesses for which condition $\langle W \rangle_{\mathrm{SEP}} = 0$, i.e., the mean value of the witness in the set of separable states is equal to 0, is achieved for one or more separable states. Within this class, for any entangled ρ_0 we can define optimal witness W_0, for which $\langle W_0 \rangle_{\mathrm{SEP}} = 0$ is attained for the closest separable state to ρ_0 in the sense of Hilbert–Schmidt norm. As a result, the distance between ρ_0 and the hyperplane $\Lambda_{W_0}(\cdot) = 0$ is the largest among all witnesses detecting entanglement of ρ_0. However, contrary to common belief, this does not correspond to the highest robustness against the noise of any universal form. Our goal is to confirm the form of such witnesses for UPB BE states.

A 5-tile bipartite UPB BE states with a fixed heliocity can be characterized by dimensions d_1, d_2 and coordinates of the central tile $(l, n), (m, o)$ with $1 < l \leq n < d_1$ and $1 < m \leq o < d_2$. We have studied cases of $d_1 = d_2 \equiv d = 3, 4, 5, 6$. This gives a single 3×3 state, nine 4×4 states, 36 5×5 states, and 100 6×6 states, giving 146 density matrices in total. In general, in dimensions $d_1 \times d_2$ we will have $\frac{1}{4}(d_1^2 - 3d_1 + 1)(d_2^2 - 3d_2 + 1)$ states.

4. The Gilbert Algorithm

According to Bertlmann–Durstberg–Heismayr–Krammer theorem [12], the optimal entanglement witness for state ρ is given by the difference of ρ and the closest separable state with respect to Hilbert–Schmidt measure, $W \propto \rho - \rho_{CSS}$. Finding ρ_{CSS} is generally a complex task, but a very good approximation can be found with the Gilbert algorithm [11]. The application of the algorithm to the classification of states as entangled was discussed in Ref. [10]. In short, the algorithm is as follows:

Input: test state ρ_0, arbitrary separable state ρ_1,

Output: approximation of a separable state closest to ρ_0

1. Choose at random a pure product state ρ_2.
2. Maximize $\mathrm{tr}(\rho_0 - \rho_1)(\rho_2 - \rho_1)$, or go to step 1 if $\mathrm{tr}(\rho_0 - \rho_1)(\rho_2 - \rho_1) \leq 0$.
3. Find the point ρ_1', which lies on line $\rho_1 d - \rho_2$, and minimizes the Hilbert–Schmidt distance $D(\sigma, \rho) = \sqrt{\mathrm{tr}(\rho - \sigma)^2}$.
4. Update $\rho_1 \leftarrow \rho_1'$.
5. Go to point 1 until a given HALT condition is met.

Let us note that local states constituting ρ_2 are drawn with the Haar measure, according to an algorithm by Życzkowski and Sommers [16]. The Hilbert–Schmidt measure has been chosen as the algorithm then requires only solving a quadratic equation, although it was shown that it is not monotonous under LOCC and thus is not a true entanglement measure [17,18]. A typical choice of the HALT condition is simply a time constrain, number of trial states (ρ_2), or number of corrections.

The Gilbert algorithm can provide an accurate approximation of the closest separable state, but, in fact, we cannot reach that state. For this reason, we only partially succeeded in finding almost optimal entanglement witness. The procedure is the following. We run the algorithm, which halts at some number of corrections of ρ_1. We then optimize the mean value of $\rho_0 - \rho_1$ for separable states. When this mean value is lower than $\mathrm{tr}(\rho_0 - \rho_1)\rho_0$, the witness is found. For We have found a witness for the 3×3 state after 25,100 corrections (time constrain), for 4×4 and 5×5 states we run the algorithm for up to 4000 corrections, while for 6×6 states we have conducted only 3500 corrections (number of corrections).

The first criterion of a state being entangled or separable is to investigate the limit of distance between ρ and the closest separable state found at the given step. This distance was registered every 50 corrections. As argued in Ref. [10] there is a strong linear dependence between number of corrections c and $|\log(\mathrm{tr}(\rho_1 - \rho_0)^2 - a)|^b$ for some values of a and b. We hence maximize the linear regression coefficient between these two quantities. The found value of b is irrelevant, while a is an approximation of the limit. The state can be considered entangled if a is above the precision of the algorithm, estimated to be less than 10^{-5}.

5. Numerical Results

All 146 UPB BE states have been shown to have the estimated distance to the set of separable sates between 0.09 and 0.06, therefore all of them can be classified as entangled as they indeed are from the construction. On the other hand, it is difficult to argue that any one of them is particularly strongly nonclassical.

In the next step, we will attempt to construct an entanglement witness for each state. In a perfect case of the algorithm actually reaching the closest separable states ρ_{CSS}, the witness would be simply proportional to $\rho_{CSS} - \rho_0 - \mathrm{tr}\rho_{CSS}(\rho_{CSS} - \rho_0)$. However, as mentioned above, we reach only an approximation ρ_1 of ρ_{CSS}, which causes the hyperplane of the witness to be tilted. We thus need to conduct an optimization over the set of separable states, $|A\rangle \otimes |B\rangle$, so that $\lambda = \max_{|A\rangle,|B\rangle} \langle A|\langle B|\rho_1 - \rho_0|A\rangle|B\rangle$. Then the witness reads:

$$W(\rho_0) = \rho_1 - \rho_0 - \mathbb{1}\lambda. \tag{2}$$

It turns out that in almost all cases ρ_1 yielded a valid entanglement witness. When transformed to a traceless form, $\rho_1 - \rho_0$, each of these witnesses has a similar structure: there are four negative eigenvalues of roughly the same magnitude and the corresponding eigenspace is approximately the support of ρ_0, and hence $W(\rho_0)$s can be compared with witnesses in Ref. [9]. A significant difference is that positive eigenvalues are not uniform.

It now remains to compare the strength of the witnesses. Because the witnesses found with the Gilbert algorithm are approximations of the witnesses $W'(\rho_0)$, constructed accordingly to Ref. [9], we would focus on the latter. The simplest quantizer of nonclassicality of UPB BE states would be the Hilbert–Schmidt distance of the state to the hyperplane defining the witness. To establish this quantity, we need vector $\vec{M}$ tangent to the hyperplane defining the witness, which in our case is the traceless part of the witness operator, $\vec{M} = W'(\rho) - \mathbb{1}\text{tr}(W'(\rho_0))$. The trace part is removed from the witness operator to ensure that the corresponding vector starts at the origin of the coordinates system. We will also need state ρ' saturating $\text{tr}\rho' W'(\rho_0) = 0$. This state is found in the process of constructing the witness. Let $\vec{R} = \rho_0 - \rho'$. Then, our distance to separable states is the length of the projection of $\vec{R}$ onto unit vector $\vec{M}/\sqrt{\vec{M}\cdot\vec{M}}$ is given by:

$$D(\rho_0, \text{Sep}) = \frac{|\vec{R}\vec{M}|}{\sqrt{|\vec{M}|^2}} = \frac{|\text{tr}W'(\rho_0)(\rho_0 - \rho')|}{\sqrt{\text{tr}W'^2(\rho_0)}}, \tag{3}$$

where Sep denotes the set of separable states.

Figure 2 presents comparison of four estimates of the distance between five-tile UPB BE states and separable states. Blue points correspond to the smallest distance found by the Gilbert algorithm after the mentioned number of corrections, green points show the estimate of the distance from the linear regression, red represent the distance computed from Equation (3) for the BGR witnesses, while black points illustrate the distance given by the witnesses found by our algorithm (again, computed from Equation (3)). The first quantity is an upper bound of the actual distance, the second is just an estimate, while the distances computed from the witnesses are lower bound.

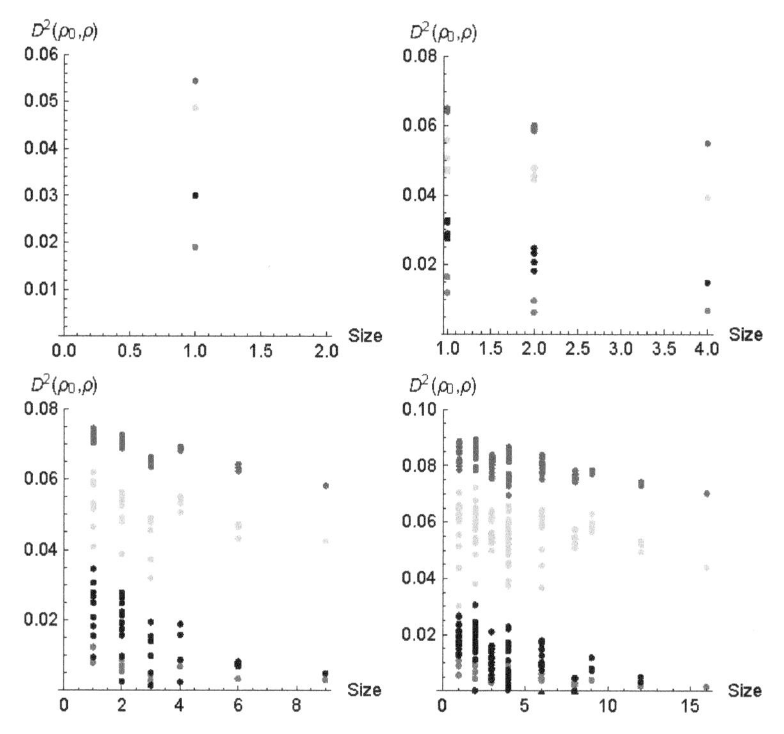

Figure 2. The comparison between last distance between a UPB BE state found by the Gilbert algorithm after 25,100 corrections (3×3, top left), 4000 corrections (4×4, top right, and 5×5, bottom left), and 3500 corrections (6×6, bottom right). Red points correspond to distances from Equation (3) with BGR witnesses, the black points show the distance to the hyperplanes of the witnesses found by the Gilbert algorithm. The last distance found by the algorithm is marked with blue points and the extrapolation of its decay is represented with green points. The data are segregated by the size of the central tile.

It should be noted that in all four plots, the red points are significantly below all other groups. This clearly indicates that we have found more optimal witnesses, than those given in Ref. [9]. Another observation, for every state, the green point is above the black one, i.e., the decay estimation is between the upper and the lower bound. However, our algorithm did not yield a valid witness for 13 states in 6×6 Hilbert space. Also, for 6 states in 6×6 and 4 states in 5×5 Hilbert spaces, the witness yielded by our algorithm was weaker than the BGR witnesses. Still, our algorithm is successful for most tested states. Furthermore, our results show that BGR construction does not lead to an optimal witness for UPB BE states, as we find more optimal ones.

6. Conclusions

In this contribution, we have studied quantifiable entanglement of bound entangled states derived from unextendible product bases in dimension $d \times d$, where $d = 3, 4, 5, 6$. As a prime tool, we have used the Gilbert algorithm to find an approximation of a separable state closest to the given entangled state. Knowledge of this approximation leads to a close-to-optimal entanglement witness. Our method succeeded to yield entanglement witnesses in 133 of the 146 studied cases. Moreover, witnesses found by our algorithm were in 123 cases more optimal than those proposed by Bandyopadhyay, Ghosh, and Roychowdhury [9]. Both the decay of the distance with successive corrections, and its extrapolation by a linear fit indicates a presence of entanglement. Also, these results show that construction of BGR witnesses is not optimal. There exist convex combinations of ρ_0 and ρ_{CSS} or ρ_1, which are not recognized as entangled by BGR witnesses, whereas our algorithm reveals quantum correlations in such cases.

We have calculated the distance between PPT states and the sets of separable states. UPB BE states turned out to be relatively close to the set of separable states, with all distances below 0.1. While no clear relationship between the structure of the state and the distance was established, one correlation appears, i.e., states with smaller central tiles lay closer to separable states.

Our results therefore have multifold scientific aspects. First, we have partially found the order of degree of entanglement of UPB BE states, by recognizing that in the given dimension, those farthest from the set of separable states have a smaller central tile. Second, the witnesses yielded by the algorithm are (in most cases) more optimal than the BGR construction. As a consequence, our witnesses allow for larger imperfections in an experimental realization. This could be relevant in quantum randomness amplification and cryptographic key distribution protocols.

Author Contributions: Conceptualization and Data Analysis: M.W.; Software and Data Acquisition: M.W., P.P. and O.S.; Writing—Original Draft Preparation M.W.; Writing—Review and Editing: B.W. All authors have read and agreed to the published version of the manuscript.

Funding: This research was funded by NCN Grants 2017/26/E/ST2/01008, 2015/19/B/ST2/01999 (MW), 2014/14/E/ST2/00020 (PP), and the ICTQT IRAP project of FNP, financed by structural funds of EU.

Acknowledgments: We thank Remigiusz Augusiak for useful insight on this publication.

Conflicts of Interest: The authors declare no conflict of interest.

References

1. Brukner, Č.; ukowski, M.Z.; Pan, J.-W.; Zeilinger, A. Bell's inequalities and quantum communication complexity. *Phys. Rev. Lett.* **2004**, *92*, 127901. [CrossRef] [PubMed]
2. Bennett, C.H.; Popescu, S.; Rohrlich, D.; Smolin, J.A.; Thapiyal, A.V. Exact and asymptotic measures of multipartite pure-state entanglement. *Phys. Rev. A* **2000**, *63*, 012307. [CrossRef]
3. Horodecki, M.; Horodecki, P.; Horodecki, R. Mixed-State Entanglement and Distillation: Is there a "Bound" Entanglement in Nature? *Phys. Rev. Lett.* **1998**, *80*, 5239. [CrossRef]
4. Horodecki, P. Separability criterion and inseparable mixed states with positive partial transposition. *Phys. Lett. A* **1997**, *232*, 333–339. [CrossRef]
5. Horodecki, M.; Horodecki, P.; Horodecki, R. Separability of n-particle mixed states: necessary and sufficient conditions in terms of linear maps. *Phys. Let. A* **2001**, *283*, 1–7. [CrossRef]

6. Jamiołkowski, A. Linear transformations which preserve trace and positive semidefiniteness of operators. *Rep. Math. Phys.* **1972**, *3*, 275–278. [CrossRef]
7. Choi, M.D. Completely positive linear maps on complex matrices. *Linear Algebra Appl.* **1975**, *10*, 285. [CrossRef]
8. Terhal, B.M. Bell inequalities and the separability criterion. *Phys. Lett. A* **2000**, *271*, 319. [CrossRef]
9. Bandyopadhyay, S.; Ghosh, S.; Roychowdhury, V. Non-full-rank bound entangled states satisfying the range criterion. *Phys. Rev. A* **2005**, *71*, 012316. [CrossRef]
10. Pandya, P.; Sakarya, O.; Wieśniak, M. Hilbert-Schmidt distance and entanglement witnessing. *arxiv* **2018**, arxiv:1811.06599.
11. Gilbert, E.G. An iterative procedure for computing the minimum of a quadratic form on a convex set. *SIAM J. Control.* **1966**, *4*, 61–80. [CrossRef]
12. Bertlmann, R.; Durstberg, K.; Heismayr, B.C.; Krammer, P. Optimal entanglement witnesses for qubits and qutrits. *Phys. Rev. A* **2005**, *72*, 052331. [CrossRef]
13. DiVincenzo, D.P.; Mor, T.; Shor, P.W.; Smolin, J.A.; Terhal, B.M. Unextendible Product Bases and Bound Entanglement. *Phys. Rev. Lett.* **1999**, *82*, 5385.
14. DiVincenzo, D.P.; Mor, T.; Shor, P.W.; Smolin, J.A.; Terhal, B.M. Unextendible product bases, uncompletable product bases and bound entanglement. *Commun. Math. Phys.* **2003**, *238*, 379. [CrossRef]
15. Brukner, Č.; Vedral, V. Macroscopic Thermodynamical Witnesses of Quantum Entanglement, preprint arxiv:quant.ph/0406040; Tóth, G. Entanglement witnesses in spin models. *Phys. Rev. A* **2005**, *71*, 010301.
16. Życzkowski, K.; Sommers, H.-J. Induced measures in the space of mixed quantum states. *J. Phys. A Math. Theor.* **2001**, *34*, 7111. [CrossRef]
17. Witte, C.; Trucks, M. A new entanglement measure induced by the Hilbert–Schmidt norm. *Phys. Lett. A* **1999**, *257*, 14. [CrossRef]
18. Ozawa, M. Entanglement measures and the Hilbert–Schmidt distance. *Phys. Lett. A* **2000**, *286*, 158. [CrossRef]

Article

Nonclassical States for Non-Hermitian Hamiltonians with the Oscillator Spectrum

Kevin Zelaya [1,2], Sanjib Dey [3], Veronique Hussin [1,4,*] and Oscar Rosas-Ortiz [2]

1 Centre de Recherches Mathématiques, Université de Montréal, Montréal, QC H3C 3J7, Canada; zelayame@crm.umontreal.ca
2 Physics Department, Cinvestav, AP 14-740, México City 07000, Mexico; orosas@fis.cinvestav.mx
3 Indian Institute of Science Education and Research Mohali, Knowledge City, Sector 81, SAS Nagar (Mohali), PO Manauli, Punjab 140306, India; dey@iisermohali.ac.in
4 Département de Mathématiques et de Statistique, Université de Montréal, Montréal, QC H3C 3J7, Canada
* Correspondence: hussin@dms.umontreal.ca

Received: 29 November 2019; Accepted: 24 December 2019; Published: 27 December 2019

Abstract: In this paper, we show that the standard techniques that are utilized to study the classical-like properties of the pure states for Hermitian systems can be adjusted to investigate the classicality of pure states for non-Hermitian systems. The method is applied to the states of complex-valued potentials that are generated by Darboux transformations and can model both non-PT-symmetric and PT-symmetric oscillators exhibiting real spectra.

Keywords: non-hermitian operators; real spectrum; nonlinear algebras; coherent states; nonclassical states

1. Introduction

In quantum mechanics one finds two important classes of states of the radiation field: Fock and coherent states. The former, introduced by Fock at the dawn of quantum theory, contain a precise number of photons and produce average fields equal to zero. With the exception of the vacuum, the Wigner distribution [1] exhibits negative values in some regions of the phase-space when it is evaluated with the Fock states. Thus, the properties of the quantum states of radiation fields that are occupied by a finite number of photons are far from the Maxwell theory. On the other hand, the coherent states were formally introduced in quantum optics by Glauber [2], although the first antecedents can be traced back to the Schrödinger papers on quantization, see, e.g., [3]. They produce average fields different from zero as well as nonnegative symmetrical Wigner distributions. While the coherent states are constructed as superpositions of Fock states, it is remarkable that their properties are very close to the Maxwell theory. This is because such states satisfy the notion of full coherence introduced by Glauber, while the Fock states (for $n \neq 0$) lack second (and higher) order coherence, and thus they are nonclassical. The set of nonclassical states includes also squeezed states [4], even and odd coherent states (also called Schrödinger cats) [5], binomial states [6,7], photon-added coherent states [8], etc. The difference between the classical and nonclassical properties of a given quantum state is strongly linked to the notion of entanglement [9], which is a fundamental concept required in the development of quantum computation and quantum information [10]. Noticeably, while investigating the quantum properties of a beam splitter, it has been found that the "entanglement of the output states is strongly related to the nonclassicality of the input fields" [11]. That is, the nonclassicality of an arbitrary (input) state can be tested by detecting entanglement in the output of a quantum beam splitter [11,12].

In the present work we study the nonclassical properties of the states of non-Hermitian Hamiltonians with real spectrum that are generated by Darboux transformations [13,14]. In general, the systems associated with non-Hermitian Hamiltonians are subject of investigation in many branches of contemporary physics [15]. The applications include the study of unstable (decaying) systems [16], light propagating in materials with a complex refractive index [17], PT-symmetric [18,19] and PTC-symmetric [20] interactions, multi-photon transition processes [21], diverse measurement techniques [22], and coherent states [3,23,24], among others. In the case of the Darboux deformed non-Hermitian systems, the eigenfunctions obey a series of oscillation theorems [25] that permit their study as if they were associated to Hermitian systems. Indeed, the representation space of such non-Hermitian systems can be equipped with a bi-orthogonal structure that provides complete sets of orthonormal states [13]. We shall focus on non-Hermitian oscillators since their bases of eigenfunctions can be used to construct optimized binomial states [26] and generalized coherent states [24], which in turn give rise to a wide diversity of pure states as particular cases. One of the main results reported in this paper is to show that the techniques used to study classical-like properties for Hermitian systems can be adjusted to investigate the classicality of the states of Darboux deformed non-Hermitian systems.

The outline of the paper is as follows: In Section 2 we revisit the main properties of the non-Hermitian oscillators we are interested in. The main differences between the conventional orthogonality and the bi-orthogonality are discussed in detail. In Section 3 we introduce the different bi-orthogonal superpositions of states that are to be analyzed, these include the optimized binomial states and diverse forms of generalized coherent states. Section 4 deals with the nonclassicality of the states defined in Section 3. Final comments are given in Section 5. We have added three appendices with relevant information that is used throughout the manuscript concerning operator algebras (Appendix A), and criteria of nonclassicality (Appendix B).

2. Non-Hermitian Oscillators

The solution to the eigenvalue problem of the dimensionless (mathematical) oscillator $H_{osc} = -\frac{d^2}{dx^2} + x^2$ is given by the discrete eigenvalues $E_n^{osc} = 2n+1$ and normalized eigenfunctions

$$\varphi_n(x) = \frac{e^{-x^2/2}}{\sqrt{2^n n! \sqrt{\pi}}} H_n(x), \quad n = 0, 1, \ldots, \tag{1}$$

where $H_n(x)$ stands for the Hermite polynomials [27]. Using the linearly independent solutions of the eigenvalue equation $H_{osc} u = -u$ and a set of real numbers $\{a, b, c, \lambda\}$ fulfilling $4ac - b^2 = 4\lambda^2$, it can be shown that

$$\alpha(x) = e^{x^2/2} \left[\frac{a\pi}{4} \text{Erf}^2(x) + \frac{b\sqrt{\pi}}{2} \text{Erf}(x) + c \right]^{1/2} \tag{2}$$

is a real-valued function which is free of zeros in $\mathbb{R}$, and defines the dimensionless potential

$$V_\lambda(x) = x^2 - 2 - 4\frac{d}{dx} \left[\frac{a\sqrt{\pi}\,\text{Erf}(x) + b - i\frac{\lambda}{2}}{\alpha^2(x)} \right] \tag{3}$$

as a Darboux transformation of $V_{osc}(x) = x^2$ [13]. Here $\text{Erf}(x)$ defines the error function. If $\lambda \neq 0$, the Hamiltonian $H_\lambda = -\frac{d^2}{dx^2} + V_\lambda(x)$ is not self-adjoint since $V_{\lambda \neq 0}(x)$ is a complex-valued function. Nevertheless, it may be shown that the imaginary part of such potential is continuous in $\mathbb{R}$ and satisfies the condition of zero total area [25]:

$$\int_{\mathbb{R}} \text{Im} V_\lambda(x) dx = -\left. \frac{2\lambda}{\alpha^2(x)} \right|_{-\infty}^{+\infty} = 0. \tag{4}$$

Equation (4) implies a balanced interplay between gain and loss of probability that does not depend on any other symmetry of either $\text{Im}V_\lambda(x)$ or $\text{Re}V_\lambda(x)$. For instance, besides fulfilling Equation (4), the profile of the real and imaginary parts of the oscillator depicted in Figure 1a is not particularly special (this has been obtained from Equation (3) with $a=5$, $b=8$, $c=5$ and $\lambda=3$). Looking for concrete symmetries we can take $b=0$ to get even and odd forms of $\text{Im}V_\lambda(x)$ and $\text{Re}V_\lambda(x)$, respectively. The result is a non-Hermitian oscillator which is invariant under PT-transformations [18]. This case is illustrated in Figure 1b for $a = c = \lambda = 1$, and $b = 0$. On the other hand, for $b = \pm 2\sqrt{ac}$, with a and c positive numbers, the parameter λ is equal to zero and Equation (3) decouples into the pair of real-valued functions

$$V^{\pm}_{\lambda=0}(x;\gamma) = x^2 - 2 - 2\frac{d}{dx}\left(\frac{e^{-x^2}}{\int^x e^{-y^2}dy \pm \gamma}\right), \quad \gamma = \sqrt{\frac{c}{a}}, \tag{5}$$

where the condition $\gamma > \sqrt{\pi}/2$ must be satisfied to avoid singularities. Figure 1c shows $V^{-}_{\lambda=0}(x;\gamma)$ with $a = 1$ and $c = 0.88$. The concomitant $V^{+}_{\lambda=0}(x;\gamma)$ corresponds to the specular reflection of $V^{-}_{\lambda=0}(x;\gamma)$. The Hermitian oscillators of Equation (5) were first found by Abraham and Moses through the systematic use of the Gelfand-Levitan equation [28], and then recovered by Mielnik as an application of his generalized factorization method [29].

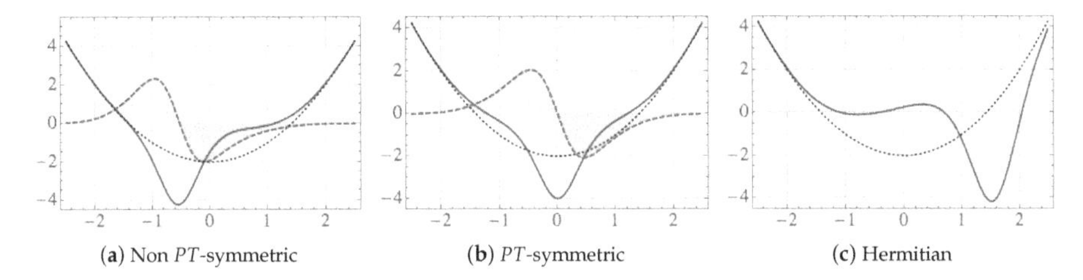

(**a**) Non PT-symmetric (**b**) PT-symmetric (**c**) Hermitian

Figure 1. (Color online) the oscillators $V_\lambda(x)$ defined in Equation (3) can be categorized in three different classes: (**a**) Non PT-symmetric, (**b**) PT-symmetric, and (**c**) Hermitian. In (**a**) and (**b**) the imaginary part (dashed-red) satisfies the condition of zero total area, Equation (4). The zones of gain and loss of probability are shadowed in orange and red, respectively. In all the cases the curve in blue corresponds to the real part of $V_\lambda(x)$ and the dotted-black curve represents the shifted oscillator $V_{osc}(x) - 2$, the latter is obtained from Equation (3) as a particular case.

Additionally, Equations (5) represent a family of systems that converge to the conventional oscillator in the limit $|\gamma| \to \infty$. Indeed, it may be verified that

$$\lim_{|\gamma|\to\infty} V_{\lambda=0}(x;\gamma) \to x^2 - 2. \tag{6}$$

The oscillator $V_{osc}(x) - 2$ is thus a member of the family of complex-valued potentials $V_\lambda(x)$ introduced in Equation (3). This is depicted in the plots of Figure 1 as a reference.

- Fundamental solutions. Another remarkable profile of the complex-valued oscillators of Equation (3) is that the functions

$$\psi_{n+1}(x) = \left[\frac{d}{dx} - \frac{\alpha'(x)}{\alpha(x)} + i\frac{\lambda}{\alpha^2(x)}\right]\varphi_n(x), \quad n = 0, 1, \ldots, \tag{7}$$

are normalizable solutions of the related eigenvalue equation $H_\lambda \psi_{n+1}(x) = E_{n+1}\psi_{n+1}(x)$ for $E_{n+1} = 2n + 1$. The additional normalizable function

$$\psi_0(x) = \frac{1}{\alpha(x)} \exp\left[i \arctan\left(\frac{a\sqrt{\pi}\,\mathrm{Erf}(x) + b}{2\lambda}\right)\right] \tag{8}$$

also solves the eigenvalue equation defined by H_λ, but it is not derivable from the set $\varphi_n(x)$ and belongs to the energy $E_0 = -1$. The spectrum of the potential $V_\lambda(x)$ is therefore composed of the equidistant energies $E_n = 2n - 1$, $n \geq 0$.

- Bi-orthogonality. While the functions $\psi_n(x)$ are normalizable, they form a peculiar set since $\psi_0(x)$ is orthogonal to all the $\psi_{n+1}(x)$ while the latter are not mutually orthogonal. As a result, in contrast with the Hermitian case, the norm of any superposition of states $\psi_{n+1}(x)$ depends not only on the modulus of the related coefficients but also on the phase shift between them. For instance, the norm of

$$\Psi(x) = \frac{\theta}{||\psi_{n+1}||}\psi_{n+1}(x) + \frac{\mu}{||\psi_{m+1}||}\psi_{m+1}(x), \quad \theta, \mu \in \mathbb{C}, \tag{9}$$

satisfies

$$||\Psi||^2 = |\theta|^2 + |\mu|^2 + 2|\theta\mu|\,\zeta_{n+1,m+1}\cos(\delta + \gamma_{n+1,m+1}), \quad \delta = \arg(\theta^*\mu), \tag{10}$$

where z^* is the complex-conjugate of $z \in \mathbb{C}$. The numbers $\zeta_{n+1,m+1}$ and $\gamma_{n+1,m+1}$ are, respectively, the modulus and argument of the product between $\psi_{n+1}(x)$ and $\psi_{m+1}(x)$. Thus, depending on the phase-shift δ, the non-orthogonality of the set $\psi_{n+1}(x)$ produces the oscillations of $||\Psi||$. The complexity of such a dependence increases with the number of elements in the superposition.

One can face the above difficulties by introducing a bi-orthogonal system (see relevant information in Reference [24]), formed by the eigenfunctions $\psi_n(x)$ of H_λ and those of its Hermitian-conjugate $H_\lambda^\dagger \equiv \overline{H}_\lambda$, written as $\overline{\psi}_m(x)$. The main point is that the bi-product $(\overline{\psi}_m, \psi_n) = \int_{\mathbb{R}} \overline{\psi}_m^*(x)\psi_n(x)dx$ is equal to zero if $n \neq m$, and serves to define the bi-norm $||\psi_n||_B = ||\overline{\psi}_n||_B$ if $n = m$.

Therefore, besides the conventional normalization $\psi_n(x)/||\psi_n||$, we have at hand the bi-normalization $\psi_n(x)/||\psi_n||_B$, with $||\psi_{n+1}||_B = \sqrt{2(n+1)}$ for $n \geq 0$. The bi-norm $||\psi_0||_B$ of the ground state depends on the set $\{a, b, c, \lambda\}$ [24]. The real and imaginary parts of $\psi_n(x)$, as well as the probability density $|\psi_n(x)|^2$, behave qualitatively equal in both normalizations but their bi-normalized values are usually larger than those obtained with the conventional normalization. Such a difference is reduced as n increases. This property is illustrated in Figure 2 for the first three eigenfunctions $\psi_n(x)$, and the corresponding probability densities $|\psi_n(x)|^2$, of the complex-valued oscillator depicted in Figure 1a.

The bi-orthogonal approach avoids the interference produced by the non-orthogonality. For instance, if the states in Equation (9) are substituted by their bi-normalized versions, we obtain the bi-orthogonal superposition

$$\Psi_B(x) = \frac{\theta}{\sqrt{2(n+1)}}\psi_{n+1}(x) + \frac{\mu}{\sqrt{2(m+1)}}\psi_{m+1}(x). \tag{11}$$

The bi-norm of the latter state does not depend on the phase-shift

$$||\Psi||_B^2 = \int_{\mathbb{R}} \overline{\Psi}_B^*(x)\Psi_B(x)dx = |\theta|^2 + |\mu|^2, \tag{12}$$

so that $\Psi_B(x)$ is uniquely bi-normalized to 1 by the constant $(|\theta|^2 + |\mu|^2)^{-1}$. An additional property of the bi-orthogonal superpositions is that the values of their real and imaginary parts, as well as the values of the corresponding probability densities, are shorter than those obtained from the conventional approach. We show this property in Figure 3 for the superpositions of Equations (9) and (11) with $n = 0$, $m = 1$, $|\theta| = |\mu| = 1/\sqrt{2}$, and three different values of the phase-shift δ. Formally, one may say that $\Psi(x)$ and $\Psi_B(x)$ represent two different superpositions of the states $\psi_{n+1}(x)$ and $\psi_{m+1}(x)$. In the sequel we shall take full advantage of the mathematical simplifications offered by the bi-orthogonal approach.

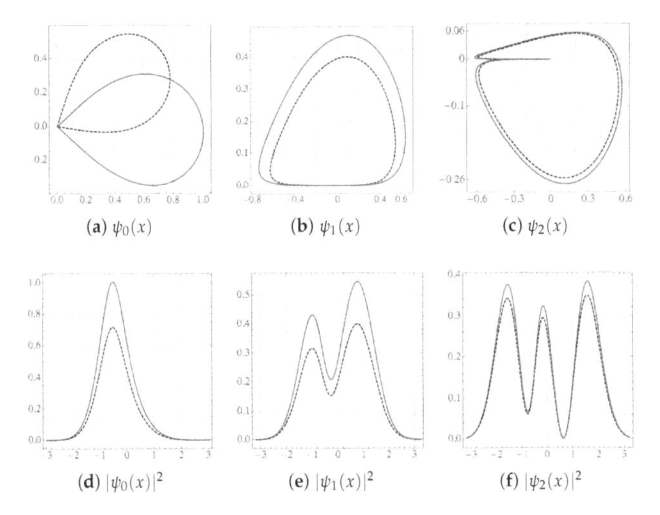

Figure 2. (Color online) bi-normalization (continuous-purple) against conventional normalization (dotted-black) for the first three eigenfunctions $\psi_n(x)$—upper row—and probability densities $|\psi_n(x)|^2$—lower row—of the complex-valued oscillator of Equation (3) shown in Figure 1a. The eigenfunctions are depicted in the Argand–Wessel representation with Re $\psi_n(x)$ and Im $\psi_n(x)$ in the horizontal and vertical axes, respectively. They have been constructed from Equations (7) and (8) with $a = 5$, $b = 8$, $c = 5$ and $\lambda = 3$.

- Operator algebras. It can be shown that there exist at least two different algebras of operators associated with the eigenstates of the complex-valued oscillator $V_\lambda(x)$ [30]. They are generated by two different pairs of ladder operators (see Appendix A for details). The first pair, $\mathcal{A}$ and $\mathcal{A}^+$, together with the Hamiltonian H_λ, satisfy the quadratic polynomial (Heisenberg) algebra

$$[\mathcal{A}, \mathcal{A}^+] = 2\,(3H_\lambda + 1)\,(H_\lambda + 1)\,, \quad [H_\lambda, \mathcal{A}] = -2\mathcal{A}, \quad [H_\lambda, \mathcal{A}^+] = 2\mathcal{A}^+. \tag{13}$$

The second pair of ladder operators, denoted by $\mathcal{C}_w$ and $\mathcal{C}_w^+$, together with the Hamiltonian H_λ, and an additional operator I_w, satisfy the distorted (Heisenberg) algebra

$$[\mathcal{C}_w, \mathcal{C}_w^+] = I_w, \quad [H_\lambda, \mathcal{C}_w] = -2\mathcal{C}_w, \quad [H_\lambda, \mathcal{C}_w^+] = 2\mathcal{C}_w^+, \tag{14}$$

where w is a non-negative parameter that defines the 'distortion' suffered by the oscillator algebra when one substitutes the operators $\mathcal{C}_w$, $\mathcal{C}_w^+$, and H_λ for the conventional boson operators, $\hat{a}$, $\hat{a}^\dagger$, and $\hat{n}$, respectively. The algebras of Equations (13) and (14) will serve to the analysis of classicality of the states associated with the complex-valued oscillators of Equation (3).

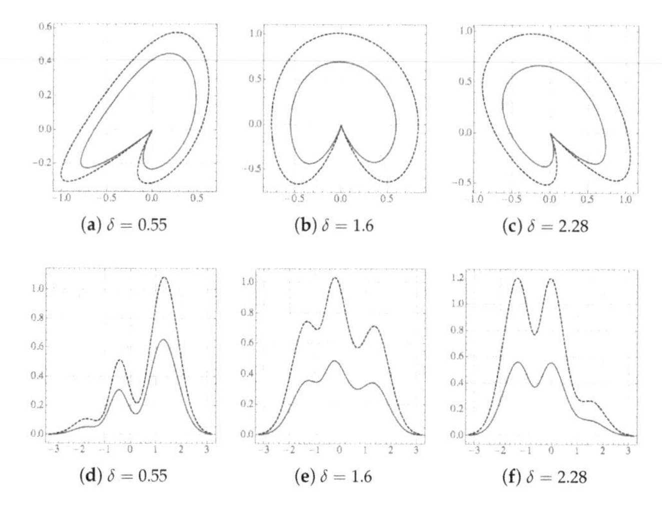

Figure 3. (Color online) the upper row shows the Argand–Wessel diagrams of the superpositions $\Psi(x)$ and $\Psi_B(x)$, Equations (9) and (11), respectively, for $n = 0$, $m = 1$, $|\theta| = |\mu| = 1/\sqrt{2}$, and the indicated values of the phase-shift δ. The corresponding probability densities are shown in the lower row. The code of colors and plot-style is the same as in Figure 2.

3. Bi-Orthogonal Superpositions

We shall study the properties of diverse superpositions of states $\psi_n(x)$ by using

(i) Bi-normalization to obtain regular probability densities.
(ii) Bi-orthogonality to avoid the interference associated with non-orthogonality.

The space of states of H_λ, denoted $\mathcal{H}_\lambda$, consists of all the bi-orthogonal superpositions

$$\phi(x) = \sum_{k=0}^{\infty} c_k \psi_k(x), \quad c_k = (\overline{\psi}_k, \phi) \in \mathbb{C}, \tag{15}$$

such that

$$||\phi||_B^2 = (\overline{\phi}, \phi) = \sum_{k=0}^{\infty} |c_k|^2 < \infty. \tag{16}$$

In turn, $\overline{\mathcal{H}}_\lambda$ denotes the space spanned by the states $\overline{\psi}_n(x)$. The series decomposition of $\overline{\phi}(x)$, which is concomitant to $\phi(x)$ in Equation (16), is obtained from Equation (15) by changing $\psi_k(x) \to \overline{\psi}_k(x)$. To simplify the notation, hereafter we use no label to denote the bi-orthogonal properties of the system $\{\overline{\psi}_n(x), \psi_m(x)\}$.

The approach we are going to use identifies a threefold partnership between H_λ, $\overline{H}_\lambda$, and the conventional oscillator H_{osc}, where the eigenvalues play a main role. That is, the triad $\{\overline{H}_\lambda, H_{osc}, H_\lambda\}$ represents a system for which the energies $E_{n+1} = E_n^{osc} = 2n + 1$ are threefold degenerate while the energy $E_0 = -1$ is only twofold degenerate. In this sense, our model is a generalization of the supersymmetric approach [31] since H_λ and $\overline{H}_\lambda$ are treated as two different faces of the same system, which can be studied in much the same way as in the Hermitian approaches [24].

We are interested in two different classes of superpositions. The optimized superpositions are such that their nonclassicality can be manipulated by tuning the number of elements as well as the state of lowest energy in the sum. The generalized coherent states are constructed in terms of the algebras underlying the complex-valued oscillators of Equation (3) and form an over-complete set in $\mathcal{H}_\lambda$. Next, we analyze these classes in detail.

3.1. Optimized Binomial States

Let us consider a superposition of $N+1$ adjacent states

$$\phi_b(x) = \sum_{k=0}^{N} c_k^b \psi_{k+r}(x), \tag{17}$$

where the non-negative integer $r \geq 0$ determines the state $\psi_r(x)$ of the lowest energy that is included in the packet, and the coefficients c_k^b give rise to the binomial distribution

$$|c_k^b|^2 = \begin{pmatrix} N \\ k \end{pmatrix} p^k (1-p)^{N-k}, \quad 0 \leq p \leq 1, \quad k \leq N. \tag{18}$$

That is, Equation (17) is bi-normalized and includes only the eigenstates belonging to the energies $E_r, E_{r+1}, \ldots, E_{r+N}$. Equation (18) means that the probability of finding the system in the state $\psi_{k+r}(x)$ is weighted by the probability p of having success k times in n trials. The superpositions of Equation (17) are called "optimized binomial" states [26] since (for $r=0$) they converge to the conventional binomial superposition at the oscillator limit.

To investigate the properties of $\phi_b(x)$ let us take $p \approx 0$. The modulus $|c_0^b|$ of the first coefficient in the expansion is close to 1 and the other coefficients are almost zero. Then, the state $\psi_r(r)$ is very influential in the behavior of the entire packet since it is weighted by c_0^b in the superposition. The situation changes if $p \approx 1$ since in this case the modulus $|c_N^b|$ of the last coefficient in the sum is close to 1, while the moduli of the other coefficients are almost zero. In such case, $\psi_{N+r}(x)$ is the state of major influence in the packet.

Another important feature of the optimized binomial states is that their mean energy $\mathcal{E}_b$ depends on the number r which labels the lowest of the involved energies

$$\mathcal{E}_b = \langle H_\lambda \rangle_b = 2(Np+r) - 1. \tag{19}$$

Considering $0 \leq p \leq 1$, we have

$$2r - 1 \leq \mathcal{E}_b \leq 2(N+r) - 1. \tag{20}$$

That is, $\mathcal{E}_b$ is bounded from below by r, and from above by N and r. Fixing $\mathcal{E}_b$ and r in Equation (19) we can write

$$Np = \tfrac{1}{2}\left(\mathcal{E}_b + 1 - 2r\right). \tag{21}$$

Then, from Equation (18) we arrive at the Poisson distribution:

$$\lim_{N \to +\infty} |c_k^b|^2 = \frac{e^{-\frac{\mathcal{E}_b+1-2r}{2}}}{k!} \left(\frac{\mathcal{E}_b + 1 - 2r}{2}\right)^k. \tag{22}$$

In this form, as a limit version of the optimized binomial state $\phi_b(x)$, we introduce the *optimized Poisson* state

$$\phi_P(x) = e^{-\frac{|z|^2}{2}} \sum_{k=0}^{\infty} \frac{z^k}{\sqrt{k!}} \psi_{k+r}(x), \qquad |z|^2 = \frac{\mathcal{E}_b + 1 - 2r}{2}. \tag{23}$$

In the quantum oscillator limit of Equation (6), the superposition of Equation (23) acquires the form

$$\phi_P^{osc}(x) = \lim_{|\gamma| \to +\infty} \phi_P(x)\Big|_{\lambda=0} = \sum_{k=0}^{+\infty} c_k^P \varphi_{k+r}(x), \quad c_k^P = e^{-\frac{|z|^2}{2}} \frac{z^k}{\sqrt{k!}}, \tag{24}$$

which coincides with the conventional coherent state for $r=0$, see Equation (A25) of Appendix B.

3.2. Generalized Coherent States

The ground state $\psi_0(x)$ is annihilated by both pairs of ladder operators $\mathcal{A}$, $\mathcal{A}^+$, and $\mathcal{C}_w$, $\mathcal{C}_w^+$, see Appendix A. In any case we can construct coherent states as eigenstates of the respective annihilator operator. We write

$$\phi^{(\gamma)}(x) = \sum_{k=0}^{+\infty} c_k^{(\gamma)} \psi_{k+1}(x), \quad \gamma = \mathcal{N}, w, \mathcal{N}_d, w_d. \tag{25}$$

The super-index γ stands for natural ($\mathcal{N}$), distorted (w), natural displaced ($\mathcal{N}_d$), and distorted displaced (w_d). Such nomenclature refers to the form in which the coefficients $c_k^{(\gamma)}$ have been selected, as it is explained below. For simplicity, we use Dirac notation.

If the superposition of Equation (25) is an eigenvector of the natural annihilation operator $\mathcal{A}$ with complex eigenvalue z, then it is a natural coherent state [24]. We write (A factor 1/2 of the complex eigenvalue reported in [24] has been absorbed in z.)

$$|z\rangle = |\phi^{(\mathcal{N})}\rangle, \quad c_k^{(\mathcal{N})} = \frac{1}{\sqrt{{}_0F_2(1,2,|z|^2)}} \frac{z^k}{k!\sqrt{(k+1)!}}. \tag{26}$$

Notice that also the ground state $|\psi_0\rangle$ is eigenvector of $\mathcal{A}$, but its complex eigenvalue is equal to zero. Moreover, as $|z = 0\rangle = |\psi_1\rangle$, we see that the eigenvalue $z = 0$ is twice degenerate. The straightforward calculation shows that the vectors $|\psi_0\rangle$ and $|z\rangle$ minimize the uncertainty relation (A9). In this sense $|\psi_0\rangle$ and $|z\rangle$ represent two different types of minimal uncertainty states.

Similarly, the distorted coherent states [24] are defined as eigenvectors of the distorted annihilation operator $\mathcal{C}_w$ with complex eigenvalue z. In this case we write

$$|z, w\rangle = |\phi^{(w)}\rangle, \quad c_k^{(w)} = \frac{1}{\sqrt{{}_1F_1(1,w,|z|^2)}} \frac{z^k}{\sqrt{(w)_k}}, \tag{27}$$

with $(w)_k = w(w-1)\cdots(w-k+1)$ being the Pochhammer symbol [27]. The eigenvalue $z = 0$ is also twice degenerate because $|\psi_0\rangle$ and $|z = 0, w\rangle = |\psi_1\rangle$ are annihilated by $\mathcal{C}_w$. As $|\psi_0\rangle$ and $|z, w\rangle$ minimize the uncertainty relation of Equation (A16), they represent two different types of minimal uncertainty states.

Another interesting superposition, Equation (25), is obtained by demanding that the vector $|\phi^{(\gamma)}\rangle$ be the displaced version of a fiducial state. The states $|\psi_0\rangle$ and $|\psi_1\rangle$ are invariant under the action of the operator exponentiations $e^{-z^*\mathcal{A}}$ and $e^{-z^*\mathcal{C}_w}$, because they are annihilated by both operators $\mathcal{A}$ and $\mathcal{C}_w$. The latter means that $|\psi_0\rangle$ and $|\psi_1\rangle$ can be used as fiducial states. Considering the operators

$$D(z) = e^{z\mathcal{A}^+} e^{-z^*\mathcal{A}}, \qquad D_w(z) = e^{z\mathcal{C}_w^+} e^{-z^*\mathcal{C}_w}, \tag{28}$$

which one has the 'displaced' states $|\phi^{(\mathcal{N}_d)}\rangle = D(z)|\psi_1\rangle$ and $|\phi^{(w_d)}\rangle = D_w(z)|\psi_1\rangle$. Remark that the ground state $|\psi_0\rangle$ is invariant under the action of $D(z)$ and $D_w(z)$, so it is also a displaced state.

In the rest of this work we shall omit the description of the states $|\phi^{(\mathcal{N}_d)}\rangle$ since their properties are qualitatively similar to those of the (distorted) displaced coherent states:

$$|\psi_0\rangle, \qquad |z, w\rangle_d = |\phi^{(w_d)}\rangle, \quad c_k^{(w_d)} = \frac{1}{\sqrt{{}_1F_1(w,1,|z|^2)}} \frac{z^k\sqrt{(w)_k}}{k!}. \tag{29}$$

In contrast to the previous superpositions, the states $|z, w\rangle_d$ do not minimize the uncertainty relation of Equation (A16) for arbitrary values of z, but only for $z = 0$. The latter means that $|\psi_0\rangle$ and $|\psi_1\rangle$ are the only displaced coherent states of minimal uncertainty.

4. Nonclassical States for Non-Hermitian Oscillators

The P-representation introduced by Glauber [32] and Sudarshan [33] defines a limit for the classical description of radiation fields. If the P-function is not accurately interpretable as a probability distribution then the field "will have no classical analog" [2]. This notion of nonclassicality applies immediately to the Fock states $|n\rangle$ for $n \neq 0$ because their P-functions are as singular as the derivatives of the delta function $\delta^{(2)}(z) = \delta(\text{Re}(z))\delta(\text{Im}(z))$, and are negative in some regions of the complex z-plane. The latter means that the light beams represented by any of the Fock states $|n \neq 0\rangle$ cannot be described in terms of the electrodynamics introduced by Maxwell. In turn, the Glauber states [2], usually denoted $|\alpha\rangle$, are classical because their P-function is precisely $\delta^{(2)}(z - \alpha)$. The vacuum $|0\rangle$ is also classical as it is a coherent state with complex eigenvalue equal to zero. In general, the states for which the quadrature variances obey the inequalities

$$(\Delta \hat{x})^2 \geq \tfrac{1}{2}, \quad (\Delta \hat{p})^2 \geq \tfrac{1}{2}, \tag{30}$$

which admit a non-negative P-function [2]. Here, $(\Delta A)^2 = \langle A^2 \rangle - \langle A \rangle^2$ stands for the variance of the operator A. If either of Equations (30) is not satisfied, the P-function is ill-defined and the state is called "squeezed" [4,34,35]. In these cases it is better to represent the state using the Wigner distribution [1], which is regular and always exists. Nonclassical states have Wigner functions which either are negative in some regions of the phase-space or are squeezed in one of the variables of the phase-space.

In modern days, the nonclassicality of a state can be tested by another excellent method, which follows from the use of the quantum beam splitter. It creates entangled states at its output ports while at least one of its inputs are fed with a nonclassical state. The device is used frequently both in theory and experiment, since it not only tests the nonclassicality of the state, but also quantifies the degree of nonclassicality in an efficient way [11,12]. There are some well-known nonclassical states in the literature, which include squeezed states [4], even and odd coherent states (also called Schrödinger cats) [5], binomial states [6,7], photon-added coherent states [8], etc. Apart from the nonclassical states that arise from the harmonic oscillator, there exist many other such states emerging from different generalizations of coherent states; see, for instance [36]. Other interesting nonclassical states have also been shown to originate from some mathematical frameworks, in particular, from the noncommutative systems reported in [37–39]. The striking feature of such systems is that they give rise to well-defined nonclassical states although the corresponding models are non-Hermitian.

Quite recently it has been shown that the appropriate generalizations of the oscillator algebra permit the construction of nonlinear coherent states that satisfy a closure relation which is expressed uniquely in terms of the Meijer G-function [40]. This property automatically defines the delta distribution as the corresponding P-representation. However, in the same work, it is also shown that does not exist a classical analog for such states since they lack second-order coherence and exhibit antibunching. Thus, although their P-representation is a delta function, the nonlinear coherent states studied in [40] are not classical since they are not fully coherent in the sense established by Glauber [2]. Similar results have been obtained for the para-Bose oscillator [41,42]. The algebras of Equations (13) and (14) that we have used to construct the coherent states of Section 3.2 satisfy the requirements established in [40]. Considering also that the bi-orthogonality permits to operate the non-Hermitian oscillators as in the conventional Hermitian case, it may be shown that the corresponding P-representation is proportional to the delta distribution. Nevertheless, the latter means the existence of a classical analog only if the states of Section 3.2 are fully coherent.

Instead of investigating whether our coherent states lack second (or higher) order coherence, we shall use some of the techniques described in Appendix B to study their classicalness. Namely, we are going to study the corresponding variances, Mandel parameter, Wigner distribution and purity.

4.1. Nonclassical Optimized Binomial States

Let us investigate the properties of the distorted quadratures of Equation (A15) in terms of the optimized binomial states of Equations (17) and (18). The variances $(\Delta X_w)^2$ and $(\Delta P_w)^2$ may be expressed in the form of Equation (A17) of Appendix B:

$$(\Delta X_w)^2 = \tfrac{1}{4}|\langle I_w\rangle_b| + U_1, \quad (\Delta P_w)^2 = \tfrac{1}{4}|\langle I_w\rangle_b| - U_2, \tag{31}$$

where we have used Equation (A16) and $\langle I_w\rangle_b \equiv \langle\overline{\phi}_b|I_w|\phi_b\rangle = 1 + (w-1)\left(\delta_{r,1}\, c_0^2 + \delta_{r,0}\, c_1^2\right) - \delta_{r,0}\, c_0^2$. The straightforward calculation gives $U_1 = \frac{1}{2}(M_1 + M_2) - M_3^2$ and $U_2 = \frac{1}{2}(M_1 - M_2)$, with

$$M_1 = \sum_{k=0}^{N-2} \Omega_{k,r}^{(1)} c_{k+2} c_k \sqrt{(k+r+w-1)(k+r+w)}, \tag{32}$$

$$M_2 = \sum_{k=0}^{N} (k+r+w-2)\Omega_{k,r}^{(2)} c_k^2, \quad M_3 = \sum_{k=0}^{N-1} \Omega_{k,r}^{(1)} c_{k+1} c_k \sqrt{(k+r+w-1)}. \tag{33}$$

In the above expressions $\Omega_{k,r}^{(1)} = 1 - \delta_{k,r}\delta_{r,0}$, and $\Omega_{k,r}^{(2)} = \Omega_{k,r}^{(1)} - \delta_{k,1}\delta_{r,0} - \delta_{k,0}\delta_{r,1}$. Hereafter we make $\theta_k = 0$ in $c_k = |c_k^b| e^{i\theta_k}$.

It may be shown that there is always a subset of the mean energies of Equations (19) and (20) for which $(\Delta X_w)^2 < \frac{1}{4}\langle I_w\rangle_b$. Thus, it is always possible to find a mean energy $\mathcal{E}_b$ producing the squeezing of the distorted quadrature X_w. The latter is illustrated in Figure 4 for $r = 2$, $N = 10$ and two different values of w. In turn, the squeezing of P_w occurs for $r = 0$ only, and it is not as strong as the one suffered by X_w since $(\Delta P_w)^2$ is just slightly larger than $\frac{1}{4}\langle I_w\rangle_b$. In general, the optimized binomial states of Equation (17) are of minimal uncertainty (i.e., they are classical states) only at the lowest mean energy for either (i) $r = 0$ and any value of w, or (ii) $r = 2$ and $w = 0$.

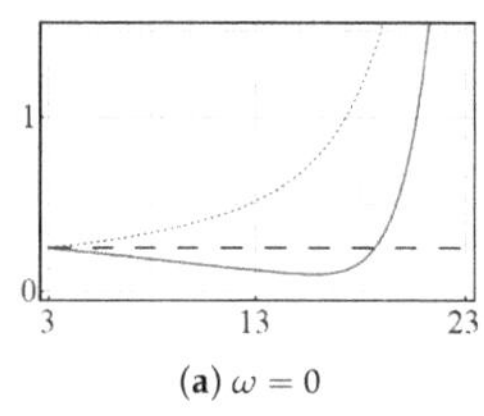

(a) $\omega = 0$

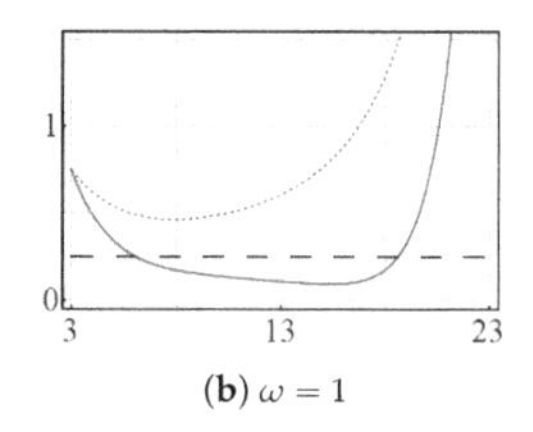

(b) $\omega = 1$

Figure 4. (Color online) the variances of Equation (31) and expectation value $\frac{1}{4}\langle I_w\rangle_b$ of the optimized binomial states of Equation (17). The functions $(\Delta X_w)^2$ and $(\Delta P_w)^2$ are depicted in solid-red and dotted-blue, respectively; the expectation value is in dashed-black. We have used $N = 10$, $r = 2$, and the indicated values of ω. The horizontal axis refers to the allowed mean energy $\mathcal{E}_b$ introduced in Equations (19) and (20).

In the quantum oscillator limit of Equation (6) one has

$$\phi_b^{osc}(x) = \lim_{|\gamma|\to\infty} \phi_b(x)\Big|_{\lambda=0} = \sum_{k=0}^{N} |c_k^b| e^{i\theta_k} \varphi_{k+r}(x). \tag{34}$$

Clearly, we recover the conventional binomial state for $r = 0$. This state is squeezed in the quadrature $\hat{x}$, for $N = 10$ and $p = 0.5$, as shown in the Wigner distribution of Figure 5a. If $r \geq 1$, besides the squeezing of $\hat{x}$, the Wigner distribution exhibits negative values in different zones of the phase-space, see Figure 5b,c. The latter results confirm that the optimized binomial states of Equation (34) are stronger nonclassical than the conventional ones.

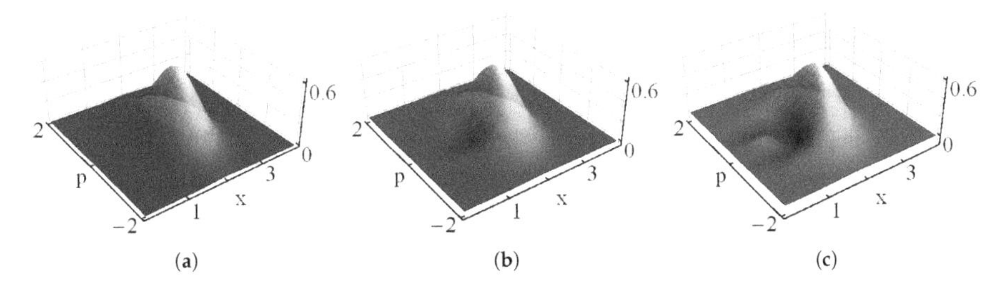

Figure 5. Wigner distribution of the optimized binomial states of Equation (34). The panel is constructed with (**a**) $p = 0.5$, $r = 0$, (**b**) $p = 0.4$, $r = 1$, and (**c**) $p = 0.3$, $r = 2$. In all cases $N = 10$ and $\mathcal{E}_b = 9$.

To get more insights about the optimized binomial states of Equation (34) we have depicted the related Mandel parameter Q, as a function of p, in Figure 6. For $p = 1$ the superposition of Equation (17) is reduced to the state $\varphi_{N+r}(x)$, so that all the curves shown in Figure 6 converge to -1 as $p \to 1$. In turn, for $r \neq 0$, all the curves take the value -1 at $p = 0$ since only the Fock state $\varphi_{r\neq 0}(x)$ is included in the sum. Thus, for $r \neq 0$ and any p, the Mandel parameter Q in this case is always negative. Then, the optimized binomial states of Equation (34) are sub-Poissonian for $r \neq 0$, and thus do not admit any classical description.

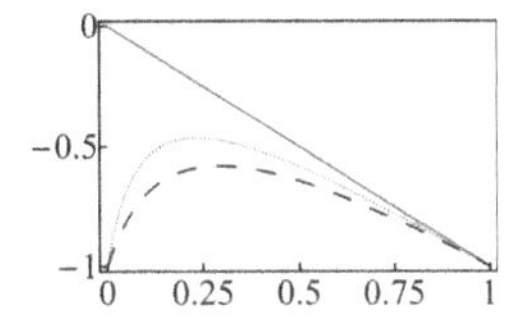

Figure 6. (Color online) Mandel parameter Q of the optimized binomial states of Equation (34) in terms of the probability p. The curves correspond to $N = 10$ with $r = 0$ (red-solid), $r = 1$ (blue-dotted) and $r = 2$ (black-dashed).

Let us apply also the beam-splitter technique (see Appendix B) to the states we are dealing with. We may study the purity (linear entropy) S_L in terms of either the transmission coefficient T of the beam-splitter or the probability p that defines the superposition of Equation (34). In the former case it is convenient to fix N and $\mathcal{E}_b$ to parameterize the purity with r and p. Then one finds that the purity is equal to zero for $T = 0$ and $T = 1$, see Figure 7. However, it is important to remark that such results give no special information since they correspond to complete reflectance (i.e., the beam-splitter is indeed a perfect mirror) and complete transparency (no beam-splitter), so the output is, respectively, the vacuum and the state which was injected in input. In general, for other values of T, the purity reaches its maximum at $T = \sqrt{0.5}$, i.e., when the testing beam-splitter is 50/50.

To analyze S_L in terms of the probability p we must fix N and T. Then, with exception of the purity for $r = 0$, we find $S_L > 0$ for any value of p (see Figure 8). The latter confirms, once again, that the optimized binomial states $\phi_b(x)$ are nonclassical. Concerning the case $r = 0$, the purity is equal to zero at $p = 0$ only. Thus, the superposition of Equation (34) only includes the vacuum state $\varphi_0(x)$, which is clearly classical.

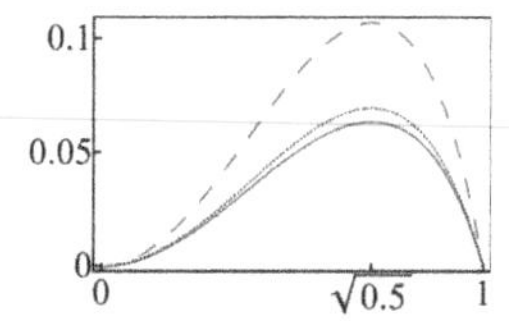

Figure 7. (Color online) purity S_L of the optimized binomial states of Equation (34) in terms of the transmission coefficient T of a testing beam-splitter. In all cases $N = 10$ and $\mathcal{E}_b = 9$. The curves correspond to $p = 0.5, r = 0$ (solid-blue), $p = 0.4, r = 1$ (dotted-black), and $p = 0.3, r = 2$ (dashed-red).

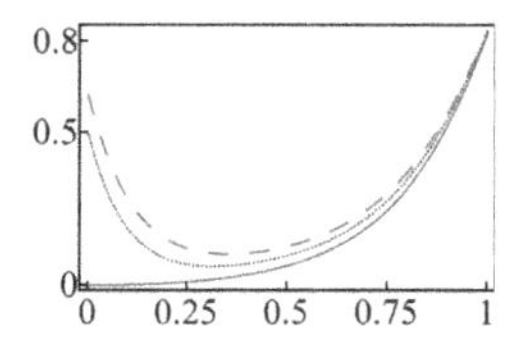

Figure 8. (Color online) purity S_L of the optimized binomial states of Equation (34) in terms of the probability p with $N = 10$. We have used a 50/50 beam-splitter ($T = \sqrt{0.5}$) for $r = 0$ (solid-blue), $r = 1$ (dotted-black) and $r = 2$ (dashed-red). Compare with Figure 7.

4.1.1. Nonclassical Optimized Poisson States

It may be shown that the optimized Poisson states of Equation (23) are minimum uncertainty states for concrete values of the parameters like either $r = 1, \omega = 1$, or $r = 2, \omega = 0$. In any case, they become states of minimum uncertainty as $|z| \to \infty$. For other values of the parameters either $(\Delta X_w)^2$ or $(\Delta P_w)^2$ is squeezed, see Figure 9.

In the quantum oscillator limit of Equation (6), the superposition of Equation (23) acquires the form of Equation (24). As indicated above, for $r = 0$ this state is reduced to the conventional coherent state, so that the Mandel parameter is equal to zero, as expected. For other values of r, the optimized Poisson states of Equation (24) are sub-Poissonian. In fact, in Figure 10 we see that the Mandel parameter becomes zero as $|z| \to \infty$, and it is equal to -1 for $r \neq 0$ at $z = 0$ (i.e. at the lowest mean energy $\mathcal{E}_b = 2r - 1$). In other words, for $r \neq 0$ and finite values of z, the optimized Poisson states $\phi_P^{osc}(x)$ are nonclassical.

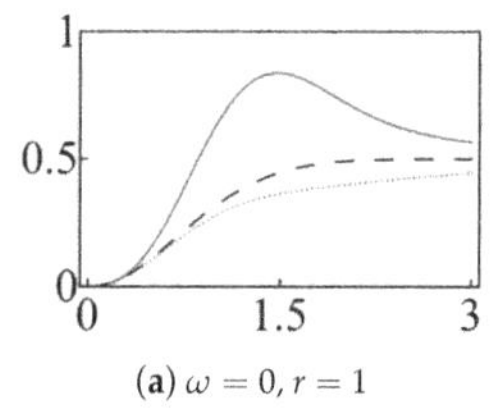

(**a**) $\omega = 0, r = 1$

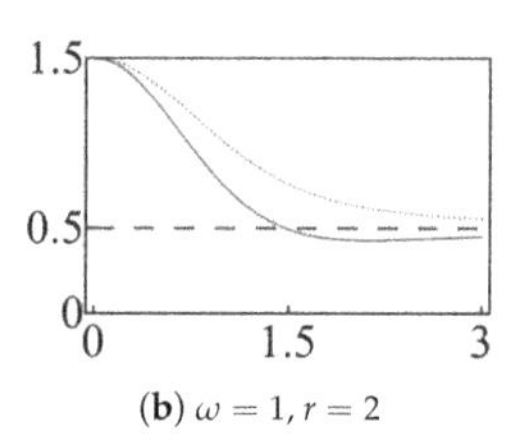

(**b**) $\omega = 1, r = 2$

Figure 9. (Color online) the variances $(\Delta X_w)^2$ and $(\Delta P_w)^2$, in solid-red and dotted-blue, respectively, of the optimized Poisson states of Equation (23) for the indicated values of ω and r. The minimum uncertainty (dashed-black) is included as a reference.

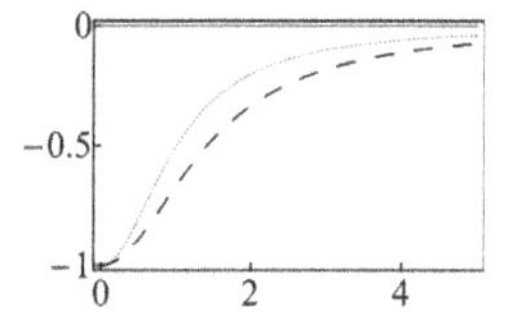

Figure 10. (Color online) Mandel parameter Q of the optimized Poisson states of Equation (24) as a function of $|z|$. The curves correspond to $r = 0$ (solid-red), $r = 1$ (dotted-blue) and $r = 2$ (dashed-black).

For completeness, we show the Wigner function of $\phi_P^{osc}(x)$ in Figure 11. Clearly, for $r = 0$ one has a conventional (classical) coherent state. Other values of r give rise to both, the squeezing of $\hat{x}$ and the negativity of the Wigner function in some regions of the phase-space.

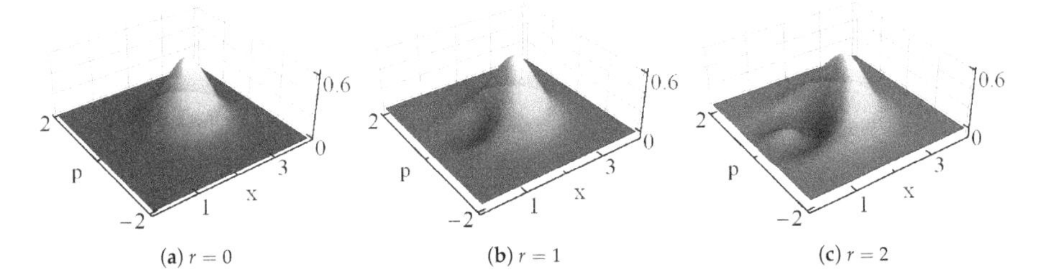

Figure 11. Wigner distribution of the optimized Poisson states of Equation (24). The panel is constructed with $|z| = \sqrt{5}$ (**a**), $|z| = 2$ (**b**), $|z| = \sqrt{3}$ (**c**), and the indicated values of r; in all cases $\mathcal{E}_b = 9$.

- **Photon added coherent states.** Equation (24) resembles the structure of the photon-added coherent states for the conventional harmonic oscillator [8], written in Dirac notation as

$$|\alpha^{(pa)}\rangle = \sum_{k=0}^{+\infty} c_k^{(pa)} |\varphi_{k+r}\rangle, \quad c_k^{(pa)} = \left[\frac{(k+r)!}{{}_1F_1(r+1,1,|\alpha|^2)\, r!} \right]^{1/2} \frac{\alpha^k}{k!}, \quad \alpha \in \mathbb{C}. \tag{35}$$

In both cases the superpositions are infinite and start with the eigenstate $|\varphi_r\rangle$. The main difference lies on the coefficients, for if we make $z = \alpha$, then

$$c_k^{(pa)} = \left[\frac{(k+1)_r}{{}_1F_1(r+1,1,|\alpha|^2)\, r!} \right]^{1/2} e^{\frac{|\alpha|^2}{2}} c_k^P, \tag{36}$$

If $r = 0$ we see that, up to the normalization constant $e^{-|\alpha|^2/2}$, the coefficients are the same. For other values of r the superpositions of Equations (24) and (35) are different in general.

Let us compare $|\alpha^{(pa)}\rangle$ with $|\phi_P^{osc}\rangle$ in order to identify which one is stronger nonclassical. Using the beam-splitter technique, we find that the nonclassicality of these states depends on different parameters. For instance, in Figure 12a we have fixed the value of the mean energy as $\mathcal{E}_b = 9$ and, after adjusting the parameters z, α, and r, the purities $S_L(\phi_P^{osc})$ and $S_L(\alpha^{(pa)})$ have been depicted in terms of the transmission coefficient T. As a global property, we see that the purity is maximum at $T = \sqrt{0.5}$ in all the cases. Besides, we can appreciate that $S_L(\phi_P^{osc}) < S_L(\alpha^{(pa)})$. Thus, for the parameters used in the figure the state $|\alpha^{(pa)}\rangle$ is stronger nonclassical than $|\phi_P^{osc}\rangle$. The situation changes if we fix T, and adjust the other parameters to plot the purity S_L in terms of $|z|$. Then $S_L(\phi_P^{osc}) > S_L(\alpha^{(pa)})$ holds for small values of $|z|$ and $|\alpha|$, and $S_L(\phi_P^{osc}) < S_L(\alpha^{(pa)})$ is valid for large values of such parameters. The latter is illustrated in Figure 12b for $T = \sqrt{0.5}$ and $|z| = |\alpha|$.

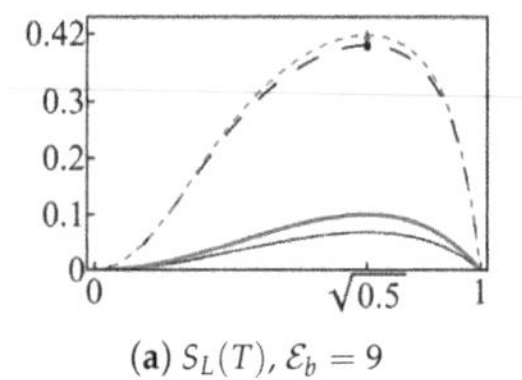

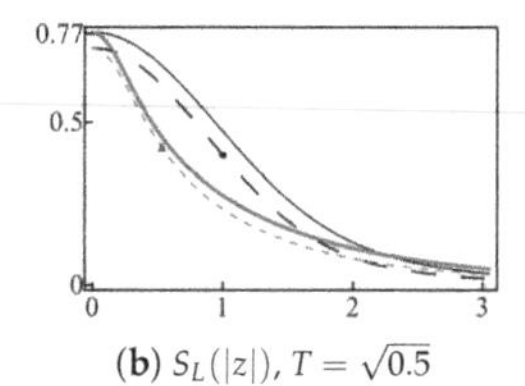

(**a**) $S_L(T)$, $\mathcal{E}_b = 9$ (**b**) $S_L(|z|)$, $T = \sqrt{0.5}$

Figure 12. (Color online) the purities of the photon-added coherent state, Equation (24), and the Poisson state, Equation (35), are, respectively, depicted in red and black. In (**a**) the mean energy is fixed, $\mathcal{E}_b = 9$, and the purities are plotted in terms of the transmission coefficient T with $r = 2$ (thick-solid for Equation (24) and solid for Equation (35)) and $r = 4$ (shot-dashed for Equation (24) and long-dashed for Equation (35)). In (**b**) the transmission coefficient is fixed, $T = \sqrt{0.5}$, and the purities are shown in terms of z with $r = 4$ (shot-dashed for Equation (24) and long-dashed for Equation (35)) and $r = 6$ (thick-solid for Equation (24) and solid for Equation (35)). For comparison, in both figures the disk and triangle mark the purity evaluated at the same parameters.

4.2. Nonclassical Natural Coherent States

By construction, the natural coherent states $\{|\psi_0\rangle, |z\rangle\}$ minimize the uncertainty relation of Equation (A9) of Appendix A to investigate their properties in the oscillator limit of Equation (6), we first calculate the purity S_L using Equation (A23) of Appendix B with $r = 1$, $c_k = c_k^{(\mathcal{N})}$, and $N \to +\infty$ in the Fock basis $|\varphi_k\rangle$. The result is depicted in Figure 13a. As we can see, this function is always different from zero and decreases as $|z| \to \infty$ for any value of the transmission coefficient T. That is, the states $|z^{osc}\rangle$ are nonclassical. The latter is confirmed by noticing that the Mandel parameter Q is always negative, as it is shown in Figure 13b.

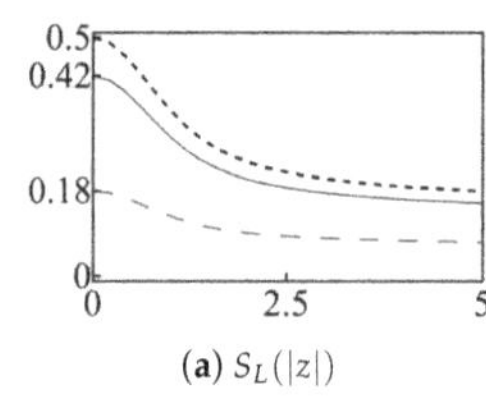

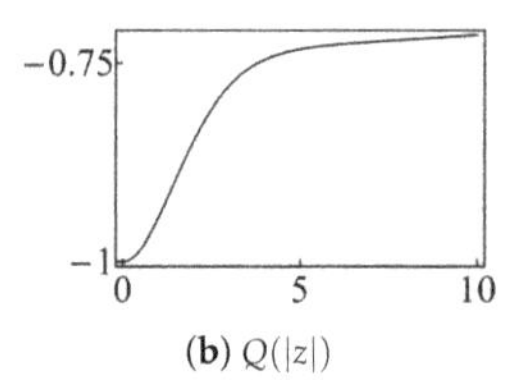

(**a**) $S_L(|z|)$ (**b**) $Q(|z|)$

Figure 13. (Color online) (**a**) purity of the natural coherent state $|z^{osc}\rangle$ in terms of $|z|$ for different values of the transmission coefficient: $T = \sqrt{0.1}$ (dashed-red), $T = \sqrt{0.3}$ (solid-blue) and $T = \sqrt{0.5}$ (dotted-black). (**b**) The corresponding Mandel parameter in terms of z.

On the other hand, the variances of the physical quadratures $\hat{x}$ and $\hat{p}$ can be expressed in the form of Equation (A17) of Appendix B, with $A = \hat{x}$, $B = \hat{p}$, and

$$U_1 = 1 + [\mathrm{Re}(z)]^2 f(z), \quad U_2 = 1 + [\mathrm{Im}(z)]^2 f(z),$$
$$f(z) = \frac{{}_0F_2(2,3;|z|^2)}{{}_0F_2(1,2;|z|^2)} - 2\left[\frac{{}_0F_2(2,2;|z|^2)}{{}_0F_2(1,2;|z|^2)}\right]^2. \tag{37}$$

In Figure 14, we show the regions of the complex z-plane where the squeezing of either $\hat{x}$ (grey zones) or $\hat{p}$ (mesh, dashed blue zones) occurs. Notice that the squeezing is maximum at the phase values $\phi = 0, \pi$, and $\phi = \frac{\pi}{2}, \frac{3\pi}{2}$, respectively. The white zones are the regions of no-squeezing, they overrun four distinguishable areas defined along the phase values $\phi = \frac{k\pi}{4}$, $k = 1,3,5,7$, and a circle of radius $|z| \approx 5$ that is centered at the origin. Comparing the Figures 13 and 14, we see that the purity and the Mandel parameter give information that is complementary to that obtained from the variances of the quadratures. Namely, S_L and Q indicate strong nonclassicality for the states $|z^{osc}\rangle$ that satisfy the condition $|z| \lesssim 5$, where no squeezing is expected for neither $\hat{x}$ nor $\hat{p}$.

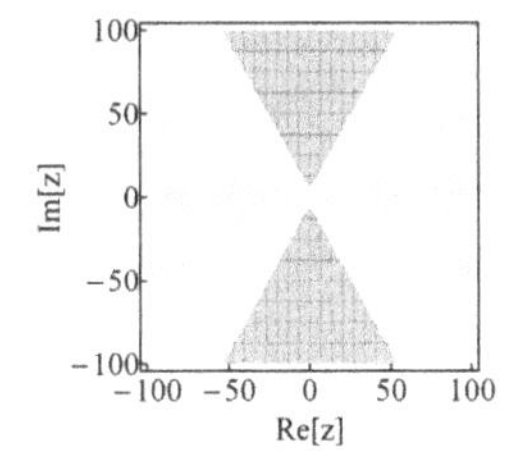

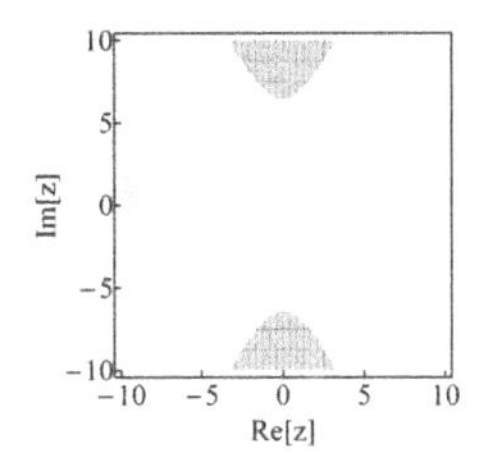

Figure 14. (Color online) regions of the complex plane where squeezing occurs. The grey (mesh, dashed blue) zones correspond to the values of $z \in \mathbb{C}$ for which $\hat{x}$ ($\hat{p}$) is squeezed. In turn, the grey and blue zones are centered at $\phi = 0, \pi$ and $\phi = \frac{\pi}{2}, \frac{3\pi}{2}$, respectively.

In Figure 15 we show the density plots of the Wigner function of $|z^{osc}\rangle$ for different values of z. In particular, for $z = 0$ the natural coherent state $|z^{osc}\rangle$ is nonclassical because it coincides with the first excited state of the oscillator $|\varphi_1\rangle$, as shown in Figure 15a. If $|z| > 5$ and $\phi = 0$ the quadrature $\hat{x}$ is squeezed; this is illustrated in Figure 15b with $z = 30$. In turn, for $z = i30$ one has the opposite result, $\hat{p}$ is squeezed; see Figure 15c.

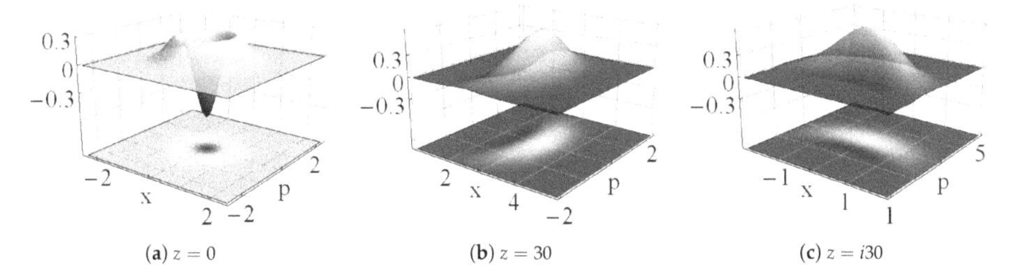

(**a**) $z = 0$ (**b**) $z = 30$ (**c**) $z = i30$

Figure 15. Wigner distribution of the natural coherent sates $|z^{osc}\rangle$ for the indicated values of z.

4.2.1. Even and Odd Natural Coherent States

We can use the natural coherent states of Equation (26) to construct additional nonclassical states. Following [5] we introduce the vectors

$$|z_{\pm}\rangle = \left[1 \pm \frac{{}_0F_2(1,2,-|z|^2)}{{}_0F_2(1,2,|z|^2)}\right]^{-1/2} \left(\frac{|z\rangle \pm |-z\rangle}{\sqrt{2}}\right), \tag{38}$$

and call them "even" and "odd" natural cat states, respectively. The related Mandel parameter Q is always negative, see Figure 16a, so that $|z_{\pm}\rangle$ are nonclassical. The variances of the natural quadratures of Equation (A8) can be expressed in the form of Equation (A17) of Appendix B, with $A = X_{\mathcal{N}}$, $B = P_{\mathcal{N}}$, and

$$\begin{aligned} U_1^{(\pm)} &= 2\left[\frac{[\text{Re}(z)]^2\,{}_0F_2(1,2;|z|^2) \mp [\text{Im}(z)]^2\,{}_0F_2(1,2;-|z|^2)}{{}_0F_2(1,2;|z|^2) \pm {}_0F_2(1,2;-|z|^2)}\right], \\ U_2^{(\pm)} &= -2\left[\frac{[\text{Im}(z)]^2\,{}_0F_2(1,2;|z|^2) \mp [\text{Re}(z)]^2\,{}_0F_2(1,2;-|z|^2)}{{}_0F_2(1,2;|z|^2) \pm {}_0F_2(1,2;-|z|^2)}\right]. \end{aligned} \tag{39}$$

In this case the quadratures are squeezed in very localized regions that are distributed along the real and imaginary axes of the complex z-plane. This is illustrated in Figure 16b, where we can appreciate a 'discretization' of the eigenvalue z as follows. The quadrature $X_{\mathcal{N}}$ is maximally squeezed along the real axis, in intervals $|z_k^{(j)}| \pm \ell_k^{(j)}$ that are defined by the points $|z_k^{(j)}|$, $k = 1, 2, \ldots$, with $\ell_k^{(j)} > 0$, and j denoting either even (e) or odd (o). For odd natural cats we make $j = o$, with

$|z_1^{(o)}|$ defining the green zones that are closest to the origin in Figure 16b. For the even natural cats we write $j = e$, with $|z_1^{(e)}|$ defining the blue zones that are closest to the origin in the same figure. The intersection of the above intervals is empty, and they are such that the points $|z_k^{(j)}|$ interlace. That is $|z_k^{(o)}| < |z_k^{(e)}| < |z_{k+1}^{(o)}| < |z_{k+1}^{(e)}| < \cdots$. The phases constraining the green and blue zones are defined in intervals $\phi_k^{(j)} \pm d_k^{(j)}$, with $d_k^{(j)} > 0$, that are shorter as k increases. A similar description holds for the squeezing of $P_{\mathcal{N}}$ along the imaginary axis.

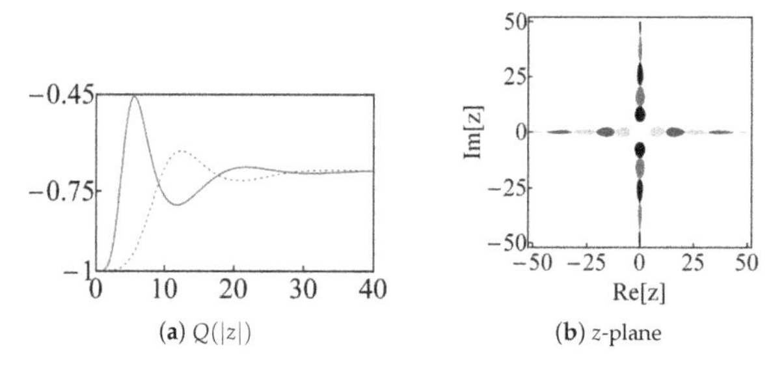

(a) $Q(|z|)$ (b) z-plane

Figure 16. (Color online) (**a**) the Mandel parameter Q of the even (solid-red) and odd (dahsed-blue) natural cat states of Equation (38) as a function of $|z|$. (**b**) Regions of the complex z-plane for which squeezing occurs. The red (even) and black (odd) regions correspond to the values of $z \in \mathbb{C}$ for which $P_{\mathcal{N}}$ is squeezed. Blue (even) and green (odd) regions correspond to the squeezing of $X_{\mathcal{N}}$.

In the oscillator limit of Equation (6), the natural cat states $|z_{\pm}^{osc}\rangle$ preserve their nonclassicality. In Figure 17 the Wigner function becomes negative in diverse zones of the complex z-plane. Such regions are less evident for $|z| \to \infty$.

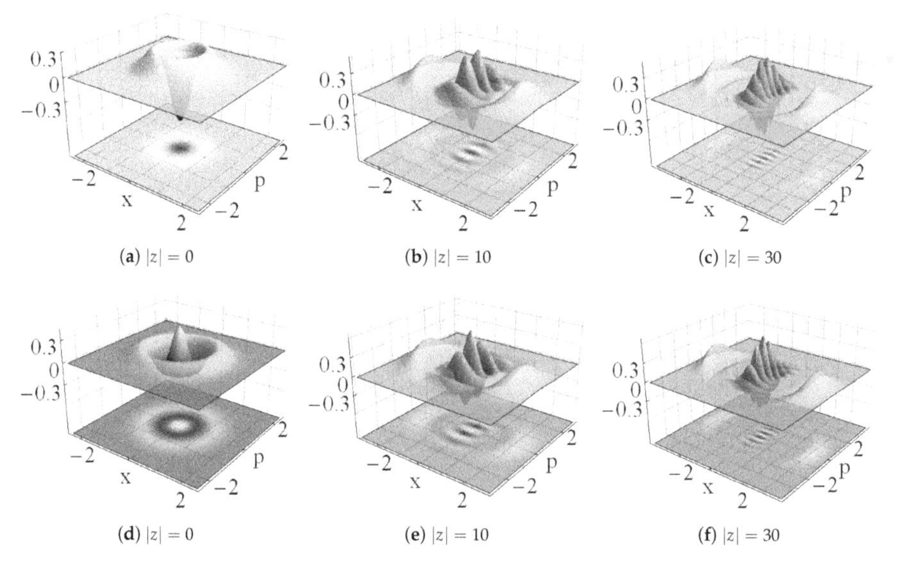

(**a**) $|z| = 0$ (**b**) $|z| = 10$ (**c**) $|z| = 30$

(**d**) $|z| = 0$ (**e**) $|z| = 10$ (**f**) $|z| = 30$

Figure 17. Wigner distributions of the even (upper row) and odd (lower row) natural cat states $|z_{\pm}^{osc}\rangle$ for $z = |z|e^{i\theta}$, with $\theta = 0$ and the indicated values of $|z|$.

4.3. Nonclassical Distorted Coherent States

By construction, the distorted coherent states $\{|\psi_0\rangle, |z,w\rangle\}$ are minimal uncertainty states with respect to the uncertainty relation of Equation (A16). In the oscillator limit, the Mandel parameter Q depends on the parameter of distortion. In Figure 18a, we see that the states $|z^{osc}, w\rangle$ are sub-Poissonian for any value of $|z|$ whenever $w = 1$. For arbitrary values of $w > 1$, they are nonclassical only in the interval $(0, |z_w|)$. Here the value of $|z_w|$ is defined by w: it is larger if the value of w is increased.

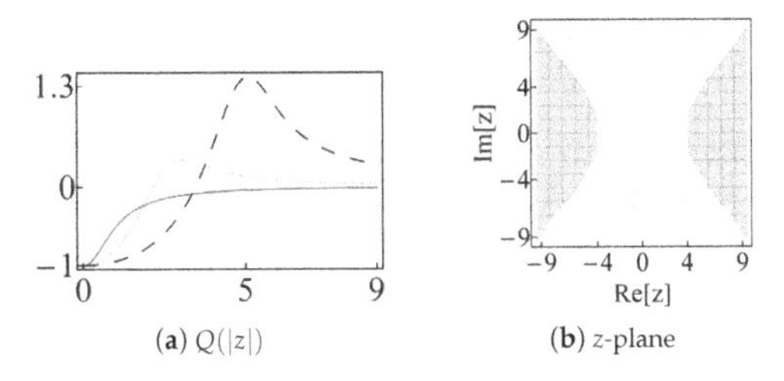

Figure 18. (Color online) the Mandel parameter Q of the distorted coherent state $|z^{osc}, w\rangle$ is shown in (**a**) as a function of $|z|$ for $w = 1$ (solid-red), $w = 5$ (dotted-blue), and $w = 20$ (dashed-black). In (**b**) we show the regions of squeezing in the complex z-plane for $w = 20$. The code of colors is the same as in Figure 14b.

The variances of the physical quadratures $\hat{x}$ and $\hat{p}$ can be expressed in the form given by Equation (A17) of Appendix B, with $A = \hat{x}$, $B = \hat{p}$, and

$$
\begin{aligned}
U_1 &= \frac{\mathrm{Re}(z^2) f_2(|z|^2, \omega) + {}_1F_1\,(2, \omega, |z|^2)}{{}_1F_1(1, \omega; |z|^2)} - 2\left[\frac{\mathrm{Re}(z) f_1(z, \omega)}{{}_1F_1(1, \omega; |z|^2)}\right]^2, \\
U_2 &= \frac{\mathrm{Re}(z^2) f_2(|z|^2, \omega) - {}_1F_1\,(2, \omega, |z|^2)}{{}_1F_1(1, \omega; |z|^2)} + 2\left[\frac{\mathrm{Im}(z) f_1(z, \omega)}{{}_1F_1(1, \omega; |z|^2)}\right]^2,
\end{aligned}
\tag{40}
$$

where

$$
f_1(z, \omega) = \sum_{n=0}^{+\infty} |z|^{2n} \sqrt{\frac{n+2}{(\omega)_n (\omega)_{n+1}}}, \quad f_2(|z|^2, \omega) = \sum_{n=0}^{+\infty} |z|^{2n} \sqrt{\frac{(n+2)(n+3)}{(\omega)_n (\omega)_{n+2}}}. \tag{41}
$$

For $w \approx 1$, the quadratures $\hat{x}$ and $\hat{p}$ are squeezed along the real and imaginary axes, respectively. However, for larger values of w, they are squeezed along the imaginary and real axes. Thus, for $w >> 1$ the squeezing is rotated by $\frac{\pi}{2}$, as shown in Figure 18b.

4.3.1. Even and Odd Distorted Coherent States

For the even and odd distorted cats

$$
|z_{\pm}, w\rangle = \left[1 \pm \frac{{}_1F_1(1, w, -|z|^2)}{{}_1F_1(1, w, |z|^2)}\right]^{-1/2} \left(\frac{|z, w\rangle \pm |-z, w\rangle}{\sqrt{2}}\right), \tag{42}
$$

the variances of the distorted quadratures of Equation (A15) can be expressed in the form given by Equation (A17) of Appendix B, with $A = X_w$, $B = P_w$, and

$$
\begin{aligned}
U_1^{(\pm)} &= 2\left[\frac{[\mathrm{Re}(z)]^2\, {}_1F_1(1, w; |z|^2) \mp [\mathrm{Im}(z)]^2\, {}_1F_1(1, w; -|z|^2)}{{}_1F_1(1, w; |z|^2) \pm {}_1F_1(1, w; -|z|^2)}\right], \\
U_2^{(\pm)} &= -2\left[\frac{[\mathrm{Im}(z)]^2\, {}_1F_1(1, w; |z|^2) \mp [\mathrm{Re}(z)]^2\, {}_1F_1(1, w; -|z|^2)}{{}_1F_1(1, w; |z|^2) \pm {}_1F_1(1, w; -|z|^2)}\right].
\end{aligned}
\tag{43}
$$

In Figure 19, we show the regions of the complex z-plane where X_w and P_w are squeezed for the even distorted cat $|z_+, w\rangle$ with the indicated value of w. Depending on the phase ϕ, the squeezing is present in zones defined by the interval $0 < |z| \lesssim 7$ and aligned with either the real or the imaginary axes. Such zones are of maximum width at $|z| \approx 3.5$ and are slenderer as either $|z| \to 0$ or $|z| \to \infty$.

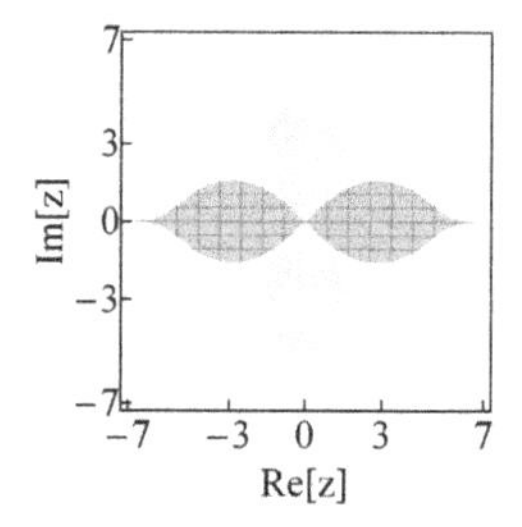

Figure 19. (Color online) regions of squeezing in the complex z-plane for the even distorted cat $|z_+, w\rangle$ with $w = 20$. The grey (mesh, dashed blue) zones correspond to the values of $z \in \mathbb{C}$ for which $X_w(P_w)$ is squeezed. The white zones indicate the values of z for which both variances are bigger than the average uncertainty.

In the oscillator limit of Equation (6), the variances of $\hat{x}$ and $\hat{p}$ are calculated using Equation (A17) of Appendix B, with

$$U_1^{(\pm)} = \frac{\mathrm{Re}(z^2)\left[f_2(|z|^2, w) \pm f_2(-|z|^2, w)\right] + \left[{}_1F_1\left(2, \omega; |z|^2\right) \pm {}_1F_1\left(2, \omega; -|z|^2\right)\right]}{{}_1F_1(1, \omega; |z|^2) \pm {}_1F_1(1, \omega; -|z|^2)},$$
$$U_2^{(\pm)} = \frac{\mathrm{Re}(z^2)\left[f_2(|z|^2, w) \pm f_2(-|z|^2, w)\right] - \left[{}_1F_1\left(2, \omega; |z|^2\right) \pm {}_1F_1\left(2, \omega; -|z|^2\right)\right]}{{}_1F_1(1, \omega; |z|^2) \pm {}_1F_1(1, \omega; -|z|^2)}, \tag{44}$$

and $f_2(|z|^2, w)$ given in Equation (41). The quadratures $\hat{x}$ and $\hat{p}$ are squeezed in narrow zones of the complex z-plane that are defined along the real and imaginary axis, respectively. As in the previous cases, there is no squeezing for small values of $|z|$. Moreover, the squeezing regions are slenderer as $|z| \to \infty$. To illustrate the phenomenon, in Figure 20 we show the behavior of the U-functions that define the variances $(\Delta\hat{x})^2$ and $(\Delta\hat{p})^2$ of the even distorted cat $|z_+^{osc}, w\rangle$. As $U_1^{(+)}$ is positive definite along the real axis, see Figure 20c, the quadrature $\hat{p}$ is squeezed in the intervals of the real axis where the function $U_2^{(+)}$ is non-negative. The latter can be identified in Figure 20a. Similarly, from Figure 20c we see that $U_2^{(+)}$ is negative along the imaginary axis. Then, $\hat{x}$ is squeezed in the intervals of the imaginary axis for which $U_1^{(+)}$ is negative, see Figure 20b.

In Figures 21 and 22 we show the density plot of the Wigner distribution of the even and odd cat states of Equation (42), respectively. In both cases we have used $|z_\pm^{osc}, w\rangle$ for $w = 1$, $w = 5$, and different values of z. The Wigner distribution is negative in different regions of the complex z-plane, and it exhibits oscillations as $|z|$ increases. The distortion parameter w is used to control such oscillations by reducing their number as w increases.

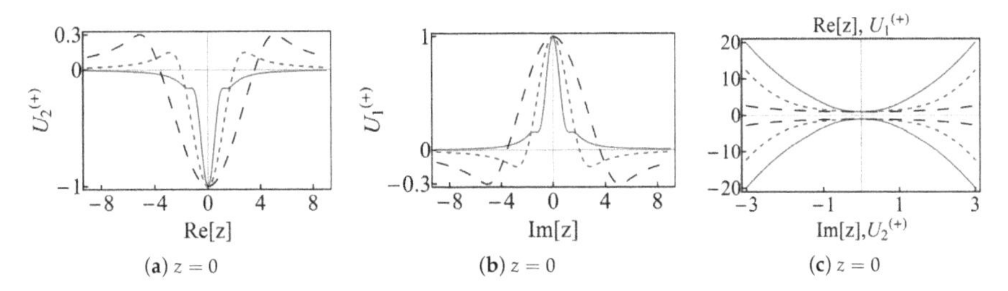

(**a**) $z = 0$ (**b**) $z = 0$ (**c**) $z = 0$

Figure 20. (Color online) the parameters $U_1^{(+)}$ and $U_2^{(+)}$ defining the variances of the physical quadratures $\hat{x}$ and $\hat{p}$ for the even distorted cat state $|z_+^{osc}, w\rangle$. In (**a**) we show the behavior of $U_2^{(+)}$ along the real axis of the complex z-plane. In (**b**) it is shown $U_1^{(+)}$ along the imaginary axis. The graphic (**c**) shows $U_1^{(+)}$ and $U_2^{(+)}$ along the imaginary and the real axis, respectively. In all the graphics $w = 1$ is in solid-red, $w = 5$ in dotted-blue, and $w = 20$ in dashed-black curves.

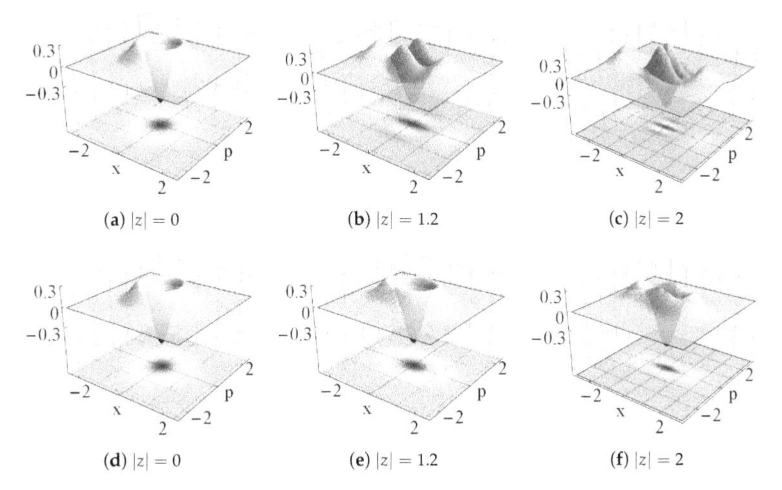

(**a**) $|z| = 0$ (**b**) $|z| = 1.2$ (**c**) $|z| = 2$

(**d**) $|z| = 0$ (**e**) $|z| = 1.2$ (**f**) $|z| = 2$

Figure 21. Wigner distribution of the distorted even cat states $|z_+^{osc}, w\rangle$ for $w = 1$ (upper row), $w = 5$ (lower row). Here $z = |z|e^{i\phi}$, with $\phi = 0$, and the indicated values of $|z|$.

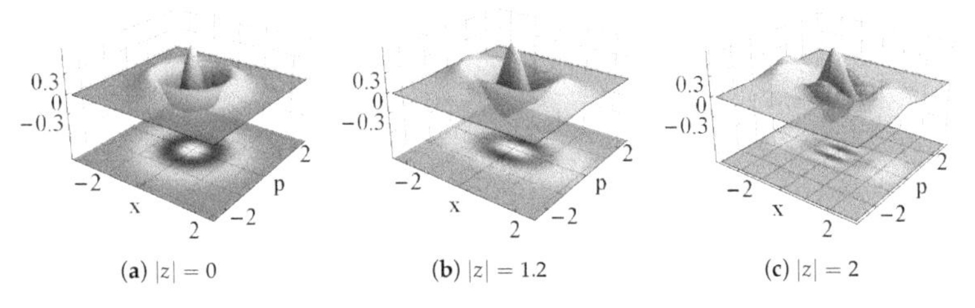

(**a**) $|z| = 0$ (**b**) $|z| = 1.2$ (**c**) $|z| = 2$

Figure 22. *Cont.*

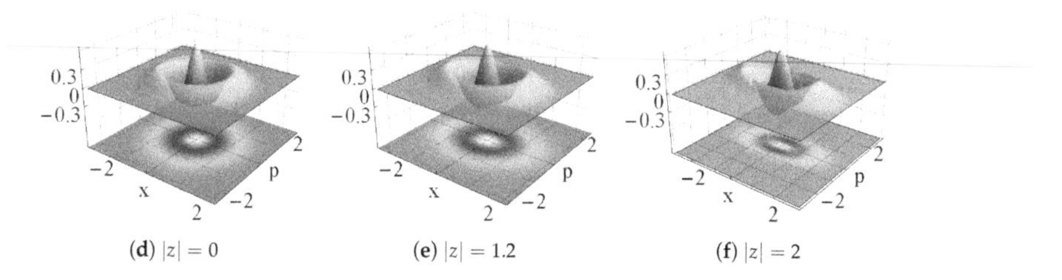

(**d**) $|z| = 0$ (**e**) $|z| = 1.2$ (**f**) $|z| = 2$

Figure 22. Wigner distribution of the distorted odd cat states $|z_-^{osc}, w\rangle$ for $w = 1$ (upper row), $w = 5$ (lower row). Here $z = |z|e^{i\phi}$, with $\phi = 0$, and the indicated values of $|z|$.

On the other hand, the straightforward calculation shows that neither X_w nor P_w are squeezed for the odd distorted cats $|z_-, w\rangle$. The same holds for the physical quadratures $\hat{x}$ and $\hat{p}$, evaluated in the oscillator limit state $|z_-^{osc}, w\rangle$.

4.4. Nonclassical Displaced Coherent States

Let us analyze the properties of the displaced coherent states $|\psi_0\rangle$ and $|z, w\rangle_d$ introduced in Equation (29). In the oscillator limit, the variances of the physical quadratures $\hat{x}$ and $\hat{p}$ are derived from Equation (A17) of Appendix B, with

$$\begin{aligned} U_1 &= \frac{\mathrm{Re}(z^2)h_2(|z|^2,\omega) + {}_2F_2\left(2,\omega;1,1;|z|^2\right)}{{}_1F_1(1,\omega;|z|^2)} - 2\left[\frac{\mathrm{Re}(z)h_1(z,\omega)}{{}_1F_1(1,\omega;|z|^2)}\right]^2, \\ U_2 &= \frac{\mathrm{Re}(z^2)h_2(|z|^2,\omega) - {}_2F_2\left(2,\omega;1,1;|z|^2\right)}{{}_1F_1(1,\omega;|z|^2)} + 2\left[\frac{\mathrm{Im}(z)h_1(z,\omega)}{{}_1F_1(1,\omega;|z|^2)}\right]^2, \end{aligned} \tag{45}$$

and

$$\begin{aligned} h_1(z,\omega) &= \sum_{n=0}^{+\infty} \frac{|z|^{2n}}{n!} \frac{\sqrt{(\omega)_n(\omega)_{n+1}(n+2)}}{(n+1)!}, \\ h_2(|z|^2,\omega) &= \sum_{n=0}^{+\infty} \frac{|z|^{2n}}{n!} \frac{\sqrt{(\omega)_n(\omega)_{n+2}(n+2)(n+3)}}{(n+2)!}. \end{aligned} \tag{46}$$

The squeezing of $\hat{x}$ and $\hat{p}$ for $|z^{osc}, w\rangle_d$ is quite similar to that obtained for the distorted coherent states of the previous section.

On the other hand, for the even and odd displaced coherent states

$$|z_\pm, w\rangle_d = \left[1 \pm \tfrac{{}_1F_1(w,1,-|z|^2)}{{}_1F_1(w,1,|z|^2)}\right]^{-1/2} \left(|z,w\rangle_d \pm |-z,w\rangle_d\right), \tag{47}$$

where the variances of the distorted quadratures of Equation (A15) are calculated from Equation (A17) of Appendix B, with

$$U_1^{(\pm)} = \frac{\mathrm{Re}(z^2)\left[h_2(|z|^2,w) \pm h_2(-|z|^2,w)\right] + \left[{}_2F_2\left(2,\omega;1,1;|z|^2\right) \pm {}_2F_2\left(2,\omega;1,1;-|z|^2\right)\right]}{{}_1F_1(1,\omega;|z|^2) \pm {}_1F_1(1,\omega;-|z|^2)},$$

$$U_2^{(\pm)} = \frac{\mathrm{Re}(z^2)\left[h_2(|z|^2,w) \pm h_2(-|z|^2,w)\right] - \left[{}_2F_2\left(2,\omega;1,1;|z|^2\right) \pm {}_2F_2\left(2,\omega;1,1;-|z|^2\right)\right]}{{}_1F_1(1,\omega;|z|^2) \pm {}_1F_1(1,\omega;-|z|^2)},$$

and $h_2(|z|^2, w)$ given in Equation (46). From Figure 23 we see that, along the real axis, the function $U_1^{(+)}$ is positive definite for any w while $U_2^{(+)}$ is non-negative for $w = 1$ only. Similarly, along the imaginary axis, $U_2^{(+)}$ is always negative but only for $w = 1$ the function $U_1^{(+)}$ is negative. The conclusion is that only for $w = 1$ we obtain the squeezing of either X_w or P_w. Remarkably, for this value of the parameter of distortion, the distorted and the displaced coherent states coincide.

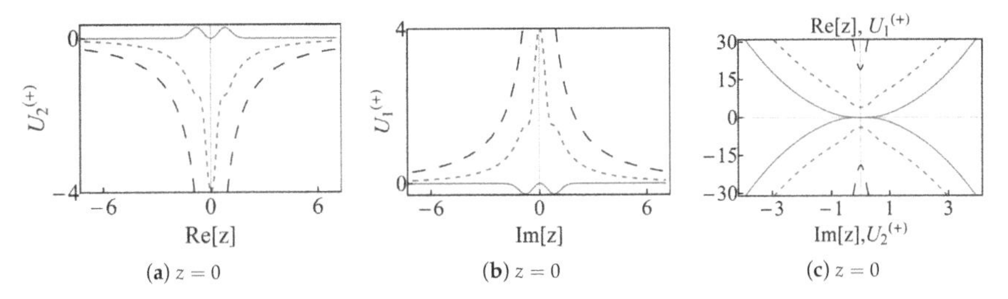

(**a**) z = 0 (**b**) z = 0 (**c**) z = 0

Figure 23. (Color online) the parameters $U_1^{(+)}$ and $U_2^{(+)}$ defining the variances of the distorted quadratures X_w and P_w according to Equation (A17) of Appendix B for the even displaced cat state $|z_+, w\rangle_d$. In (**a**) we show the behavior of $U_2^{(+)}$ along the real axis of the complex z-plane. In (**b**) it is shown $U_1^{(+)}$ along the imaginary axis. The graphic (**c**) shows $U_1^{(+)}$ and $U_2^{(+)}$ along the imaginary and the real axis, respectively. In all the graphics $w = 1$ is in solid-red, $w = 5$ in dotted-blue, and $w = 20$ in dashed-black curves.

5. Conclusions

We have studied the nonclassical properties of some pure states of the non-Hermitian Hamiltonians with real spectrum that are generated by Darboux (supersymmetric) transformations. Our interest has been focused on non-Hermitian oscillators since their eigenfunctions form bi-orthogonal bases of states in a straightforward form [24]. Depending on the parameters that define the Darboux transformation, these oscillators may be either parity-time-symmetric or non-parity-time-symmetric [13]. The pure states studied in this work are constructed as superpositions of the related bi-orthogonal basis, they include the so-called optimized Binomial and optimized Poisson states [26,30], together with the even and odd versions of the generalized coherent states that are associated with the dynamical algebras of the non-Hermitian oscillators [24,30].

One of the main results reported in this paper is to show that the techniques used to study classical-like properties for Hermitian systems can be adjusted to investigate the classicality of the states of non-Hermitian systems. The latter is relevant considering that, although the non-Hermitian systems are widely used in contemporary physics [16–22], the classicality of the related quantum states is rarely studied. The main difficulty is that non-Hermiticity of a given operator implies that the orthonormal properties of the related eigenfunctions are not granted a priori. Since most of the approaches implemented to analyze classicality assume orthonormality of states, it is not clear how to translate such methods to the non-Hermitian case in general. Such a difficulty can be overpassed by considering a bi-orthogonal basis for the space of states of the non-Hermitian system, just as we have shown throughout this work. It is expected that our approach can be applied in the study of resonances [16] (where the non-Hermiticity of the Hamiltonian defining the interaction of a scatterer with a projectile leads to the temporal trapping of the later by the former) and in the propagation of light in materials with complex refractive index [17] (where the imaginary part of the refractive index may model absorption or amplification of electromagnetic signals).

Concerning the non-Hermitian oscillators we have shown that the corresponding bi-orthogonal bases converge to the Fock basis in the appropriate limit (i.e., after canceling the imaginary part of the potential and properly selecting the parameters), so that the bi-orthogonal superpositions of such states are reduced to pure states of the conventional harmonic oscillator. In particular, we have found that the generalized coherent states of the non-Hermitian oscillators are not minimum uncertainty states for the physical position and momentum quadratures in the oscillator limit. Nevertheless, they exhibit some interesting nonclassical behavior. For instance, depending on the module and complex-phase of the coherence parameter α, the physical quadratures can be squeezed for the class of generalized coherent states that we have called "natural" (see Section 3.2). The squeezing is clearly exhibited in the corresponding Wigner distribution and revealed by the Mandel parameter for which it is shown that the photon number distribution is sub-Poissonian. The other class of generalized coherent states, called "distorted", have similar quadrature squeezing properties. However, the Mandel parameter reveals a different behavior since the photon distribution becomes sub-Poissonian for some values of $|\alpha|$, but it converges to a Poissonian distribution at the limit $|\alpha| \to \infty$. The latter implies a nonclassical-to-classical transition for the distorted coherent states (see [8] for the results on the matter for photon-added coherent states). Such a transition can be also manipulated with the parameter w that defines the related dynamical algebra.

Our method can be extended to construct and study intelligent states associated with the dynamical algebras of the non-Hermitian oscillators. The model involves finite difference equations, which deserve special treatment. Work in this direction is in progress.

Author Contributions: Each author contributed equally to this article. All authors have read and agree to the published version of the manuscript.

Funding: This research received no external funding.

Acknowledgments: The financial support from CONACyT (Mexico), grant number A1-S-24569, MINECO (Spain), project MTM2014- 57129-C2-1-P, and Junta de Castilla y León (Spain), project VA057U16, is acknowledged. K. Zelaya is supported by a postdoctoral fellowship from the Mathematical Physics Laboratory, Centre de Recherches Mathématiques. S. Dey acknowledges the support of research grant (DST/INSPIRE/04/2016/001391) from the DST, Govt. of India. V. Hussin acknowledges the support of research grants from NSERC of Canada.

Conflicts of Interest: The authors declare no conflict of interest.

Appendix A. Operator Algebras

The properties of the algebras used throughout this work are summarized below. For the sake of simplicity, hereafter we use the Dirac's notation to represent the states of the system.

- **Quadratic polynomial Heisenberg algebra.** The first pair of ladder operators, $\mathcal{A}$ and $\mathcal{A}^+$, together with the Hamiltonian H_λ, satisfy the *quadratic polynomial* (Heisenberg) algebra introduced in Equation (13):

$$[\mathcal{A},\mathcal{A}^+] = 2\,(3H_\lambda+1)\,(H_\lambda+1), \quad [H_\lambda,\mathcal{A}] = -2\mathcal{A}, \quad [H_\lambda,\mathcal{A}^+] = 2\mathcal{A}^+. \tag{A1}$$

The action of $\mathcal{A}$ and $\mathcal{A}^+$ on the eigenvectors of H_λ is as follows

$$\begin{gathered}\mathcal{A}|\psi_{n+1}\rangle = 2n\sqrt{2(n+1)}|\psi_n\rangle, \quad \mathcal{A}^+|\psi_{n+1}\rangle = 2(n+1)\sqrt{2(n+2)}|\psi_{n+2}\rangle, \\ \mathcal{A}|\psi_0\rangle = \mathcal{A}^+|\psi_0\rangle = 0, \quad n\geq 0.\end{gathered} \tag{A2}$$

Thus, $\mathcal{A}$ annihilates the vectors $|\psi_1\rangle$ and $|\psi_0\rangle$, and $\mathcal{A}^+$ annihilates the vector $|\psi_0\rangle$. In the Hermitian case ($\lambda = 0$), the above operators coincide with the generators of the *natural* SUSY algebra reported in [43]

$$\mathcal{A}|_{\lambda=0} \equiv a_{\mathcal{N}}^-, \quad \mathcal{A}^+|_{\lambda=0} \equiv a_{\mathcal{N}}^+. \tag{A3}$$

In the harmonic oscillator limit (6), they are reduced to the following f-oscillator [44] ladder operators:

$$\mathcal{A} \to \hat{a}_f = 2\hat{N}\hat{a}, \quad \mathcal{A}^+ \to \hat{a}_f^\dagger = 2\hat{a}^\dagger\hat{N}, \tag{A4}$$

where $\hat{N}$, $\hat{a}$, and $\hat{a}^\dagger$ are, respectively, the number, annihilation and creation operators of the (mathematical) harmonic oscillator

$$\hat{x} = \tfrac{1}{2}(\hat{a}^\dagger + \hat{a}), \quad \hat{p} = \tfrac{i}{2}(\hat{a}^\dagger - \hat{a}), \quad H_{osc} = \hat{p}^2 + \hat{x}^2 = \hat{a}^\dagger \hat{a} + 1 = 2\hat{N} + 1, \tag{A5}$$

with

$$[\hat{x}, \hat{p}] = i, \qquad \Delta\hat{x}\Delta\hat{p} \geq \tfrac{1}{2}, \tag{A6}$$

and

$$[\hat{a}, \hat{a}^\dagger] = 2, \quad [\hat{N}, \hat{a}] = -\hat{a}, \quad [N, \hat{a}^\dagger] = \hat{a}^\dagger, \quad \hat{N} = \tfrac{1}{2}\hat{a}^\dagger \hat{a}. \tag{A7}$$

Note that, $\hat{a}_f$ and $\hat{a}_f^\dagger$ operate on the set $\{|\varphi_n\rangle\}_{n\geq 0}$ quite similar to the form in which $\mathcal{A}$ and $\mathcal{A}^+$ operate on $\{|\psi_n\rangle\}_{n\geq 0}$. That is, $\hat{a}_f$ annihilates the vectors $|\varphi_1\rangle$ and $|\varphi_0\rangle$, and $\hat{a}_f^\dagger$ annihilates $|\varphi_0\rangle$.

One may introduce the quadrature operators corresponding to $\mathcal{A}$ and $\mathcal{A}^+$:

$$X_{\mathcal{N}} - \frac{1}{2}\left(\mathcal{A}^+ + \mathcal{A}\right), \quad P_{\mathcal{N}} = \frac{i}{2}\left(\mathcal{A}^+ - \mathcal{A}\right), \tag{A8}$$

which we call the 'natural quadratures'. They satisfy

$$\begin{aligned} [X_{\mathcal{N}}, P_{\mathcal{N}}] &= \tfrac{i}{2}[\mathcal{A}, \mathcal{A}^+] = i\,(3H_\lambda + 1)\,(H_\lambda + 1), \\ \Delta X_{\mathcal{N}} \Delta P_{\mathcal{N}} &\geq \tfrac{1}{2}\left|\langle (3H_\lambda + 1)(H_\lambda + 1)\rangle\right|. \end{aligned} \tag{A9}$$

- **Distorted Heisenberg algebra.** The second pair of ladder operators, denoted by $\mathcal{C}_w$ and $\mathcal{C}_w^+$, together with the Hamiltonian H_λ, and an additional operator I_w, satisfy the *distorted* (Heisenberg) algebra introduced in Equation (14):

$$[\mathcal{C}_w, \mathcal{C}_w^+] = I_w, \quad [H_\lambda, \mathcal{C}_w] = -2\mathcal{C}_w, \quad [H_\lambda, \mathcal{C}_w^+] = 2\mathcal{C}_w^+.$$

The action of $\mathcal{C}_w$, $\mathcal{C}_w^+$ and I_w on the eigenvectors of H_λ is as follows

$$\mathcal{C}_w|\psi_n\rangle = (1 - \delta_{n,0} - \delta_{n,1})\sqrt{n - 2 + w}\,|\psi_{n-1}\rangle, \tag{A10}$$

$$\mathcal{C}_w^+|\psi_n\rangle = (1 - \delta_{n,0})\sqrt{n - 1 + w}\,|\psi_{n+1}\rangle, \tag{A11}$$

$$I_w|\psi_n\rangle = [1 - \delta_{n,0} + \delta_{n,1}(w - 1)]\,|\psi_n\rangle, \tag{A12}$$

where $n \geq 0$ and w is a non-negative parameter that defines the 'distortion' of the oscillator algebra (A7). In the Hermitian case, these operators coincide with the generators of the *distorted* SUSY algebra reported in [45,46],

$$\mathcal{C}_w|_{\lambda=0} = C_w, \quad \mathcal{C}_w^+|_{\lambda=0} = C_w^\dagger. \tag{A13}$$

In the harmonic oscillator limit one has the f-oscillator ladder operators

$$\mathcal{C}_w \to \hat{c}_w = \frac{1}{2N}\sqrt{\frac{N + w - 1}{N + 1}}\,\hat{a}_{\mathcal{N}}, \quad \mathcal{C}_w^+ \to \hat{c}_w^\dagger = \frac{1}{2(N-1)}\sqrt{\frac{N + w - 2}{N}}\,\hat{a}_{\mathcal{N}}^\dagger. \tag{A14}$$

As in the previous case, $\hat{c}_w$ annihilates the vectors $|\varphi_1\rangle$ and $|\varphi_0\rangle$ while $\hat{c}_w^\dagger$ annihilates $|\varphi_0\rangle$. The corresponding quadrature operators are given by

$$X_w = \frac{1}{2}\left(\mathcal{C}_w^+ + \mathcal{C}_w\right), \quad P_w = \frac{i}{2}\left(\mathcal{C}_w^+ - \mathcal{C}_w\right), \tag{A15}$$

which will be called 'distorted quadratures' and they satisfy

$$[X_w, P_w] = \tfrac{i}{2}[\mathcal{C}_w, \mathcal{C}_w^+] = \tfrac{i}{2} I_w, \qquad \Delta X_w \Delta P_w \geq \tfrac{1}{4}|\langle I_w \rangle|. \tag{A16}$$

Appendix B. Nonclassicality Criteria

The criteria used to analyze the nonclassicality of the superpositions throughout this work are the following.

- **Squeezing.** For any two operators A and B with commutator $[A, B] = iC$, the variances can be expressed as

$$(\Delta A)^2 = \langle A^2 \rangle - \langle A \rangle^2 = \tfrac{1}{2}|\langle C \rangle| + U_1, \quad (\Delta B)^2 = \langle B^2 \rangle - \langle B \rangle^2 = \tfrac{1}{2}|\langle C \rangle| - U_2. \tag{A17}$$

If U_1 and U_2 are both equal to zero then the root-mean-square deviations become equal $\Delta A = \Delta B = \sqrt{\frac{1}{2}|\langle C \rangle|}$, and the uncertainty relationship between A and B is minimized, $\Delta A \Delta B = \frac{1}{2}|\langle C \rangle|$. If $U_1 \neq U_2 \neq 0$ we have two different cases (i) U_1 and U_2 are both positive, then $\Delta A > \Delta B$ and we say that B is squeezed (ii) U_1 and U_2 are both negative, then $\Delta A < \Delta B$ and we say that A is squeezed. This criterion is used along the paper to analyze the inequalities (A9) and (A16) in their respective state spaces.

On the other hand, it is well known that the Mandel parameter [47]

$$Q = \frac{(\Delta N)^2}{\langle N \rangle} - 1, \tag{A18}$$

dictates a sub-Poissonian, Poissonian and super-Poissonian photon number distribution for $-1 \leq Q < 0$, $Q = 0$ and $Q > 0$, respectively. Nonclassicality corresponds to the sub-Poissonian distributions, which is associated with the squeezing of the photon number $0 \leq (\Delta N)^2 < \langle N \rangle$.
- **Beam-splitter technique.** The action of a beam splitter on a given state $|\text{in}\rangle = |\Phi_1\rangle \otimes |\Phi_2\rangle$ is that it produces non-separable outputs $|\text{out}\rangle$ in general [11]. If $|\text{out}\rangle$ is entangled, then the signal $|\Phi_1\rangle$ is nonclassical, even if the ancilla $|\Phi_2\rangle$ is a classical state. The latter criterion is used with $|\Phi_2\rangle = |\varphi_0\rangle$, which is classical, and $|\Phi_1\rangle$ being any of the bi-orthogonal superpositions at the oscillator limit (6). Thus, we may write

$$|\text{in}^{osc}\rangle = \sum_{k=0}^{K} c_k |\varphi_{k+r}\rangle \otimes |\varphi_0\rangle, \tag{A19}$$

where the super-label "osc" means that the oscillator limit (6) has been applied. The action of the beam-splitter is represented by the unitary operator [11]

$$BS = \exp\left[\frac{\theta}{2}\left(a_1^\dagger \otimes a_2 e^{i\phi} - a_1 \otimes a_2^\dagger e^{-i\phi}\right)\right], \tag{A20}$$

where a_k and $a_k^\dagger$ are the ladder operators acting on the input states $|\Phi_k^{osc}\rangle$, with $k = 1, 2$. Up to a phase, the reflection and transmission coefficients of the beam-splitter are, respectively, given by $R = \sin\frac{\theta}{2}$ and $T = \cos\frac{\theta}{2}$, with $\theta \in [0, \pi)$. Using $|\text{in}^{osc}\rangle$, the straightforward calculation shows that the output state is of the form

$$|\text{out}^{osc}\rangle = \left(\sum_{k=0}^{K+r} \sum_{p=0}^{K+r-k} - \sum_{k=0}^{r-1} \sum_{p=0}^{r-1-k} \right) \Gamma_{k,p} |\varphi_k\rangle \otimes |\varphi_p\rangle, \tag{A21}$$

where

$$\Gamma_{k,p} = e^{-i(\phi-\pi)p} R^p T^k \sqrt{\frac{(k+p)!}{k!p!}}\, c_{k+p-r}. \tag{A22}$$

The *purity* (linear entropy) $S_L(\rho_1) = 1 - \mathrm{Tr}(\rho_1^2)$ of the signal state $\rho_1 = \mathrm{Tr}_2(|\mathrm{out}^{osc}\rangle\langle \mathrm{out}^{osc}|)$ is given by

$$\begin{aligned} S_L(\rho_1) = 1 - \mathrm{Tr}(\rho_1^2) &= \sum_{n,m=0}^{K} |F_{n+r,m+r}|^2 + \sum_{n,m=0}^{r-1} |F_{n,m} + G_{n,m} - H_{n,m} - H^*_{m,n}|^2 \\ &\quad + 2\sum_{n=0}^{r-1}\sum_{m=0}^{K} |F_{n,m+r} - H^*_{m+r,n}|^2, \end{aligned} \tag{A23}$$

with

$$\begin{aligned} F_{n,m} &= \sum_{p=0}^{K+r-\mathrm{Max}\{n,m\}} \Gamma_{n,p}\Gamma^*_{m,p}, \qquad G_{n,m} = \sum_{p=0}^{r-1-\mathrm{Max}\{n,m\}} \Gamma_{n,p}\Gamma^*_{m,p}, \\ H_{n,m} &= \sum_{p=0}^{\mathrm{Max}\{K+r-n, r-1-m\}} \Gamma_{n,p}\Gamma^*_{m,p}. \end{aligned} \tag{A24}$$

Classical states satisfy the separability condition $S_L = 0$ while the maximal entanglement is obtained for $S_L = 1$. Then, the nonclassicality is associated with $0 < S_L \leq 1$.

The results for $|\phi^{(\gamma)}\rangle$ are obtained by making $r = 0$, at the limit $K \to +\infty$. In such case, the last two additive terms in (A23) are equal to zero, so that $S_L(\rho_1)$ acquires the form of the linear entropy studied in, e.g., [37,39].

- **Wigner function.** In the basis of the Glauber states [2]:

$$|\alpha_G\rangle = e^{-|\alpha|^2/2} \sum_{k=0}^{+\infty} \frac{\alpha^k}{\sqrt{k!}} |\varphi_k\rangle, \quad \alpha \in \mathbb{C}, \tag{A25}$$

the Wigner function [1] of the state $\rho_\phi = |\phi\rangle\langle\phi|$, with $|\phi\rangle$ given either by (17) or (25), is expressed as

$$W(\alpha,\phi) = \tfrac{2}{\pi} e^{2|\alpha|^2} \int d^2\beta\, \langle -\beta|\rho_\phi|\beta\rangle e^{2(\beta^*\alpha - \beta\alpha^*)}, \quad \beta \in \mathbb{C}. \tag{A26}$$

If the Wigner function $W(\alpha,\phi)$ is negative in at least a definite region of the phase-space, then the state ρ_ϕ is nonclassical.

References

1. Wigner, E. On the Quantum Correction For Thermodynamic Equilibrium. *Phys. Rev.* **1932**, 40, 749–759. [CrossRef]
2. Glauber, R.J. *Quantum Theory of Optical Coherence*; Selected Papers and Lectures; Wiley: Hoboken, NJ, USA, 2007.
3. Rosas-Ortiz, O. Coherent and Squeezed States: Introductory Review of Basic Notions, Properties and Generalizations. In *Integrability, Supersymmetry and Coherent States*; CRM Series in Mathematical Physics; Kuru, S., Negro, J., Nieto, L.M., Eds.; Springer: Berlin/Heidelberg, Germany, 2019.
4. Walls, D.F. Squeezed states of light. *Nature* **1983**, *306*, 141–146. [CrossRef]
5. Dodonov, V.V.; Malkin, I.A.; Man'ko, V.I. Even and odd coherent states and excitations of a singular oscillator. *Physica* **1974**, *72*, 597–615. [CrossRef]
6. Stoler, D.; Salek, B.E.A.; Teich, M.C. Binomial states of the quantized radial field. *Opt. Acta* **1985**, *32*, 345–355. [CrossRef]
7. Lee, C.T. Photon antibunching in a free-electron laser. *Phys. Rev. A* **1985**, *31*, 1213–1215. [CrossRef]
8. Agarwal, A.S.; Tara, K. Nonclassical properties of states generated by the excitations on a coherent state. *Phys. Rev. A* **1991**, *43*, 492–497. [CrossRef]

9. Aczel, A.D. *Entanglement*; Plume: New York, NY, USA, 2013.
10. Nielsen, M.A.; Chuang, I.L. *Quantum Theory and Quantum Information*; Cambridge University Press: Cambridge, UK, 2000.
11. Kim, M.S.; Son, W.; Buzek, V.; Knight, P.L. Entanglement by a beam splitter: Nonclassicality as a prerequisite for entanglement. *Phys. Rev. A* **2002**, *65*, 032323. [CrossRef]
12. Scheel, S.; Welsch, D.G. Entanglement generation and degradation by passive optical devices. *Phys. Rev. A* **2001**, *64*, 063811. [CrossRef]
13. Rosas-Ortiz, O.; Castanos, O.; Schuch, D. New supersymmetry-generated complex potentials with real spectra. *J. Phys. A: Math. Theor.* **2015**, *48*, 445302. [CrossRef]
14. Blanco-Garcia, Z.; Rosas-Ortiz, O.; Zelaya, K. Interplay between Riccati, Ermakov and Schrödinger equations to produce complex-valued potentials with real energy spectrum. *Math. Meth. Appl. Sci.* **2019**, *42*, 4925–4938. [CrossRef]
15. Bagarello, F.; Passante, R.; Trapani, C. (Eds.) *Non-Hermitian Hamiltonians in Quantum Physics*; Springer: Cham, Switzerland, 2016.
16. Moiseyev, N. *Non-Hermitian Quantum Mechanics*; Cambridge University Press: New York, NY, USA, 2011.
17. Mosk, A.P.; Lagendijk, A.; Lerosey, G.; Fink, M. Controlling waves in space and time for imaging and focusing in complex media. *Nat. Photonics* **2012**, *6*, 282–292. [CrossRef]
18. Bender, C.M. Making sense of non-Hermitian Hamiltonians. *Rep. Prog. Phys.* **2007**, *70*, 947–1018. [CrossRef]
19. Fernandez, H.M.; Guardiola, R.; Ros, J.; Znojil, M. Strong-coupling expansions for the $\mathcal{PT}$-symmetric oscillators $V(x) = a(ix) + b(ix)^2 + c(ix)^3$. *J. Phys. A: Math. Gen.* **1998**, *31*, 10105–10112. [CrossRef]
20. Fakhri, H.; Mojaveri, B.; Dehghani, A. Coherent states and Schwinger models for pseudo generalization of the Heisenberg algebra. *Mod. Phys. Lett. A* **2009**, *24*, 2039–2051. [CrossRef]
21. Faisal, F.H.M. *Theory of Multiphoton Processes*; Springer: New York, NY, USA, 1987.
22. Simon, D.S.; Jaeger, G.; Sergienko, A.V. *Quantum Metrology, Imaging, and Communication*; Springer: New York, NY, USA, 2017.
23. Dey, S.; Fring, A.; Hussin, V. A Squeezed Review on Coherent States and Nonclassicality for Non-Hermitian Systems with Minimal Length. In *Coherent States and Their Applications*; Springer Proceedings in Physics; Antoine, J.P., Bagarello, F., Gazeau, J.P., Eds.; Springer: Cham, Switzerland, 2018; Volume 205.
24. Rosas-Ortiz, O.; Zelaya, K. Bi-Orthogonal Approach to Non-Hermitian Hamiltonians with the Oscillator Spectrum: Generalized Coherent States for Nonlinear Algebras. *Ann. Phys.* **2018**, *388*, 26–53. [CrossRef]
25. Jaimes-Nájera, A.; Rosas-Ortiz, O. Interlace properties for the real and imaginary parts of the wave functions of complex-valued potentials with real spectrum. *Ann. Phys.* **2017**, *376*, 126–149. [CrossRef]
26. Zelaya, K.D.; Rosas-Ortiz, O. Optimized Binomial Quantum States of Complex Oscillators with Real Spectrum. *J. Phys. Conf. Ser.* **2016**, *698*, 012026. [CrossRef]
27. Olver, F.W.J.; Lozier, D.W.; Boisvert, R.F.; Clark, C.W. (Eds.) *NIST Handbook of Mathematical Functions*; Cambridge University Press: Cambridge, UK, 2010.
28. Abraham, P.B.; Moses, H.E. Changes in potentials due to changes in the point spectrum: Anharmonic oscillators with exact solutions. *Phys. Rev. A* **1980**, *22*, 1333–1340. [CrossRef]
29. Mielnik, B. Factorization method and new potentials with the oscillator spectrum. *J. Math. Phys.* **1984**, *25*, 3387–3389. [CrossRef]
30. Zelaya, K. Non-Hermitian and Time-Dependent Systems: Exact Solutions, Generating Algebras and Nonclassicality of Quantum States. Ph.D. Thesis, Cinvestav: Physics Department, Mexico City, Mexico, 2019.
31. Mielnik, B.; Rosas-Ortiz, O. Factorization: Little or great algorithm? *J. Phys. A Math. Gen.* **2004**, *37*, 10007–10035. [CrossRef]
32. Glauber, R.J. Coherent and incoherent states of the radiation field. *Phys. Rev.* **1963**, *131*, 2766–2788. [CrossRef]
33. Sudarshan, E.C.G. Equivalence of semiclassical and quantum mechanical descriptions of statistical light beams. *Phys. Rev. Lett.* **1963**, *10*, 277–279. [CrossRef]
34. Hollenhorst, J.N. Quantum limits on resonant-mass gravitational-radiation detectors. *Phys. Rev. D* **1979**, *19*, 1669–1679. [CrossRef]
35. Caves, C.M. Quantum-mechanical noise in an interferometer. *Phys. Rev. D* **1981**, *23*, 1693–1708. [CrossRef]
36. Dodonov, V.V.; Man'ko, V.I. (Eds.) *Theory of Nonclassical States of Light*; Taylor and Francis: New York, NY, USA, 2003.

37. Dey, S.; Hussin, V. Entangled squeezed states in noncommutative spaces with minimal length uncertainty relations. *Phys. Rev. D* **2015**, *91*, 124017. [CrossRef]
38. Dey, S. Q-deformed noncommutative cat states and their nonclassical properties. *Phys. Rev. D* **2015**, *91*, 044024. [CrossRef]
39. Dey, S.; Fring, A.; Hussin, V. Nonclassicality versus entanglement in a noncommutative space. *Int. J. Mod. Phys. B* **2017**, *31*, 1650248. [CrossRef]
40. Zelaya, K.; Rosas-Ortiz, O.; Blanco-Garcia, Z.; Cruz, S. Completeness and Nonclassicality of Coherent States for Generalized Oscillator Algebras. *Adv. Math. Phys.* **2017**, *2017*, 7168592. [CrossRef]
41. Mojaveri, B.; Dehghani, A.; Jafarzadeh, Bahrbeig, R. Excitation on the para-Bose states: Nonclassical properties. *Eur. Phys. J. Plus* **2018**, *133*, 346. [CrossRef]
42. Mojaveri, B.; Dehghani, A.; Jafarzadeh, Bahrbeig, R. Nonlinear coherent states of the para-Bose oscillator and their non-classical features. *Eur. Phys. J. Plus* **2018**, *133*, 529. [CrossRef]
43. Fernandez, D.J.; Hussin, V.; Rosas-Ortiz, O. Coherent states for Hamiltonians generated by supersymmetry. *J. Phys A: Math. Theor.* **2007**, 40, 6491–6511. [CrossRef]
44. Man'ko, V.I.; Marmo, G.; Sudarshan, E.C.G.; Zaccaria, F. F-oscillators and nonlinear coherent states. *Phys. Scr.* **1997**, *55*, 528–541. [CrossRef]
45. Fernandez, C.D.J.; Nieto, L.M.; Rosas-Ortiz, O. Distorted Heisenberg Algebra and Coherent States for Isospectral Oscillator Hamiltonians. *J. Phys. A Math. Gen.* **1995**, *28*, 2693–2708. [CrossRef]
46. Rosas-Ortiz, J.O. Fock-Bargman Representation of the Distorted Heisenberg Algebra. *J. Phys. A Math. Gen.* **1996**, *29*, 3281–3288. [CrossRef]
47. Mandel, L. Sub-Poissonian photon statistics in resonance fluorescence. *Opt. Lett.* **1979**, *4*, 205–207. [CrossRef]

Article

Remembering George Sudarshan

Alberto Ibort [1,2,*] and Giuseppe Marmo [3]

1 ICMAT, Instituto de Ciencias Matematicas (CSIC-UAM-UC3M-UCM), Nicolás Cabrera, 1315, Campus de Cantoblanco, UAM, 28049 Madrid, Spain
2 Dipartimento de Matemáticas, Universitario Carlos III de Madrid, Avda. de la Universidad 30, Leganés, 28911 Madrid, Spain
3 Dipartimento di Fisica "E. Pancini", Università di Napoli Federico II and Sezione INFN di Napoli. Complesso Universitario di Monte S. Angelo, via Cintia, 80126 Naples, Italy; marmo@na.infn.it
* Correspondence: albertoi@math.uc3m.es

Received: 16 November 2019; Accepted: 27 November 2019; Published: 2 December 2019

Abstract: In these brief notes we want to render homage to the memory of E.C.G. Sudarshan, adding it to the many contributions devoted to preserve his memory from a personal point of view.

Keywords: Sudarshan; apology; no-interaction theorem

---

1. Sudarshan and the Foundations of Quantum Mechanics

E.C.G. Sudarshan has been one of the most innovative theoretical physicists of Indian origin in the second part of the 20th century. He has made epoch-making contributions to many areas of theoretical physics. We may quote, for instance, the V-A theory of weak interactions (formulated with his advisor R. Marshak [1]), which was the key for the formulation of electro-weak interactions. Indeed, as stressed by S. Weinberg [2], Sudarshan's formulation was essential for the development of a unified model of electromagnetic and weak interactions.

Following A. Einstein's famous quote: *Physical concepts are free creations of the human mind, and are not, however it may seem, uniquely determined by the external world (A. Einstein, The Evolution of Physics (1938) (co-written with Leopold Infeld))* we may agree that the intrinsic freedom of the researcher to provide interpretations of physical reality drinks heavily from the social, cultural and economical background of the individual. In this sense, Sudarshan's scientific work is a rare combination of western logic and precision with eastern imagination; Greek-type philosophy with an ancient Indian way of thought. Who else could have thought of what nowadays we call tachyons? [3,4].

We may characterize him as iconoclast, with marked absence of all-pervading cultural prejudices. These unique features of him, also make that his seminal work is often quoted but likely unread, often not really understood until considerable time has passed.

The deep common thread of his contribution has been an attempt to resolve the fundamental conflict between *being* and *becoming*, that is, between objects and processes: *All Natural Philosophy is characterized by one central problem: that is, to understand and describe the meaning of existence and the nature of change.*

An example of this was provided by the important contribution of Sudarshan to the formulation of quantum Zeno effect [5], by now experimentally verified, that finds applications in quantum information processing. It is difficult to find a better example of a physical situation where the tension between *being* and *becoming* is more apparent. But this peculiar way of thinking by Geoge has had an enormous impact in other areas of physics. For instance, modern quantum optics has build on the concept of coherent states, again a counter-intuitive quantum mechanical notions whose discovery constituted a leap forward in the understanding of Quantum Mechanics. In a few years, it is clearly realized the connection between coherent states and group theory. It is remarkable that George's ideas are constantly followed up by experimentalists thanks to the developments in experimental techniques.

2. Contributions to Physics

Apart from the already mentioned contributions, many other contributions where discussed by leading scientists in the IOP open-access Journal Conferences in occasion of his 75th birthday:

- *Sudarshan: Seven Science Quest*, Austin, Texas, US 6–7 November 2006. Journal of Physics: Conference Series, Volume 196, Number 1.
- *Particle and Fields: Classical and Quantum*, Jaca, Huesca, Spain 18–21 September 2006. Journal of Physics: Conference Series, Volume 87.

The Seven Quest that were referred to in the first paper are:

- V-A Symmetry.
- Spin and Statistics.
- Quantum Coherence.
- Quantum Zeno Effect.
- Tachyons.
- Open systems.

In the ten years elapsed since other topics were investigated, we would add: Entanglement and Tomography (see, for instance, [6–9] and referencers therein).

All of the Seven Quest are commented by Sudarshan himself in the two quoted open-access journal for conferences, therefore we would refer to these issues for additional information. As for Open Systems, we refer the reader to the article: *A brief History of the GKLS Equation*, by Darek Chruscinski and Saverio Pascazio [10].

George wrote over 500 papers, alone or with his students and collaborators all over the world. We list here the books written by ECG Sudarshan in collaboration with a number of collaborators:

- *Introduction to Elementary Particle Physics* with R. Marshak. Wiley Interscience, 1962.
- *Introduction to Quantum Optics* with John Klauder, W.A. Benjamin 1968.
- *Classical Dynamics: A Modern Perspective* with N. Mukunda, J. Wiley 1974.
- *100 Years of Planck Quantum* with Ian M. Duck. World Scientific, 1996.
- *W. Pauli and the Spin-Statistic Theorem* with Ian M. Duck. World Scientific, 1997.
- *Doubt and Certainty, conversations with Tony Rothman*. Perseus Book, 2000.
- *From Classical to Quantum Mechanics* with G. Esposito, G. Marmo. Cambridge UP, 2004.
- *Advanced Concepts in Quantum Mechanics* with G. Esposito, G. Marmo, G. Miele. Cambridge UP, 2015.

3. Some Biographical Traits

Ennackal Chandy George Sudarshan was born in Pallam, Kottayam District in the State of Kerala, India, on 16 September 1931, from Achamma, his mother Achamma, a school teacher, and his father, E.I. Chandy, who was revenue supervisor for the Kerala Government Service. He had two brothers and three sons. George attended the CMS College of Kottayam (1946–48), graduated B.Sc. at Madras, Christian College (1948–51) and got the Master of Arts Degree from Madras University (1952).

George Sudarshan joined the Tata Institute of Fundamental Research (TIFR) in Bombay in the spring of 1952 as a research student. He started collaborating with B. Peters in cosmic rays research, the greatest discovery he made by that time was that he could be a theoretician inspired by experimental research rather than being an experimentalist himself. The presence of a strong group in Mathematics at the Tata Institute allowed George Sudarshan to become acquainted with topology, modern algebra, theory of spinors and functional analysis.

Two years later, in 1954, Paul A.M. Dirac, a teacher of Bhabha at Cambridge in the 1930s, visited TIFR and gave a course of lectures on quantum mechanics. George collaborated with K.K. Gupta, a student of Bhabha, in the preparation of the lecture notes. This work gave him the occasion to be in close contact with Dirac, a formidable and unimaginably fortunate opportunity to learn the subject from the master himself. This experience shaped George's attitude to quantum mechanics all along his life, giving him the daring and courage to push and test the principles of the subject in many directions. ECG greatly admired Dirac, and the two remained lifelong friends.

Later George, in September 1955, went to Rochester for his Ph.D. Thesis with Bob Marshak (a former student of Hans A. Bethe). Here he started a life-long friendship with Susumo Okubo. Immediately after he went to Harvard University for two years as a post-doc with J. Schwinger.

With hindsight, one can say that some inspiration by Schwinger on George's later work may be found in the action principle (as enunciated by P. Weiss, is a variational principle which includes variations of the boundaries for the action integral); his presentation of classical mechanics with group theory, and, more generally, his pervasive use of group theory in many important papers. In his own words, in the opening of his book Classical Dynamics: A Modern Perspective, he says: This book is the public declaration of an "affair of the heart".

We owe much inspiration to three masters: Edmund Whittaker Paul Dirac, whose writings and words are an inspiration in all other endeavours in physics as well as in this; and Julian Schwinger, mentor to the senior author and the most extraordinary teacher of the spirit of dynamics.

In Harvard, George was in very good terms with Sheldon Glashow, another lifelong friend. From Boston, he went regularly back to Rochester, to continue the collaboration with Marshak and Okubo. Later, the group was joined by Steven Weinberg.

From Harvard, George moved back to Rochester (1959) as Assistant Professor, to become Full Professor in 1961. In 1963 he conceived a way to describe quantum states close to classical beams of light, the already mentioned coherent states, when he pointed out the importance of using the coherent states of the photon field [11].

After a one-year leave in Bern, he went to Syracuse as full professor in 1964. Here he gave birth to a group of particle physics (Syracuse, with the presence of Peter Bergmann, was mainly general relativity). In 1969, George moved again to join the University of Texas at Austin where he taught and continued his research until he passed away in 2017.

4. Sudarshan and Classical Dynamics

There is yet a seminal contribution by George related to relativistic invariance and hamiltonian theories of interacting particles. George always had a deep love for the structure of classical mechanics, continuing with his own words in the preface of his book on classical mechanics together with N. Mukunda [12]:

This book is the public declaration of an "affair of the heart" that we have had with classical dynamics for most of our adult lives. Indian tradition recognizes ten stages for love, beginning with the beauty of form, through stages of closeness and of agony, to ultimate bliss. It is our belief that the beauty of form of classical dynamics is only her calling card and recognition of this beauty but the first step. We see classical dynamics not as part of physics, but as physics itself. We see her form pervading all of physics...

Consider things in context. From 1960, and for almost thirty years, George kept returning to the canonical description of relativistic particles. In the eighties there was a renewed interest in the canonical description of relativistic interacting particles (many specialists in relativity contributed to the field, among others Arthur Komar, Joshua N. Goldberg, Fritz Rohrlich). The activity was aimed at "evading" the no-interaction theorem which George, with his collaborators, had established much earlier. We shall try to sketch the meaning and results of the theorem.

When addressing the description of the dynamics of interacting particles together with the requirements of special relativity, there are new features with respect to the Newtonian description. The traditional specification of the instantaneous state of a system of N interacting particles is to give

N triplets of Cartesian coordinates $\mathbf{q}$ and their conjugate momenta $\mathbf{p}$. The time evolution would be described by a Hamiltonian function of the $3N$ classical pairs $\mathbf{q}_a, \mathbf{p}_a$.

In the Hamiltonian formalism, the irrelevance of time origin implies that $\partial H/\partial t = 0$. Moreover, irrelevance of orientation of the coordinate frame in 3-space implies the immutability of the angular momentum $\sum_a \mathbf{q}_a \wedge \mathbf{p}_a$; which, in turn, implies the rotation invariance of the Hamiltonian function H. The irrelevance of the space origin implies the invariance of the Hamiltonian under common transformation of the Cartesian coordinates and the corresponding invariance of the total momentum $\sum_a \mathbf{p}_a$.

Thus, we have a seven parameter set of canonical transformations implementing time translations, space translations and space rotations:

$$\begin{aligned} \{H,\mathbf{P}\} &= 0, \qquad \{H,\mathbf{J}\} = 0, \\ \{\mathbf{P}\cdot\mathbf{a},\mathbf{P}\cdot\mathbf{b}\} &= 0, \qquad \{\mathbf{P}\cdot\mathbf{a},\mathbf{J}\cdot\boldsymbol{\theta}\} = (\boldsymbol{\theta}\wedge\mathbf{a})\cdot\mathbf{P}, \\ \{\mathbf{J}\cdot\boldsymbol{\theta},\mathbf{J}\cdot\boldsymbol{\varphi}\} &= (\boldsymbol{\varphi}\wedge\boldsymbol{\theta})\cdot\mathbf{J}. \end{aligned}$$

Poisson brackets are defined at a single time. For any dynamical variable F,

$$\frac{dF}{dt} = \{F,H\}.$$

All previous expressions look very non-relativistic. They employ a clock time t and dynamical variables determined simultaneously at the same time t. But special relativity tells us that distant simultaneity is not independent of the frame of reference.

Dirac showed that the invariance under changes of the relativistic frames may be implemented by adding boosts transformations to moving inertial frames. Thus, in any Hamiltonian relativistic theory there are ten generators which are canonically realized. The dynamical evolution is described by a Hamiltonian which is one of the ten generators in Dirac's generator formalism. To previous commutation relations we add:

$$\begin{aligned} \{\mathbf{K}\cdot\mathbf{u},H\} &= -\mathbf{P}\cdot\mathbf{u}, \\ \{\mathbf{K}\cdot\mathbf{u},\mathbf{P}\cdot\mathbf{a}\} &= -(\mathbf{u}\cdot\mathbf{a})\cdot H, \\ \{\mathbf{K}\cdot\mathbf{u},\mathbf{K}\cdot\mathbf{v}\} &= (\mathbf{u}\wedge\mathbf{v})\cdot\mathbf{J}. \end{aligned}$$

This gives Dirac's "instant form" of the relativistic dynamics for any canonical system (see [13] for a discussion of Dirac's forms of relativistic dynamics in this issue). From these relations, we learn that if the Hamiltonian changes (because of interactions) also the boosts are obliged to change.

Interacting systems may be constructed by choosing $H = H_0 + V$, where H_0 is appropriate for a collection of free particles, and V is a rotationally and translationally invariant momentum-dependent "potential". Existence of invariant world lines for each particle can be encoded in the following relations:

$$\{\mathbf{K}\cdot\mathbf{u},(\mathbf{q}_a)_j\} = (\mathbf{q}_a\cdot\mathbf{u})(\mathbf{v}_a)_j = \{(\mathbf{q}_a)_j,H\}(\mathbf{q}_a\cdot\mathbf{u}).$$

If we try to find out the interactions which are compatible with the world line condition (WLC) expressed above, we find that there are none, all accelerations must vanish if we have canonical realizations of the relativistic transformation group satisfying the WLC.

In the early eighties, the collaboration of Giuseppe Marmo with George Sudarshan started. Marmo joined Sudarshan, Mukunda, Balachandran, Nilsson and Zaccaria in the analysis of this problem and they provided a new proof of the no-interaction theorem based on a Lagrangian approach [14].

The Lagrangian proof of the no-interaction theorem. When going from the Hamiltonian to the Lagrangian formalism we face some novel problems. Hamiltonian equations, which are explicit differential equations:

$$\frac{dq_a}{dt} = \frac{\partial H}{\partial p_a}, \qquad \frac{dp_a}{dt} = -\frac{\partial H}{\partial q_a}$$

where q_a denote now a generic set of configuration coordinates and p_a the corresponding conjugate momenta, are replaced by Euler–Lagrange equations:

$$\frac{dq_a}{dt} = v_a, \qquad \frac{d}{dt}\frac{\partial L}{\partial v_a} = \frac{\partial L}{\partial q_a},$$

which are implicit differential equations. Spelling them out we get:

$$\frac{\partial^2 L}{\partial v_r \partial v_s}\frac{dv_s}{dt} + \frac{\partial^2 L}{\partial v_r \partial q_s}\frac{dq_s}{dt} = \frac{\partial L}{\partial q_r},$$

To put them into normal form we need:

$$\det\left(\frac{\partial^2 L}{\partial v_r \partial v_s}\right) \neq 0\,.$$

Whenever we have to deal with implicit differential equations, we will face difficulties with the notion of symmetries and constants of the motion (a convenient presentation of these issues will be done in [15]). However to implement the action of space translations and rotations we have no problems because they are implemented by point transformations. We encounter problems with the implementation of boosts because they cannot be realized as "point transformation". They do not respect the tangent bundle structure of the carrier space. Moreover, when the Lagrangian function is degenerate, there are no Poisson brackets directly available. By handling these various problems, it was found that the no-interaction theorem applies also for degenerate Lagrangians, as long as there are no second-class constraints.

From a better understanding of the proof it was possible to see how to evade the no-interaction theorem. However, by requiring a notion of separability for far apart particles, a novel and more profound no-interaction theorem was found [16]. The curious result was reminiscent of the EPR "paradox" in quantum theory [17,18] but the indicated circle of ideas seemed to point out that correlations between distant objects need not always involve transport of material influences. It might rather depend upon the indecomposable nature of the dynamical system itself. This was brought about by the imposition of the apparently innocent WLC. We have the feeling that a deeper analysis of the conditions and the reasons for the theorem to work would enlighten substantially the current problem of causality in quantum mechanics.

As a conclusion from the results of the new version of the no-interaction theorem we may take alternative points of view at this juncture. One can accept the result and proceed to a field formalism as the sole vehicle for particle interactions (this is the road taken in [15]). One could abandon differential equations of motion and replace them by integro-differential equations instead (perhaps these two methods are not essentially different).

A third way is to seek a more geometrical and explicitly invariant formalism from the start. In these descriptions one uses an explicitly invariant set of configuration variables to describe the system and guarantee that the system does describe particles interacting by imposing suitable constraints. In this geometrical constraint approach, one has an enlargement of Dirac's original framework: It is necessary to consider dynamical choices of temporal evolution variables rather than the kinematical choices offered by Dirac in his various forms of relativistic dynamics (see [13] for further discussion on these issues).

Author Contributions: Conceptualization, G.M.; writing-review and editing, G.M. and A.I.; funding acquisition, G.M. and A.I.

Funding: This research was funded by the Spanish Ministry of Economy and Competitiveness, through the Severo Ochoa Programme for Centres of Excellence in RD (SEV-2015/0554). A.I. and F.C. would like to thank partial support provided by the MINECO research project MTM2017-84098-P and QUITEMAD++, S2018/TCS-A4342. G.M. would like to thank the support provided by the Santander/UC3M Excellence Chair Programme 2019/2020.

Conflicts of Interest: The authors declare no conflict of interest.

References

1. Sudarshan, E.C.G.; Marshak, R.E. The Nature of the Four-Fermion Interaction. In Proceedings of the Conference on Mesons and Newly Discovered Particles, Padua, Venice, Italy, 22–27 September 1957; reprinted in *The Development of Weak Interaction Theory*; Kabir, P.K., Ed.; Gordon and Breach: New York, NY, USA, 1963; pp. 119–128.
2. Weinberg, S. Essay: Half a Century of the Standard Model. *Phys. Rev. Lett.* **2018**, *121*, 220001. [CrossRef] [PubMed]
3. Bilaniuk, O.M.P.; Deshpande, V.K.; Sudarshan, E.C.G. "Meta" relativity. *Am. J. Phys.* **1962**, *30*, 718. [CrossRef]
4. Arons, M.E.; Sudarshan, E.C.G. Lorentz invariance, local field theory, and faster-than-light particles. *Phys. Rev.* **1968**, *173*, 1622. [CrossRef]
5. Misra, B.; Sudarshan, E.C.G. The Zeno's paradox in quantum theory. *J. Math. Phys.* **1977**, *18*, 756. [CrossRef]
6. Asorey, M.; Facchi, P.; Manko, V.I.; Marmo, G.; Pascazio, S.; Sudarshan, E.C.G. Generalized Tomographic Maps. *Phys. Rev.* **2008**, *A 77*, 042115. [CrossRef]
7. Modi, K.; Sudarshan, E.C.G. Role of preparation in quantum process tomography. *Phys. Rev.* **2010**, *A 81*, 052119. [CrossRef]
8. Ibort, A.; Lopez-Yela, A.; Manko, V.I.; Marmo, G.; Simoni, A.; Ventriglia, F.; Sudarshan, E.C.G. On the tomographic description of classical fields. *Phys. Lett.* **2012**, *A 376*, 1417–1425. [CrossRef]
9. Manko, V.I.; Marmo, G.; Sudarshan, E.C.G.; Zaccaria, F. Positive maps of density matrix and a tomographic criterion of entanglement. *Phys. Lett.* **2004**, *A 327*, 353–364. [CrossRef]
10. Chruscinski, D.; Pascazio, S. Gorini-Kossakowski-Lindblad-Sudarshan Master Equation 40 years after. In Proceedings of the 48th Symposium on Mathematical Physics, Torun, Poland, 10–12 June 2016.
11. Sudarshan, E.C.G. Equivalence of Semiclassical and Quantum Mechanical Descriptions of Statistical Light Beams. *Phys. Rev. Lett.* **1963**, *10*, 277. [CrossRef]
12. Sudarshan, E.C.G.; Mukunda, N. *Classical Dynamics: A Modern Perspective*; John Wiley and Sons: New York, NY, USA, 1974.
13. Ciaglia, F.M.; di Cosmo, F.; Ibort, A.; Marmo, G. Description of Relativistic Dynamics with World Line Condition. *Quantum Rep.* **2019**, *1*, 181–192. [CrossRef]
14. Marmo, G.; Mukunda, N.; Sudarshan, E.C.G. Relativistic Particle Dynamics—Lagrangian Proof of the No-Interaction Theorem. *Phys. Rev. D* **1984**, *30*, 2110. [CrossRef]
15. Asorey, M.; Falceto, F.; Ibort, A.; Marmo, G. *Classical Field Theory: A Geometric Approach*; Springer: Berlin/Heidelberg, Germany, 2020.
16. Balachandran, A.; Dominici, D.; Marmo, G.; Mukunda, N.; Nilsson, J.; Samuel, J.; Sudarshan, E.C.G.; Zaccaria, F. Separability in Relativistic Hamiltonian Particle Dynamics. *Phys. Rev. D* **1982**, *26*, 3492. [CrossRef]
17. Bohm, D. *Quantum Theory*; Prentice Hall: New York, NY, USA, 1951.
18. Einstein, A.; Podolsky, B.; Rosen, N. Can Quantum-Mechanical Descripton of Physical Reality be Considered Complete? *Phys. Rev.* **1935**, *47*, 777–780. [CrossRef]

 *quantum reports*

Article

Coherent States for the Isotropic and Anisotropic 2D Harmonic Oscillators

James Moran [1,2,*] and Véronique Hussin [2,3,*]

1 Département de Physique, Université de Montréal, C. P. 6128, Succ. Centre-ville, Montréal, QC H3C 3J7, Canada
2 Centre de Recherches Mathématiques, Université de Montréal, C. P. 6128, Succ. Centre-ville, Montréal, QC H3C 3J7, Canada
3 Département de Mathématiques et de Statistique, Université de Montréal, C. P. 6128, Succ. Centre-ville, Montréal, QC H3C 3J7, Canada
* Correspondence: james.moran@umontreal.ca (J.M.); hussin@dms.umontreal.ca (V.H.)

Received: 13 September 2019; Accepted: 13 November 2019; Published: 15 November 2019

Abstract: In this paper we introduce a new method for constructing coherent states for 2D harmonic oscillators. In particular, we focus on both the isotropic and commensurate anisotropic instances of the 2D harmonic oscillator. We define a new set of ladder operators for the 2D system as a linear combination of the x and y ladder operators and construct the $SU(2)$ coherent states, where these are then used as the basis of expansion for Schrödinger-type coherent states of the 2D oscillators. We discuss the uncertainty relations for the new states and study the behaviour of their probability density functions in configuration space.

Keywords: coherent states; harmonic oscillator; $SU(2)$ coherent states; 2D coherent states; resolution of the identity; uncertainty principle, isotropic harmonic oscillator, anisotropic harmonic oscillator

1. Introduction

Degeneracy in the spectrum of the Hamiltonian is one of the first problems we encounter when trying to define a new type of coherent state for the 2D oscillator. Klauder described coherent states of the hydrogen atom [1] which preserved many of the usual properties required by coherent state analysis [2]. Fox and Choi proposed the Gaussian Klauder states [3], an alternative method for producing coherent states for more general systems with degenerate spectra. An analysis of the connection between the two definitions was studied in [4].

When labeling energy eigenstates of a 2D system, $|n, m\rangle$, there exist several representations of the state space. In this paper, we present a motivation for an $SU(2)$ representation of the state space. Discussions of alternate state-space representations, as well as its application to 2D magnetism, may be found in [5,6]. When generalising beyond 2D, there exist many more state-space representations, leading to many definitions of coherent states in higher dimensions.

In this work, we aim to develop an approach for constructing coherent states for 2D oscillators in both isotropic and commensurate anisotropic settings. We aim to minimally extend the standard definitions of coherent states in the 1D setting, and we determine new properties of the constructed coherent states for the 2D system.

In the first part of the paper, we address the degeneracy in the energy spectrum by constructing non-degenerate states, the $SU(2)$ coherent states. We define a generalised ladder operator formed from a linear combination of the 1D ladder operators with complex coefficients. The $SU(2)$ coherent states are then used as a basis of expansion to describe the Schrödinger-type coherent states for the 2D system.

In the second part of the paper, we modify the $SU(2)$ coherent states according to Chen [7] to produce coherent states for the commensurate anisotropic oscillator, and we discuss the emergent properties and their correspondence to Lissajous figures in configuration space. Finally, we suggest some future directions the work can take, as well as problems that may arise in more complicated systems than the oscillator.

2. Coherent States of the 1D Harmonic Oscillator

The very well-known coherent states of the 1D harmonic oscillator, labelled by $z \in \mathbb{C}$, satisfy

$$a^- |z\rangle = z\, |z\rangle\, ; \tag{1}$$

$$|z\rangle = e^{(za^+ - \bar{z}a^-)}\, |0\rangle \equiv D(z)\, |0\rangle\, ; \tag{2}$$

$$|z\rangle = e^{-\frac{|z|^2}{2}} \sum_{n=0}^{\infty} \frac{z^n}{\sqrt{n!}}\, |n\rangle\, ; \tag{3}$$

$$\Delta\, \hat{x} \Delta\, \hat{p} = \frac{1}{2}, \forall\, |z\rangle \quad \text{with} \quad \Delta\, \hat{x} = \Delta\, \hat{p}. \tag{4}$$

Equations (1)–(4) describe some of the basic definitions of coherent states. These definitions were formalised by Glauber and Sudarshan [8,9], but these minimal uncertainty wave-packets were first studied by Schrödinger [10], and so we will refer to them as Schrödinger-type coherent states throughout.

Furthermore, these properties can be used to show that the states $|z\rangle$ form an over-complete basis, and they resolve the identity in the following way:

$$\int \frac{d^2 z}{\pi}\, |z\rangle \langle z| = \sum_{n=0}^{\infty} |n\rangle \langle n| = \mathbb{I}_{\mathcal{H}}. \tag{5}$$

Here, $d^2 z = d\Re z\, d\Im z$. The basis is over-complete because the states $|z\rangle$ are not orthogonal, $\langle z'|z\rangle \neq 0$. In the theory of coherent states, the resolution of the identity is often taken as a basic requirement. This allows one to use the coherent states as a basis for describing other states in the space.

3. The 2D Oscillator

For a 2D isotropic oscillator, we have the quantum Hamiltonian

$$\hat{H} = -\frac{1}{2}\frac{d^2}{dx^2} - \frac{1}{2}\frac{d^2}{dy^2} + \frac{1}{2}x^2 + \frac{1}{2}y^2, \tag{6}$$

where we have set $\hbar = 1$, the mass $m = 1$, and the frequency $\omega = 1$. We solve the time-independent Schrödinger equation $H\,|\Psi\rangle = E\,|\Psi\rangle$ and obtain the usual energy eigenstates (or Fock states) labelled by $|\Psi\rangle = |n, m\rangle$ with eigenvalue $E_{n,m} = n + m + 1$ and $n, m \in \mathbb{Z}^{\geq 0}$. These states may all be generated by the action of raising and lowering the operators in the following way [11]:

$$\begin{aligned} a_x^- \,|n,m\rangle &= \sqrt{n}\,|n-1,m\rangle\,,\ a_x^+\,|n,m\rangle = \sqrt{n+1}\,|n+1,m\rangle\,; \\ a_y^- \,|n,m\rangle &= \sqrt{m}\,|n,m-1\rangle\,,\ a_y^+\,|n,m\rangle = \sqrt{m+1}\,|n,m+1\rangle\,. \end{aligned} \tag{7}$$

In configuration space, the states $|n, m\rangle$ have the following wave-function:

$$\langle x, y|n, m\rangle = \psi_n(x)\psi_m(y) = \frac{1}{\sqrt{2^{n+m} n! m!}} \sqrt{\frac{1}{\pi}} e^{-\frac{x^2}{2} - \frac{y^2}{2}} H_n(x)\, H_m(y)\,, \tag{8}$$

where $\psi_n(x) = \frac{1}{\sqrt{2^n n!}}\left(\frac{1}{\pi}\right)^{\frac{1}{4}} e^{-\frac{x^2}{2}} H_n(x)$ is the wave-function of the 1D oscillator, and $H_n(x)$ are the Hermite polynomials. For the physical position and momentum operators, $\hat{X}_i = \frac{1}{\sqrt{2}}(a_i^+ + a_i^-)$, $\hat{P}_i = \frac{1}{\sqrt{2}i}(a_i^- - a_i^+)$, respectively, and in the i direction, the states $|n,m\rangle$ satisfy the following

$$(\Delta\,\hat{X})^2_{|n,m\rangle} = (\Delta\hat{P}_x)^2_{|n,m\rangle} = \frac{1}{2} + n; \tag{9}$$

$$(\Delta\,\hat{Y})^2_{|n,m\rangle} = (\Delta\hat{P}_y)^2_{|n,m\rangle} = \frac{1}{2} + m, \tag{10}$$

where $(\Delta\,\hat{O})^2_{|\psi\rangle} \equiv \langle\psi|\,\hat{O}^2\,|\psi\rangle - \langle\psi|\,\hat{O}\,|\psi\rangle^2$ is the variance of the operator $\hat{O}$ in the state $|\psi\rangle$. They satisfy the Heisenberg uncertainty relation $(\Delta\;\hat{X})_{|n,m\rangle}(\Delta\hat{P}_x)_{|n,m\rangle} = \frac{1}{2} + n$, which grows linearly in n, and similarly for the Y quadratures.

In what follows, we will construct two new ladder operators as linear combinations of the operators in (7) and proceed to define a single indexed Fock state for the 2D system which yields the $SU(2)$ coherent states, as well as extend the definitions in Section 2 to obtain Schrödinger-type coherent states for the 2D system.

4. $SU(2)$ Coherent States

We extend the definitions of the ladder operators presented in Section 3 to apply to the 2D oscillator. Introducing a set of states $\{|\nu\rangle\}$, and defining a new set of ladder operators through their action on the set,

$$A^-\,|\nu\rangle = \sqrt{\nu}\,|\nu-1\rangle\,, \qquad A^+\,|\nu\rangle = \sqrt{\nu+1}\,|\nu+1\rangle\,, \qquad \langle\nu|\nu\rangle = 1, \qquad \nu = 0,1,2,\ldots. \tag{11}$$

These states have a linear increasing spectrum, $E_\nu = \nu + 1$. We may build the states by hand, starting with the only non-degenerate state, the ground state, $|0\rangle \equiv |0,0\rangle$, and we take simple linear combinations of the 1D ladder operators:

$$\begin{aligned} A^+_{\alpha,\beta} &= \alpha\, a_x^+ \otimes \mathbb{I}_y + \mathbb{I}_x \otimes \beta\, a_y^+; \\ A^-_{\alpha,\beta} &= \bar{\alpha} a_x^- \otimes \mathbb{I}_y + \mathbb{I}_x \otimes \bar{\beta} a_y^-; \\ [A^-_{\alpha,\beta}, A^+_{\alpha,\beta}] &= (|\alpha|^2 + |\beta|^2)\mathbb{I}_x \otimes \mathbb{I}_y \equiv \mathbb{I}, \end{aligned} \tag{12}$$

for $\alpha,\beta \in \mathbb{C}$ and $\mathbb{I}_x \otimes \mathbb{I}_y = \mathbb{I}_y \otimes \mathbb{I}_x \equiv \mathbb{I}$. Equation (12) defines the normalisation condition, $|\alpha|^2 + |\beta|^2 = 1$. Constructing the states $\{|\nu\rangle\}$ starting with the ground state gives us the following table:

Table 1. Construction of the states $|\nu\rangle$ using the relation $A^+\,|\nu\rangle = \sqrt{\nu+1}\,|\nu+1\rangle$.

$\|\nu\rangle$	$\|n,m\rangle$
$\|0\rangle$	$\|0,0\rangle$
$\|1\rangle$	$\alpha\;\|1,0\rangle + \beta\;\|0,1\rangle$
$\|2\rangle$	$\alpha^2\,\|2,0\rangle + \sqrt{2}\alpha\beta\,\|1,1\rangle + \beta^2\,\|0,2\rangle$
$\vdots$	$\vdots$
$\|\nu\rangle$	$\sum_{n,m}^{n+m=\nu} \alpha^n\beta^m\sqrt{\binom{\nu}{n}}\,\|n,m\rangle$

The states, $|\nu\rangle$, in Table 1 depend on α,β and may be expressed as

$$|\nu\rangle_{\alpha,\beta} = \sum_{n=0}^{\nu} \alpha^n\beta^{\nu-n}\sqrt{\binom{\nu}{n}}\,|n,\nu-n\rangle\,. \tag{13}$$

The states $|\nu\rangle_{\alpha,\beta}$ are precisely the $SU(2)$ coherent states in the Schwinger boson representation [2]. This makes sense from our construction, where the degeneracy present in the spectrum $E_{n,m}$ is an $SU(2)$ degeneracy, and so we created states which averaged out the degenerate contributions to a given ν.

These states have the following orthogonality relations

$$\langle\mu|_{\gamma,\delta}\,|\nu\rangle_{\alpha,\beta} = (\bar{\gamma}\alpha + \bar{\delta}\beta)^{\nu}\delta_{\mu,\nu}, \tag{14}$$

which reduces to a more familiar relation when $\gamma = \alpha$ and $\delta = \beta$,

$$\langle\mu|_{\alpha,\beta}\,|\nu\rangle_{\alpha,\beta} = \delta_{\mu,\nu}, \tag{15}$$

using the normalization condition $|\alpha|^2 + |\beta|^2 = 1$. The states $|\nu\rangle_{\alpha,\beta}$ have the configuration space wave function expressed in terms of (8)

$$\langle x, y|\nu\rangle_{\alpha,\beta} = \sum_{n=0}^{\nu} \alpha^n \beta^{\nu-n} \sqrt{\binom{\nu}{n}}\,\psi_n(x)\psi_{\nu-n}(y). \tag{16}$$

In Figure 1, there are two plots of the probability density functions $\left|\langle x, y|\nu\rangle_{\alpha,\beta}\right|^2$. In the picture on the left, there is an imaginary component to the relative phase between α and β, and this causes the emergence of an elliptical shape to the density. Conversely, on the right, when α and β are exactly in phase (or out of phase), the probability density is concentrated on a line, and the angle of the line to the x axis is determined by $\tan\theta = \frac{|\beta|}{|\alpha|}$. The probability densities of the quantum $SU(2)$ coherent states mimic the spatial distribution of a classical 2D isotropic oscillator—that is, ellipses in the (x, y) plane.

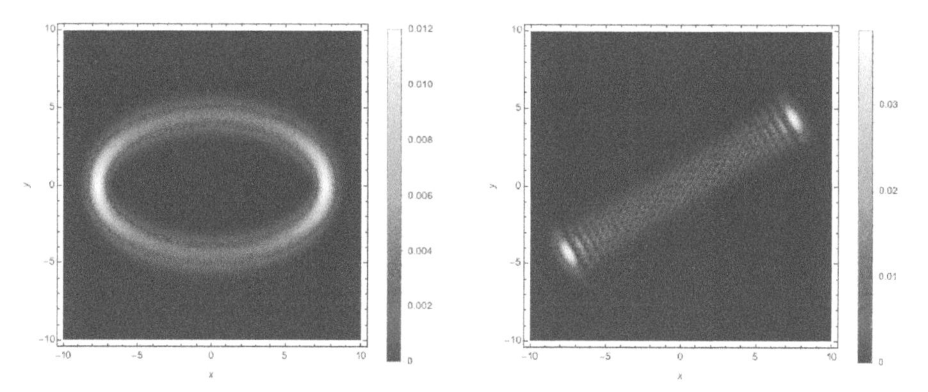

Figure 1. Density plots of $\left|\langle x, y|\nu\rangle_{\alpha,\beta}\right|^2$ for $\alpha = \frac{\sqrt{3}}{2}e^{i\frac{\pi}{2}}, \beta = \frac{1}{2}$ (**left**) and $\alpha = \frac{\sqrt{3}}{2}, \beta = \frac{1}{2}$ (**right**), both at $\nu = 40$.

The $SU(2)$ coherent states have the following variances for the physical position and momentum operators $\hat{X}_i = \frac{1}{\sqrt{2}}(a_i^+ + a_i^-)$, $\hat{P}_i = \frac{1}{\sqrt{2}i}(a_i^- - a_i^+)$, respectively, in the i direction:

$$(\Delta\,\hat{X})^2_{|\nu\rangle_{\alpha,\beta}} = (\Delta\hat{P}_x)^2_{|\nu\rangle_{\alpha,\beta}} = \frac{1}{2} + |\alpha|^2\nu; \tag{17}$$

$$(\Delta\,\hat{Y})^2_{|\nu\rangle_{\alpha,\beta}} = (\Delta\hat{P}_y)^2_{|\nu\rangle_{\alpha,\beta}} = \frac{1}{2} + |\beta|^2\nu. \tag{18}$$

The results are essentially the same as those in (9) and (10), but they are tuned by the continuous parameters α, β introduced in (12).

5. Schrödinger-Type 2D Coherent States

Using the $SU(2)$ coherent states $|\nu\rangle_{\alpha,\beta}$ as a Fock basis for defining 2D coherent states in the same vein as Section 2, we write down the following:

$$|\Psi\rangle_{\alpha,\beta} = e^{-\frac{|\Psi|^2}{2}} \sum_{\nu=0}^{\infty} \frac{\Psi^\nu}{\sqrt{\nu!}} |\nu\rangle_{\alpha,\beta}. \tag{19}$$

These states have the following inner product relation:

$$\langle \Psi'|_{\gamma,\delta} |\Psi\rangle_{\alpha,\beta} = e^{-\frac{|\Psi'|^2+|\Psi|^2}{2}} e^{\bar{\Psi}'\Psi(\bar{\gamma}\alpha+\bar{\delta}\beta)}. \tag{20}$$

Because these states are constructed so as to be analogous with the 1D definitions, we also find that they are eigenstates of the generalised lowering operator A^-

$$A^-_{\alpha,\beta} |\Psi\rangle_{\alpha,\beta} = \Psi |\Psi\rangle_{\alpha,\beta}. \tag{21}$$

The expansion in (19) also implies the existence of a displacement operator, as in the 1D case:

$$\begin{aligned} |\Psi\rangle_{\alpha,\beta} &= e^{-\frac{|\Psi|^2}{2}} \sum_{\nu=0}^{\infty} \frac{\Psi^\nu}{\sqrt{\nu!}} |\nu\rangle_{\alpha,\beta} \\ &= e^{-\frac{|\Psi|^2}{2}} \sum_{\nu=0}^{\infty} \frac{\Psi^\nu}{\sqrt{\nu!}} \frac{A^{+\,\nu}_{\alpha\beta}}{\sqrt{\nu!}} |0\rangle_{\alpha,\beta} \\ &= e^{-\frac{|\Psi|^2}{2}+\Psi A^+_{\alpha\beta}} |0\rangle_{\alpha,\beta} \equiv D(\Psi)\,|0\rangle_{\alpha,\beta}. \end{aligned} \tag{22}$$

A Baker-Campbell-Haussdorf identity, along with the annihilation of the 2D vacuum, $A^-_{\alpha,\beta}|0\rangle_{\alpha,\beta} = 0$ allows us to rewrite $D(\Psi)$ in the following way:

$$\begin{aligned} D(\Psi) &= e^{\Psi A^+_{\alpha,\beta} - \bar{\Psi} A^-_{\alpha,\beta}} \\ &= e^{(\alpha\Psi a_x^+ - \bar{\alpha}\bar{\Psi} a_x^-)+(\beta\Psi a_y^+ - \bar{\beta}\bar{\Psi} a_y^-)} \\ &= D_x(\alpha\Psi) D_y(\beta\Psi), \end{aligned} \tag{23}$$

where we have split $D(\Psi)$ into operators acting on x and y independently. The Schrödinger-type coherent states then factorise into two uncoupled 1D coherent states, $|\alpha\,\Psi\rangle_x \otimes |\beta\,\Psi\rangle_y$.

The Schrödinger-type coherent states represent an infinite sum of the elliptical, or $SU(2)$ coherent states established previously, with a Poissonian probability of being in a state $|\mu\rangle_{\alpha,\beta}$ given by

$$\left|\langle \mu|_{\alpha,\beta} |\Psi\rangle_{\alpha,\beta}\right|^2 = e^{-|\Psi|^2} \frac{|\Psi|^{2\mu}}{\mu!}, \tag{24}$$

analogous to the 1D coherent states, $|\langle n|z\rangle|^2 = e^{-|z|^2} \frac{|z|^{2n}}{n!}$.

It is clear from the factorisation of the displacement operator (23) that the wave-function of the Schrödinger-type coherent states must also factorise into the product of two 1D coherent state wave-functions. Using the general form of the 1D coherent state wave-function [2], we get the position representation of the 2D Schrödinger-type coherent states:

$$\langle x,y|\Psi\rangle_{\alpha,\beta} = \frac{1}{\sqrt{\pi}} \exp\left(-\frac{1}{2}[(x-\sqrt{2}\,\mathrm{Re}(\alpha\Psi))^2 + (y-\sqrt{2}\,\mathrm{Re}(\beta\Psi))^2]\right) e^{\left(i\sqrt{2}[x\,\mathrm{Im}(\alpha\Psi)+y\,\mathrm{Im}(\beta\Psi)]\right)}. \tag{25}$$

In Figure 2, we see the probability densities $\left|\langle x,y|\Psi\rangle_{\alpha,\beta}\right|^2$ are Gaussian in the (x,y) plane. The peak of the probability density is located at the coordinates $(x,y)=(\sqrt{2}\,\mathrm{Re}(\alpha\Psi),\sqrt{2}\,\mathrm{Re}(\beta\Psi))$.

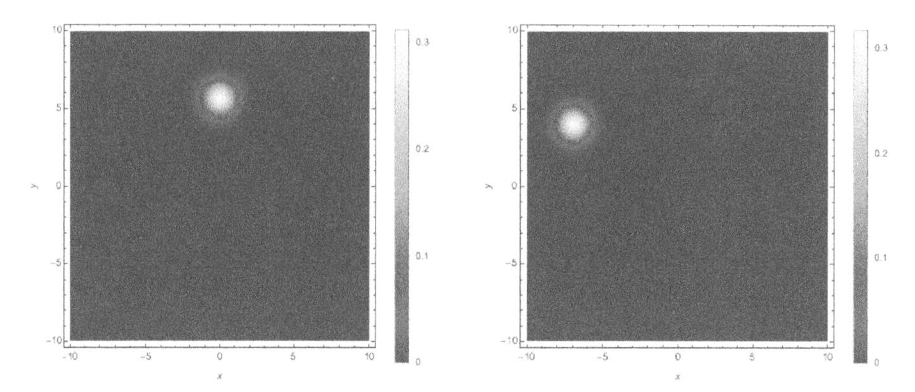

Figure 2. Density plots of $\left|\langle x,y|\Psi\rangle_{\alpha,\beta}\right|^2$ for $\Psi=8,\alpha=\frac{\sqrt{3}}{2}e^{i\frac{\pi}{2}},\beta=\frac{1}{2}$ (**left**) and $\Psi=8e^{i\frac{\pi}{4}},\alpha=\frac{\sqrt{3}}{2}e^{i\frac{\pi}{2}},\beta=\frac{1}{2}$ (**right**).

The Schrödinger-type 2D isotropic coherent states are minimal uncertainty states in both x and y, and this follows from the factorisation of the displacement operator,

$$(\Delta\hat{X})_{|\Psi\rangle_{\alpha,\beta}}(\Delta\,\hat{P}_x)_{|\Psi\rangle_{\alpha,\beta}}=\frac{1}{2},\qquad(\Delta\hat{X})_{|\Psi\rangle_{\alpha,\beta}}=(\Delta\,\hat{P}_x)_{|\Psi\rangle_{\alpha,\beta}};\tag{26}$$

$$(\Delta\hat{Y})_{|\Psi\rangle_{\alpha,\beta}}(\Delta\,\hat{P}_y)_{|\Psi\rangle_{\alpha,\beta}}=\frac{1}{2},\qquad(\Delta\hat{Y})_{|\Psi\rangle_{\alpha,\beta}}=(\Delta\,\hat{P}_y)_{|\Psi\rangle_{\alpha,\beta}}.\tag{27}$$

6. Resolution of the Identity

The $SU(2)$ coherent states resolve the identity in the following way:

$$\frac{\nu+1}{\pi^2}\int_{S^3}d^2\alpha\;d^2\beta\;\delta(|\alpha|^2+|\beta|^2-1)\,|\nu\rangle_{\alpha,\beta}\,\langle\nu|_{\alpha,\beta}=\mathbb{I}_\nu,\tag{28}$$

where $\mathbb{I}_\nu$ is the identity operator for the states $\{|\nu\rangle_{\alpha,\beta}\}$—in other words, the sum of the projectors onto states with a total occupation number of $n+m=\nu$—for example, $\mathbb{I}_2=|2,0\rangle\langle 2,0|+|1,1\rangle\langle 1,1|+|0,2\rangle\langle 0,2|$.

We retrieved the identity operator for the entire Hilbert space by summing over ν

$$\sum_{\nu=0}^{\infty}\left(\frac{\nu+1}{\pi^2}\int_{S^3}d^2\alpha\;d^2\beta\;\delta(|\alpha|^2+|\beta|^2-1)\,|\nu\rangle_{\alpha,\beta}\,\langle\nu|_{\alpha,\beta}\right)=\sum_{n=0}^{\infty}\sum_{m=0}^{\infty}|n,m\rangle\,\langle n,m|=\mathbb{I}_{\mathcal{H}}.\tag{29}$$

The resolution of the identity allowed us to express any other state in the Hilbert space in terms of the states $\{|\nu\rangle_{\alpha,\beta}\}$. The energy eigenstates were then given by

$$|n,m\rangle=\sum_{\nu=0}^{\infty}\left\{\frac{\nu+1}{\pi^2}\int_{S^3}d^2\alpha\;d^2\beta\;\delta(|\alpha|^2+|\beta|^2-1)\sqrt{\binom{\nu}{n}}\bar{\alpha}^n\bar{\beta}^m\,|\nu\rangle_{\alpha,\beta}\right\}.\tag{30}$$

The Schrödinger-type 2D coherent states resolve the identity with a slightly modified measure. It is insufficient to combine the measures used for the 1D coherent states and $SU(2)$ coherent states in Equations (5) and (28), where doing so, we would obtain

$$\frac{1}{\pi^2}\int_{S^3} d^2\alpha\, d^2\beta\, \delta(|\alpha|^2+|\beta|^2-1)\int_{\mathbb{C}}\frac{d^2\Psi}{\pi}\,|\Psi\rangle_{\alpha,\beta}\,\langle\Psi|_{\alpha,\beta}=\sum_{\nu=0}^{\infty}\frac{\mathbb{I}_\nu}{\nu+1}=\sum_{n=0}^{\infty}\sum_{m=0}^{\infty}\frac{|n,m\rangle\,\langle n,m|}{n+m+1}\not\propto\mathbb{I}_{\mathcal{H}}. \quad (31)$$

However, the identity operator for the full Hilbert space can be retrieved by the inclusion of $|\Psi|^2$ into the measure as follows:

$$\frac{1}{\pi^2}\int_{S^3} d^2\alpha\, d^2\beta\, \delta(|\alpha|^2+|\beta|^2-1)\int_{\mathbb{C}}\frac{d^2\Psi}{\pi}|\Psi|^2\,|\Psi\rangle_{\alpha,\beta}\,\langle\Psi|_{\alpha,\beta}=\mathbb{I}_{\mathcal{H}}, \quad (32)$$

thus, the Schrödinger-type coherent states for the 2D oscillator represent an over-complete basis for the full Hilbert space of the 2D oscillator. The resolution of the identity means the states could have some application in 2D coherent state quantization [2].

7. Commensurate Anisotropic $SU(2)$ Coherent States

In order to generalise coherent states to the commensurate anisotropic oscillator, we introduce two integers, p, q, in the Hamiltonian as

$$\begin{aligned}\hat{H}&=-\frac{1}{2}\frac{d^2}{dx^2}-\frac{1}{2}\frac{d^2}{dy^2}+\frac{1}{2}\omega_x^2x^2+\frac{1}{2}\omega_y^2y^2\\&=-\frac{1}{2}\frac{d^2}{dx^2}-\frac{1}{2}\frac{d^2}{dy^2}+\frac{p^2}{2}\omega^2x^2+\frac{q^2}{2}\omega^2y^2,\end{aligned} \quad (33)$$

where the frequencies are related by $\omega_x = p\omega$ and $\omega_y = q\omega$, and the ratio, $\frac{p}{q}$, represents the ratio of the two frequencies, $\frac{\omega_x}{\omega_y}$. Without loss of generality, we will set the common frequency $\omega = 1$ in what follows and choose p, q such that they are relative prime integers. A hypothesis made by Chen [12] says that the integers p, q enter the quantum $SU(2)$ coherent states in the following way:

$$|\nu\rangle^{p,q}_{\alpha,\beta}=\sum_{n=0}^{\nu}\alpha^n\beta^{\nu-n}\sqrt{\binom{\nu}{n}}\,|pn,q(\nu-n)\rangle, \quad (34)$$

where the states are normalised in the usual way: $\langle\nu|^{p,q}_{\alpha,\beta}\,|\nu\rangle^{p,q}_{\alpha,\beta}=1$ and $|\alpha|^2+|\beta|^2=1$.

Chen's hypothesis (34) suitably addresses the extension of our construction to the commensurate anisotropic oscillator. Energy eigenstates of (33) have eigenvalues $E_{n,m} = p\left(n+\frac{1}{2}\right)+q\left(m+\frac{1}{2}\right)$, which do not have the same degenerate structure as in the isotropic case where $p = q = 1$, and instead we are considering a superposition of states $|pn, qm\rangle$ such that $n + m = \nu$ for given p, q.

The energy eigenvalues of the states $|\nu\rangle^{p,q}_{\alpha,\beta}$ may be calculated from

$$\begin{aligned}\langle\nu|^{p,q}_{\alpha,\beta}\,a_x^+a_x^-+a_y^+a_y^-+1\,|\nu\rangle^{p,q}_{\alpha,\beta}&=(p-q)\left(\sum_{n=0}^{\nu}|\alpha|^{2n}|\beta|^{2(\nu-n)}\binom{\nu}{n}n\right)+q\nu+1\\&=(p-q)|\alpha|^2\nu+q\nu+1\\&=p|\alpha|^2\nu+q|\beta|^2\nu+1\\&\equiv E^{p,q}_\nu,\end{aligned} \quad (35)$$

which was computed by observing that

$$\frac{\partial}{\partial|\alpha|^2}\sum_{n=0}^{\nu}|\alpha|^{2n}|\beta|^{2(\nu-n)}\binom{\nu}{n}=\frac{\partial}{\partial|\alpha|^2}(|\alpha|^2+|\beta|^2)^{\nu}, \tag{36}$$

yielding

$$\sum_{n=0}^{\nu}|\alpha|^{2(n-1)}|\beta|^{2(\nu-n)}\binom{\nu}{n}n=\nu(|\alpha|^2+|\beta|^2)^{\nu-1}=\nu. \tag{37}$$

The states $|\nu\rangle_{\alpha,\beta}^{p,q}$ correspond to Lissajous-type probability densities in configuration space, a feature present in the classical spatial distribution of an anisotropic oscillator with commensurate frequencies [7,13].

In Figure 3, we have two types of Lissajous figures, where on the left is a closed figure and on the right, an open figure. The frequency ratio $\frac{p}{q}$ determines the type of Lissajous figure, and the relative phase between α and β deforms the figures such that when they are completely in (or out of) phase, the figure is open, and when there is an imaginary component to the relative phase, the figure is closed. Tables of Lissajous figures corresponding to different choices of p and q can be found in [14]. The correspondence of the quantum probability densities to the classical spatial distribution of a 2D commensurate anisotropic oscillator confirms Chen's definition as a suitable description of coherent states.

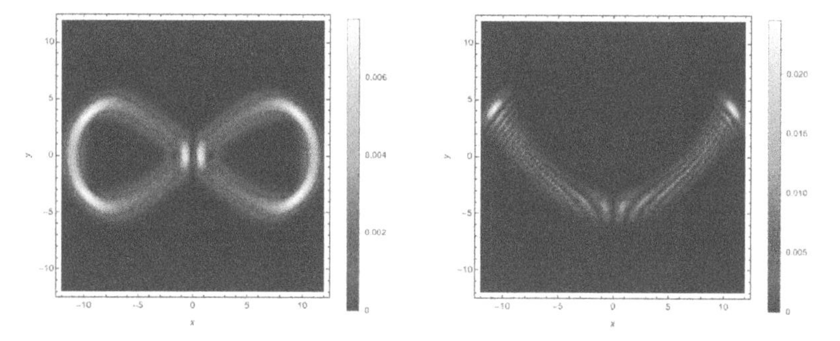

Figure 3. Density plots of $\left|\langle x,y|\nu\rangle_{\alpha,\beta}^{p,q}\right|^2$ for $\alpha=\frac{\sqrt{3}}{2}e^{i\frac{\pi}{2}},\beta=\frac{1}{2}$ (**left**) and $\alpha=\frac{\sqrt{3}}{2},\beta=\frac{1}{2}$ (**right**) for $p=2,q=1$ at $\nu=40$.

The commensurate anisotropic $SU(2)$ coherent states have slightly modified variances compared with the isotropic case

$$(\Delta\,\hat{X})^2_{|\nu\rangle_{\alpha,\beta}^{p,q}}=(\Delta\hat{P}_x)^2_{|\nu\rangle_{\alpha,\beta}^{p,q}}=\frac{1}{2}+|\alpha|^2p\nu; \tag{38}$$

$$(\Delta\,\hat{Y})^2_{|\nu\rangle_{\alpha,\beta}^{p,q}}=(\Delta\hat{P}_y)^2_{|\nu\rangle_{\alpha,\beta}^{p,q}}=\frac{1}{2}+|\beta|^2q\nu. \tag{39}$$

8. Commensurate Anisotropic 2D Schrödinger-Type Coherent States

As with the isotropic case, we can build 2D Schrödinger-type coherent states using the commensurate anisotropic $SU(2)$ coherent states as a basis, defining the states $|\Psi\rangle_{\alpha,\beta}^{p,q}$

$$|\Psi\rangle_{\alpha,\beta}^{p,q} = e^{-\frac{|\Psi|^2}{2}} \sum_{\nu=0}^{\infty} \frac{\Psi^{\nu}}{\sqrt{\nu!}} |\nu\rangle_{\alpha,\beta}^{p,q}. \tag{40}$$

These Schrödinger-type coherent states are normalised $\langle\Psi|_{\alpha,\beta}^{p,q} |\Psi\rangle_{\alpha,\beta}^{p,q} = 1$ with inner product

$$\langle\Psi'|_{\alpha,\beta}^{p,q} |\Psi\rangle_{\alpha,\beta}^{p,q} = e^{-\frac{|\Psi'|^2+|\Psi|^2}{2}} e^{\bar{\Psi}'\Psi}. \tag{41}$$

Similarly to the isotropic case, (40) may be interpreted as the infinite sum of commensurate anisotropic $SU(2)$ coherent states, determined by p, q, with a probability of being in a given coherent state, $|\mu\rangle_{\alpha,\beta}^{p,q}$, given by

$$\left|\langle\mu|_{\alpha,\beta}^{p,q} |\Psi\rangle_{\alpha,\beta}^{p,q}\right|^2 = e^{-|\Psi|^2} \frac{|\Psi|^{2\mu}}{\mu!}. \tag{42}$$

Figures 4 and 5 show four density plots for the probability density of the commensurate anisotropic 2D Schrödinger-type coherent states. We have finitely used many terms in the expansion of $|\Psi\rangle_{\alpha,\beta}^{p,q}$, and so we can see the emergence of localisation, but the pictured graphs are not properly normalised as a result. An interesting difference between the isotropic and commensurate anisotropic Schrödinger-type coherent states is that for certain values of $(\alpha, \beta, \Psi, p, q)$, the probability density can localise onto two or more separate points. This can be seen clearly in the left-most image in Figure 5, unlike the isotropic Schrödinger states which were seen to have Gaussian probability distributions in configuration space with a single maximum.

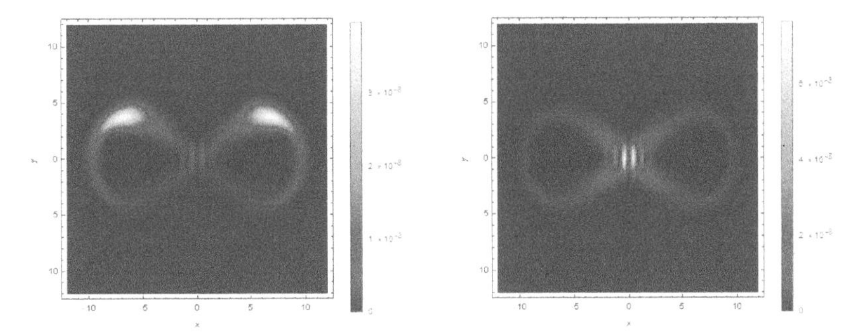

Figure 4. Density plots of $\left|\langle x, y|\Psi\rangle_{\alpha,\beta}^{p,q}\right|^2$ for $\Psi = 8, \alpha = \frac{\sqrt{3}}{2}e^{i\frac{\pi}{2}}, \beta = \frac{1}{2}$ **(left)** and $\Psi = 8e^{i\frac{\pi}{2}}, \alpha = \frac{\sqrt{3}}{2}e^{i\frac{\pi}{2}}, \beta = \frac{1}{2}$ **(right)**, with $p = 2, q = 1$ in both instances. Thirty terms are kept in the expansion of $|\Psi\rangle_{\alpha,\beta}^{p,q}$. We see the emergence of localisation onto parts of the $SU(2)$ coherent state used in the expansion.

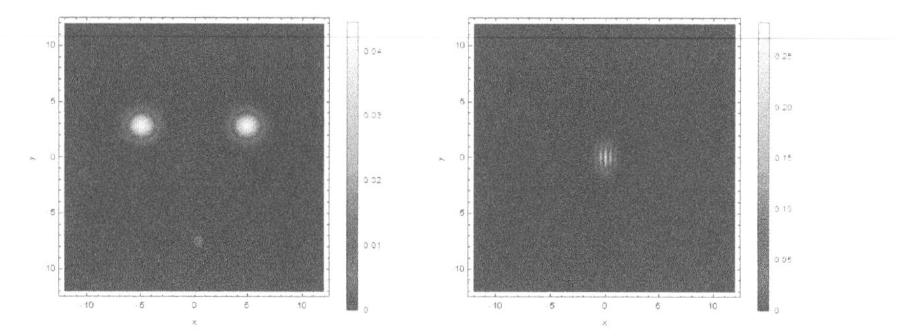

Figure 5. Density plots of $\left|\langle x,y|\Psi\rangle^{p,q}_{\alpha,\beta}\right|^2$ for $\Psi = 4, \alpha = \frac{\sqrt{3}}{2}e^{i\frac{\pi}{2}}, \beta = \frac{1}{2}$ (**left**) and $\Psi = 4e^{i\frac{\pi}{2}}, \alpha = \frac{\sqrt{3}}{2}e^{i\frac{\pi}{2}}, \beta = \frac{1}{2}$ (**right**), with $p = 2, q = 1$ in both instances. Thirty terms are kept in the expansion of $|\Psi\rangle^{p,q}_{\alpha,\beta}$.

In the right-most density plot in Figure 5 there is good localisation, but the probability distribution is fringed around the origin, and this behaviour differs from the isotropic counterparts. The graphs in Figure 4 are clearly far from normalisation (because larger Ψ was used), but they demonstrate how the first few terms in the expansion of $|\Psi\rangle^{p,q}_{\alpha,\beta}$ begin to localise onto the Lissajous figure. The parameters (α, β, p, q) determine the topology of the Lissajous figure, as described in Section 7, while $\arg\Psi$ controls the points on the Lissajous figure where the probability density will concentrate.

9. Conclusions

In this paper we have described a method for constructing coherent states for the 2D oscillator, which relies on using the minimal set of definitions used to describe the coherent states of the 1D oscillator. We found that most of the properties of the 1D coherent states were also present in their 2D isotropic Schrödinger-type counterparts: minimisation of the uncertainty principle, existence of a displacement operator, eigenstates of an annihilation operator, and correspondence to classical dynamics. A suitable measure was also found for the resolution of the identity.

Using the hypothesis of Chen, we generalised these results to the commensurate anisotropic 2D harmonic oscillator and found that their probability densities corresponded to Lissajous orbits. It is not clear at present how these results can be extended to the non-commensurate case. The relative prime integers p, q enter the $SU(2)$ coherent states in a very natural way, but it seems that a different formalism altogether would be required when dealing with non-commensurable ω_x, ω_y, where classically this would correspond to quasi-periodicity [15].

As an outlook, it would be interesting to obtain detailed results on the variances of the physical quadratures in the commensurate anisotropic Schrödinger-type coherent states. We were able to assess the localisation of the probability densities, but they lacked accurate numerical values as a result of using a finite number of terms in the expansion of $|\Psi\rangle^{p,q}_{\alpha,\beta}$. A further consideration would be to define a squeezing operator with the generalised ladder operators A^- and A^+, analogously to the 1D squeezing operator, $S(\Xi) = e^{\frac{\Xi}{2}(A^+)^2 - \frac{\bar{\Xi}}{2}(A^-)^2}$. When acting on the ground state with this operator to produce a 2D squeezed vacuum $S(\Xi)\,|0\rangle$, we obtained non-trivial interactions between the x and y oscillators due to the bilinear terms appearing in the exponent. A two-mode-like squeezing between x and y modes was found to arise.

Finally, this method could perhaps be used to describe coherent states for degenerate systems other than the harmonic oscillator, where the 2D oscillator is the simplest example of a degenerate 2D spectrum, and the next simplest example would be the particle in a 2D box. Work has been done on defining coherent states with degenerate spectra by Fox and Choi [3], and the example of the particle in a 2D box was looked at in [4]. The extension of our framework is not extremely straightforward; however, the spectrum of the particle in a box goes as $n^2 + m^2$, which contains non-algebraic degeneracies (such as $1^2 + 7^2 = 5^2 + 5^2$) and would require more careful thought when counting states in a given degenerate subgroup $|\nu\rangle$.

Author Contributions: Each author contributed equally to this article.

Funding: This research received no external funding.

Acknowledgments: V.H. acknowledges the support of research grants from NSERC of Canada.

Conflicts of Interest: The authors declare no conflict of interest.

References

1. Klauder, J.R. Coherent states for the hydrogen atom. *J. Phys. A Math. Gen.* **1996**, *29*, L293–L298, doi:10.1088/0305-4470/29/12/002. [CrossRef]
2. Gazeau, J.P. *Coherent States in Quantum Physics*; Wiley-VCH: Berlin, Germany, 2009.
3. Fox, R.F.; Choi, M.H. Generalized coherent states for systems with degenerate energy spectra. *Phys. Rev. A* **2001**, *64*, 042104, doi:10.1103/PhysRevA.64.042104. [CrossRef]
4. Dello Sbarba, L.; Hussin, V. Degenerate discrete energy spectra and associated coherent states. *J. Math. Phys.* **2007**, *48*, 012110, doi:10.1063/1.2435596. [CrossRef]
5. Novaes, M.; Gazeau, J.P. Multidimensional generalized coherent states. *J. Phys. A Math. Gen.* **2002**, *36*, 199–212, doi:10.1088/0305-4470/36/1/313. [CrossRef]
6. Li, W.; Sebastian, K. The coherent states of the two-dimensional isotropic harmonic oscillator and the classical limit of the Landau theory of a charged particle in a uniform magnetic field. *Eur. J. Phys.* **2018**, *39*, 045403, doi:10.1088/1361-6404/aab985. [CrossRef]
7. Chen, Y.F.; Huang, K.F. Vortex structure of quantum eigenstates and classical periodic orbits in two-dimensional harmonic oscillators. *J. Phys. A Math. Gen.* **2003**, *36*, 7751–7760, doi:10.1088/0305-4470/36/28/305. [CrossRef]
8. Glauber, R.J. The Quantum Theory of Optical Coherence. *Phys. Rev.* **1963**, *130*, 2529–2539, doi:10.1103/PhysRev.130.2529. [CrossRef]
9. Sudarshan, E.C.G. Equivalence of Semiclassical and Quantum Mechanical Descriptions of Statistical Light Beams. *Phys. Rev. Lett.* **1963**, *10*, 277–279, doi:10.1103/PhysRevLett.10.277. [CrossRef]
10. Schrödinger, E. Der stetige Übergang von der Mikro- zur Makromechanik. *Naturwissenschaften* **1926**, *14*, 664–666, doi:10.1007/BF01507634. [CrossRef]
11. Dirac, P.A.M. *The Principles of Quantum Mechanics*; Clarendon Press: Oxford, UK, 1930.
12. Chen, Y.F.; Huang, K.F.; Lai, H.C.; Lan, Y.P. Observation of Vector Vortex Lattices in Polarization States of an Isotropic Microcavity Laser. *Phys. Rev. Lett.* **2003**, *90*, 053904, doi:10.1103/PhysRevLett.90.053904. [CrossRef] [PubMed]
13. Doll, R.; Ingold, G.L. Lissajous curves and semiclassical theory: The two-dimensional harmonic oscillator. *Am. J. Phys.* **2007**, *75*, 208–215, doi:10.1119/1.2402157. [CrossRef]
14. Greenslade, T.B. *Adventures with Lissajous Figures*; Morgan & Claypool Publishers: Bristol, UK, 2018; pp. 2053–2571, doi:10.1088/978-1-6432-7010-4. [CrossRef]
15. Taylor, J.R. *Classical Mechanics*; University Science Books: Sausalito, CA, USA, 2005.

Article

Einstein's $E = mc^2$ Derivable from Heisenberg's Uncertainty Relations

Sibel Başkal [1], Young S. Kim [2,*] and Marilyn E. Noz [3]

[1] Department of Physics, Middle East Technical University, 06800 Ankara, Turkey; sibelbaskal@gmail.com
[2] Center for Fundamental Physics, University of Maryland, College Park, MD 20742, USA
[3] Department of Radiology, New York University, New York, NY 10016, USA; marilyne.noz@gmail.com
* Correspondence: yskim@umd.edu; Tel.: +1-301-937-1306

Received: 12 September 2019; Accepted: 7 November 2019; Published: 9 November 2019

Abstract: Heisenberg's uncertainty relation can be written in terms of the step-up and step-down operators in the harmonic oscillator representation. It is noted that the single-variable Heisenberg commutation relation contains the symmetry of the $Sp(2)$ group which is isomorphic to the Lorentz group applicable to one time-like dimension and two space-like dimensions, known as the $O(2,1)$ group. This group has three independent generators. The one-dimensional step-up and step-down operators can be combined into one two-by-two Hermitian matrix which contains three independent operators. If we use a two-variable Heisenberg commutation relation, the two pairs of independent step-up, step-down operators can be combined into a four-by-four block-diagonal Hermitian matrix with six independent parameters. It is then possible to add one off-diagonal two-by-two matrix and its Hermitian conjugate to complete the four-by-four Hermitian matrix. This off-diagonal matrix has four independent generators. There are thus ten independent generators. It is then shown that these ten generators can be linearly combined to the ten generators for Dirac's two oscillator system leading to the group isomorphic to the de Sitter group $O(3,2)$, which can then be contracted to the inhomogeneous Lorentz group with four translation generators corresponding to the four-momentum in the Lorentz-covariant world. This Lorentz-covariant four-momentum is known as Einstein's $E = mc^2$.

Keywords: $E = mc^2$ from Heisenberg's uncertainty relations; one symmetry for quantum mechanics and special relativity

---

1. Introduction

Let us start with Heisenberg's commutation relations

$$[x_i, P_j] = i\,\delta_{ij}, \tag{1}$$

with

$$P_i = -i\frac{\partial}{\partial x_i}, \tag{2}$$

where $i = 1, 2, 3$, corresponds to the x, y, z coordinates respectively.

With these x_i and P_i, we can construct the following three operators,

$$J_i = \epsilon_{ijk} x_j P_k. \tag{3}$$

These three operators satisfy the closed set of commutation relations:

$$[J_i, J_j] = i\epsilon_{ijk} J_k. \tag{4}$$

These J_i operators generate rotations in the three-dimensional space. In mathematics, this set is called the Lie algebra of the rotation group. This is a direct consequence of Heisenberg's commutation relations.

In quantum mechanics, each J_i corresponds to the angular momentum along the i direction. A remarkable fact is that it is also possible to construct the same Lie algebra with two-by-two matrices. These matrices are of course the Pauli spin matrices, leading to the observable angular momentum not seen in classical mechanics.

As the expression shows in Equation (2), each P_i generates a translation along the ith direction. Thus, the three translation generators, together with the three rotation generators constitute the Lie algebra of the Galilei group, with the additional commutation relations:

$$[J_i, P_j] = i\epsilon_{ijk}P_k. \tag{5}$$

This set of commutation relations together with those of Equation (4) constitute a closed set for both P_i and J_i. This set is called the Lie algebra of the Galilei group. This group is the basic symmetry group for the Schrödinger or non-relativistic quantum mechanics.

In the Schrödinger picture, the generator P_i corresponds to the particle momentum along the i direction. In addition, the time translation operator is

$$P_0 = i\frac{\partial}{\partial t}. \tag{6}$$

This corresponds to the energy variable.

Let us go to the Lorentzian world. Here we have to take into account the generators of the boosts. The generators thus include the time variable, and the generator of boosts along the i direction is

$$K_i = i\left(x_i\frac{\partial}{\partial t} + t\frac{\partial}{\partial x_i}\right). \tag{7}$$

These generators satisfy the commutation relations

$$[K_i, K_j] = -i\epsilon_{ijk}J_k. \tag{8}$$

Thus, these three boost generators alone cannot constitute a closed set of commutation relations (Lie algebra).

With J_i, these boost generators satisfy

$$[J_i, K_j] = i\,\epsilon_{ijk}K_k. \tag{9}$$

With P_i, they satisfy the relation

$$[P_i, K_i] = i\delta_{0i}P_0. \tag{10}$$

Thus, the commutation relations of Equations (4),(5),(8–10) constitute a closed set of the ten generators. This closed set is commonly called the Lie algebra of the Poincaré symmetry.

The three rotation and three translation generators are contained in, or are derivable from, Heisenberg's commutation relations, and the time translation operator is seen in the Schrödinger equation. They are all Hermitian operators corresponding to dynamical variables. On the other hand, the three boost generators of Equation (7) are not derivable from the Heisenberg relations. Furthermore, they do not appear to correspond to observable quantities [1].

The purpose of this paper is to show that the Lie algebra of the Poincaré symmetry is derivable from the Heisenberg commutation relations. For this purpose, we first examine the symmetry of the Heisenberg commutation relation using the Wigner function in the phase space. It is noted that the single-variable relation contains the symmetry of the Lorentz group applicable to two space-like and one time-like dimensions.

As Dirac noted in 1963 [2], two coupled oscillators lead to the symmetry of the $O(3,2)$ or the Lorentz group applicable to the three space-like directions and two time-like directions. As is illustrated in Figure 1, it is possible to contract one of those two time variables of this $O(3,2)$ group into the inhomogeneous Lorentz group, consisting of the Lorentz group applicable to the three space-like dimensions and one time-like direction, plus four translation generators corresponding to the energy-momentum four-vector. This of course leads to Einstein's energy–momentum relation of $E = mc^2$.

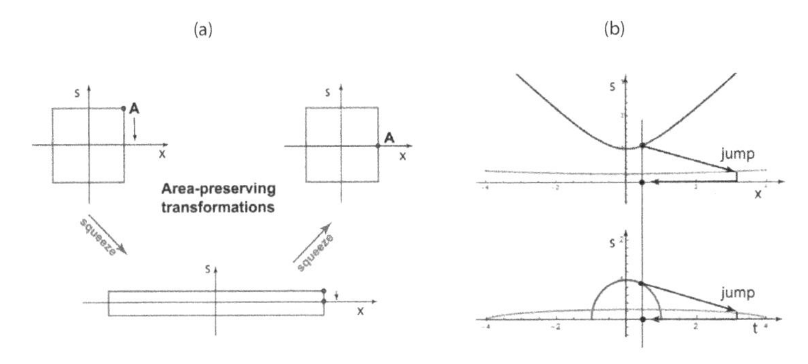

Figure 1. The Inönü–Wigner contraction procedure interpreted as squeeze transformations. In (**a**), the square becomes a narrow rectangle during the squeeze process. When the rectangle becomes narrow enough, the point A can be moved to the horizontal axis. Then, the inverse squeeze brings back the rectangle to the original shape. The point A remains on the horizontal axis. In (**b**), both the hyperbola and the circle become flattened to the horizontal axis, during the initial squeeze. The point on the curve moves to the horizontal axis. This point moves back to its finite position during the inverse squeeze.

In Section 2, it is noted that the best way to study the symmetry of the Heisenberg commutation relation is to use the Wigner function for the Gaussian function for the oscillator state. In the Wigner phase space, this function contains the symmetry for the Lorentz group applicable to two space-like dimensions and one time-like dimension. This group has three generators. This operation is equivalent to constructing a two-by-two block-diagonal Hermitian matrix with quadratic forms of the step-up and step-down operators.

In Section 3, we consider two oscillators. If these oscillators are independent, it is possible to construct a four-by-four block diagonal matrix, where each block consists of the two-by-two matrix for each operator defined in Section 2. Since the oscillators are uncoupled, this four-by-four block-diagonal Hermitian matrix contains six independent generators.

If the oscillators are coupled, then to keep the overall four-by-four block-diagonal matrix Hermitian, we need one off-diagonal block matrix, with four independent quadratic forms. Thus, the overall four-by-four matrix contains ten independent quadratic forms of the creation and annihilation operators.

It is shown that these ten independent generators can be linearly combined into the ten generators constructed by Dirac for the the Lorentz group applicable to three space-like dimensions and two time-like dimensions, commonly called the $O(3,2)$ group.

In Section 4, using the boosts belonging to one of its time-like dimensions, we contract $O(3,2)$ to produce the Lorentz group applicable to one time dimension and four translations leading to the four-momentum. This Lorentz-covariant four-momentum is commonly known as Einstein's $E = mc^2$.

This paper is essentially based on Dirac's paper published in 1949 and 1963 [1,2]. As is illustrated in Figure 2, we show here that the space-time symmetry of quantum mechanics mentioned in his 1949 paper is derivable from his two-oscillator system discussed in 1963. The route is the group contraction procedure of Inönü and Wigner [3].

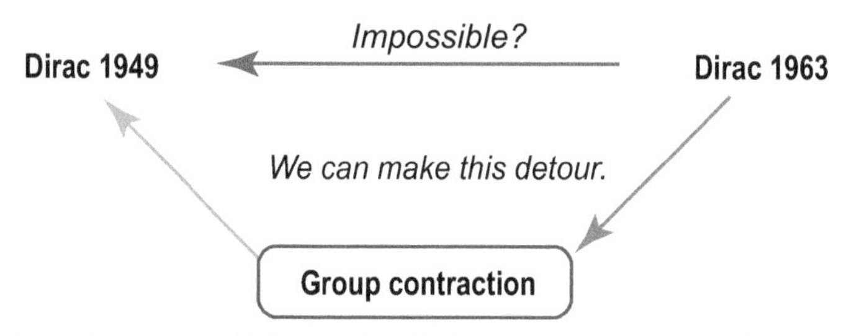

Figure 2. According to Dirac's 1949 paper, the task of constructing quantum mechanics is essentially constructing a representation of the inhomogeneous Lorentz group. In his 1963 paper, Dirac constructed the Lie algebra of the $O(3,2)$ de Sitter group from the algebra of two harmonic oscillators, which is a direct consequence of Heisenberg's uncertainty commutation relations. It is possible to derive the Lie algebra of the inhomogeneous Lorentz group from that of $O(3,2)$ using the group-contraction procedure of Inönü and Wigner [3].

Indeed, from 1927 Dirac made lifelong efforts to synthesize quantum mechanics and special relativity [4]. In 1949 and before, he treated quantum mechanics and special relativity as two separate scientific disciplines, and then in 1949 he attempted to synthesize them. Thus, it is of interest to see how Dirac's ideas evolved during the period 1929–1949. We shall give a brief review of Dirac's efforts during the period in Appendix A.

2. Symmetries of the Single-Mode States

Heisenberg's uncertainty relation for a single Cartesian variable takes the form

$$[x, p] = i \tag{11}$$

with

$$p = -i\frac{\partial}{\partial x}.$$

Very often, it is more convenient to use the operators

$$a = \frac{1}{\sqrt{2}}(x + ip), \qquad a^{\dagger} = \frac{1}{\sqrt{2}}(x - ip) \tag{12}$$

with

$$\left[a, a^{\dagger}\right] = 1. \tag{13}$$

This aspect is well known.

The representation based on a and $a^{\dagger}$ is known as the harmonic oscillator representation of the uncertainty relation and is the basic language for the Fock space for particle numbers. This representation is therefore the basic language for quantum optics.

Let us next consider the quadratic forms: $aa, a^{\dagger}a^{\dagger}, aa^{\dagger}$, and $a^{\dagger}a$. Then the linear combination

$$aa^{\dagger} - a^{\dagger}a = 1, \tag{14}$$

according to the uncertainty relation. Thus, there are three independent quadratic forms, and we are led to the following two-by-two matrix:

$$\begin{pmatrix} \left(aa^{\dagger} + a^{\dagger}a\right)/2 & aa \\ a^{\dagger}a^{\dagger} & \left(aa^{\dagger} + a^{\dagger}a\right)/2 \end{pmatrix}. \tag{15}$$

This matrix leads to the following three independent operators:

$$J_2 = \frac{1}{2}\left(aa^{\dagger} + a^{\dagger}a\right), \quad K_1 = \frac{1}{2}\left(a^{\dagger}a^{\dagger} + aa\right), \quad K_3 = \frac{i}{2}\left(a^{\dagger}a^{\dagger} - aa\right). \tag{16}$$

They produce the following set of closed commutation relations:

$$[J_2, K_1] = -iK_3, \qquad [J_2, K_3] = iK_1, \qquad [K_1, K_3] = iJ_2. \tag{17}$$

This set is commonly called the Lie algebra of the $Sp(2)$ group, locally isomorphic to the Lorentz group applicable to one time and two space coordinates.

The best way to study the symmetry property of these operators is to use the Wigner function for the ground-state oscillator which takes the form [5–8]

$$W(x,p) = \frac{1}{\pi} \exp\left[-\left(x^2 + p^2\right)\right]. \tag{18}$$

This distribution is concentrated in the circular region around the origin. Let us define the circle as

$$x^2 + p^2 = 1. \tag{19}$$

We can use the area of this circle in the phase space of x and p as the minimum uncertainty. This uncertainty is preserved under rotations in the phase space and also under squeezing. These transformations can be written as

$$\begin{pmatrix} \cos\theta & -\sin\theta \\ \sin\theta & \cos\theta \end{pmatrix}\begin{pmatrix} x \\ p \end{pmatrix}, \qquad \begin{pmatrix} e^{\eta} & 0 \\ 0 & e^{-\eta} \end{pmatrix}\begin{pmatrix} x \\ p \end{pmatrix}, \tag{20}$$

respectively. The rotation and the squeeze are generated by

$$J_2 = -i\left(x\frac{\partial}{\partial p} - p\frac{\partial}{\partial x}\right), \qquad K_1 = -i\left(x\frac{\partial}{\partial x} - p\frac{\partial}{\partial p}\right). \tag{21}$$

If we take the commutation relation with these two operators, the result is

$$[J_2, K_1] = -iK_3, \quad . \tag{22}$$

with

$$K_3 = -i\left(x\frac{\partial}{\partial p} + p\frac{\partial}{\partial x}\right). \tag{23}$$

Indeed, as before, these three generators form the closed set of commutation which form the Lie algebra of the $Sp(2)$ group, isomorphic to the Lorentz group applicable to two space and one time dimensions. This isomorphic correspondence is illustrated in Figure 3.

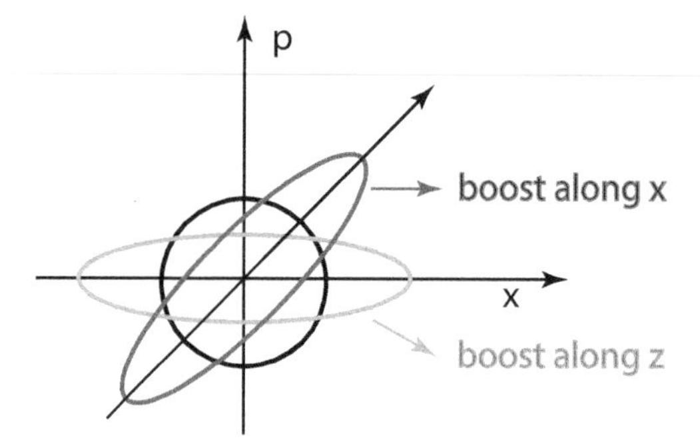

Figure 3. Rotations and squeezes in the phase space produced by the $Sp(2)$ transformations. The squeeze along the x direction corresponds to the Lorentz boost along the z direction, while the squeeze along the 45° angle corresponds to the boost along the x direction. The rotation by 45° corresponds to the rotation by 90° around the y axis.

Let us consider the Minkowski space of (x, y, z, t). It is possible to write three four-by-four matrices satisfying the Lie algebra of Equation (17). The three four-by-four matrices satisfying this set of commutation relations are:

$$J_2 = \begin{pmatrix} 0 & 0 & i & 0 \\ 0 & 0 & 0 & 0 \\ i & 0 & 0 & 0 \\ 0 & 0 & 0 & 0 \end{pmatrix}, \quad K_1 = \begin{pmatrix} 0 & 0 & 0 & i \\ 0 & 0 & 0 & 0 \\ 0 & 0 & 0 & 0 \\ i & 0 & 0 & 0 \end{pmatrix}, \quad K_3 = \begin{pmatrix} 0 & 0 & 0 & 0 \\ 0 & 0 & 0 & 0 \\ 0 & 0 & 0 & i \\ 0 & 0 & i & 0 \end{pmatrix}. \tag{24}$$

However, these matrices have null second rows and null second columns. Thus, they can generate Lorentz transformations applicable only to the three-dimensional space of (x, z, t), while the y variable remains invariant. Thus, this single-oscillator system cannot describe what happens in the full four-dimensional Minkowski space.

Yet, it is interesting that the oscillator system can produce three different representations sharing the same Lie algebra with the (2 + 1)-dimensional Lorentz group, as shown in Table 1.

Table 1. Transformation for the Gaussian function, in terms of harmonic oscillators, two-dimensional phase space, and the four-dimensional Minkowski space.

Generators	Oscillator	Phase Space	Lorentz
J_2	$\frac{1}{2}\left(aa^{\dagger} + a^{\dagger}a\right)$	$\frac{1}{2}\sigma_2$	$\begin{pmatrix} 0 & 0 & i & 0 \\ 0 & 0 & 0 & 0 \\ i & 0 & 0 & 0 \\ 0 & 0 & 0 & 0 \end{pmatrix}$
K_1	$\frac{1}{2i}\left(a^{\dagger}a^{\dagger} + aa\right)$	$\frac{i}{2}\sigma_1$	$\begin{pmatrix} 0 & 0 & 0 & i \\ 0 & 0 & 0 & 0 \\ 0 & 0 & 0 & 0 \\ i & 0 & 0 & 0 \end{pmatrix}$
K_3	$\frac{1}{2}\left(a^{\dagger}a^{\dagger} - aa\right)$,	$\frac{i}{2}\sigma_3$	$\begin{pmatrix} 0 & 0 & 0 & 0 \\ 0 & 0 & 0 & 0 \\ 0 & 0 & 0 & i \\ 0 & 0 & i & 0 \end{pmatrix}$

3. Symmetries from Two Oscillators

In order to generate Lorentz transformations applicable to the full Minkowskian space, we may need two Heisenberg commutation relations. Indeed, Paul A. M. Dirac started this program in 1963 [2]. It is possible to write the two uncertainty relations using two harmonic oscillators as

$$\left[a_i, a_j^\dagger\right] = \delta_{ij}, \tag{25}$$

with

$$a_i = \frac{1}{\sqrt{2}}\left(x_i + ip_i\right), \qquad a_i^\dagger = \frac{1}{\sqrt{2}}\left(x_i - ip_i\right), \tag{26}$$

and

$$x_i = \frac{1}{\sqrt{2}}\left(a_i + a_i^\dagger\right), \qquad p_i = \frac{i}{\sqrt{2}}\left(a_i^\dagger - a_i\right), \tag{27}$$

where i and j could be 1 or 2.

As in the case of the two-by-two matrix given in Equation (15), we can consider the following four-by-four block-diagonal matrix if the oscillators are not coupled:

$$\begin{pmatrix} \left(a_1a_1^\dagger + a_1^\dagger a_1\right)/2 & a_1a_1 & 0 & 0 \\ a_1^\dagger a_1^\dagger & \left(a_1a_1^\dagger + a_1^\dagger a_1\right)/2 & 0 & 0 \\ 0 & 0 & \left(a_2a_2^\dagger + a_2^\dagger a_2\right)/2 & a_2a_2 \\ 0 & 0 & a_2^\dagger a_2^\dagger & \left(a_2a_2^\dagger + a_2^\dagger a_2\right)/2 \end{pmatrix}. \tag{28}$$

There are six generators in this matrix.

We are now interested in coupling them by filling in the off-diagonal blocks. The most general forms for this block are the following two-by-two matrix and its Hermitian conjugate:

$$\begin{pmatrix} a_1^\dagger a_2 & a_1a_2 \\ a_1^\dagger a_2^\dagger & a_1a_2^\dagger \end{pmatrix} \tag{29}$$

with four independent generators. This leads to the following four-by-four matrix with ten (6 + 4) generators:

$$\begin{pmatrix} \left(a_1a_1^\dagger + a_1^\dagger a_1\right)/2 & a_1a_1 & a_1^\dagger a_2 & a_1a_2 \\ a_1^\dagger a_1^\dagger & \left(a_1a_1^\dagger + a_1^\dagger a_1\right)/2 & a_1^\dagger a_2^\dagger & a_1a_2^\dagger \\ a_1a_2^\dagger & a_1a_2 & \left(a_2a_2^\dagger + a_2^\dagger a_2\right)/2 & a_2a_2 \\ a_1^\dagger a_2^\dagger & a_1^\dagger a_2 & a_2^\dagger a_2^\dagger & \left(a_2a_2^\dagger + a_2^\dagger a_2\right)/2 \end{pmatrix}. \tag{30}$$

With these ten elements, we can now construct the following four rotation-like generators:

$$\begin{aligned} J_1 &= \frac{1}{2}\left(a_1^\dagger a_2 + a_2^\dagger a_1\right), & J_2 &= \frac{1}{2i}\left(a_1^\dagger a_2 - a_2^\dagger a_1\right), \\ J_3 &= \frac{1}{2}\left(a_1^\dagger a_1 - a_2^\dagger a_2\right), & S_0 &= \frac{1}{2}\left(a_1^\dagger a_1 + a_2 a_2^\dagger\right), \end{aligned} \tag{31}$$

and six squeeze-like generators:

$$K_1 = -\frac{1}{4}\left(a_1^\dagger a_1^\dagger + a_1 a_1 - a_2^\dagger a_2^\dagger - a_2 a_2\right),$$
$$K_2 = +\frac{i}{4}\left(a_1^\dagger a_1^\dagger - a_1 a_1 + a_2^\dagger a_2^\dagger - a_2 a_2\right),$$
$$K_3 = +\frac{1}{2}\left(a_1^\dagger a_2^\dagger + a_1 a_2\right), \tag{32}$$

and

$$Q_1 = -\frac{i}{4}\left(a_1^\dagger a_1^\dagger - a_1 a_1 - a_2^\dagger a_2^\dagger + a_2 a_2\right),$$
$$Q_2 = -\frac{1}{4}\left(a_1^\dagger a_1^\dagger + a_1 a_1 + a_2^\dagger a_2^\dagger + a_2 a_2\right),$$
$$Q_3 = +\frac{i}{2}\left(a_1^\dagger a_2^\dagger - a_1 a_2\right). \tag{33}$$

There are now ten operators from Equations (31)–(33), and they satisfy the following Lie algebra as was noted by Dirac in 1963 [2]:

$$[J_i, J_j] = i\epsilon_{ijk} J_k, \quad [J_i, K_j] = i\epsilon_{ijk} K_k,$$
$$[J_i, Q_j] = i\epsilon_{ijk} Q_k, \quad [K_i, K_j] = [Q_i, Q_j] = -i\epsilon_{ijk} J_k,$$
$$[K_i, Q_j] = -i\delta_{ij} S_0, \quad [J_i, S_0] = 0, \quad [K_i, S_0] = -iQ_i, \quad [Q_i, S_0] = iK_i. \tag{34}$$

Dirac noted that this set is the same as the Lie algebra for the $O(3,2)$ de Sitter group, with ten generators. This is the Lorentz group applicable to the three-dimensional space with two time variables. This group plays a very important role in space-time symmetries.

In the same paper, Dirac pointed out that this set of commutation relations serves as the Lie algebra for the four-dimensional symplectic group commonly called $Sp(4)$. For a dynamical system consisting of two pairs of canonical variables x_1, p_1 and x_2, p_2, we can use the four-dimensional phase space with the coordinate variables defined as [9]

$$\left(x_1, p_1, x_2, p_2\right). \tag{35}$$

Then the four-by-four transformation matrix M applicable to this four-component vector is canonical if [10,11]

$$MJ\tilde{M} = J, \tag{36}$$

where $\tilde{M}$ is the transpose of the M matrix, with

$$J = \begin{pmatrix} 0 & 1 & 0 & 0 \\ -1 & 0 & 0 & 0 \\ 0 & 0 & 0 & 1 \\ 0 & 0 & -1 & 0 \end{pmatrix}, \tag{37}$$

which we can write in the block-diagonal form as

$$J = i\begin{pmatrix} I & 0 \\ 0 & I \end{pmatrix}\sigma_2, \tag{38}$$

where I is the unit two-by-two matrix.

According to this form of the J matrix, the area of the phase space for the x_1 and p_1 variables remains invariant, and the story is the same for the phase space of x_2 and p_2.

We can then write the generators of the $Sp(4)$ group as [12]

$$J_1 = -\frac{1}{2}\begin{pmatrix} 0 & I \\ I & 0 \end{pmatrix}\sigma_2, \quad J_2 = \frac{i}{2}\begin{pmatrix} 0 & -I \\ I & 0 \end{pmatrix} I, \quad J_3 = \frac{1}{2}\begin{pmatrix} -I & 0 \\ 0 & I \end{pmatrix}\sigma_2, \quad S_0 = \frac{1}{2}\begin{pmatrix} I & 0 \\ 0 & I \end{pmatrix}\sigma_2, \tag{39}$$

and

$$K_1 = \frac{i}{2}\begin{pmatrix} I & 0 \\ 0 & -I \end{pmatrix}\sigma_1, \quad K_2 = \frac{i}{2}\begin{pmatrix} I & 0 \\ 0 & I \end{pmatrix}\sigma_3, \quad K_3 = -\frac{i}{2}\begin{pmatrix} 0 & I \\ I & 0 \end{pmatrix}\sigma_1,$$

$$Q_1 = -\frac{i}{2}\begin{pmatrix} I & 0 \\ 0 & -I \end{pmatrix}\sigma_3, \quad Q_2 = \frac{i}{2}\begin{pmatrix} I & 0 \\ 0 & I \end{pmatrix}\sigma_1, \quad Q_3 = \frac{i}{2}\begin{pmatrix} 0 & I \\ I & 0 \end{pmatrix}\sigma_3. \tag{40}$$

Among these ten matrices, six of them are in block-diagonal form. They are S_0, J_3, K_1, K_2, Q_1, and Q_2. In the language of two harmonic oscillators, these generators do not mix up the first and second oscillators. There are six of them because each operator has three generators for its own $Sp(2)$ symmetry. These generators, together with those in the oscillator representation, are tabulated in Table 2.

Table 2. Transformation generators for the two-oscillator system.

Generators	Two Oscillators	Phase Space
J_1	$\frac{1}{2}\left(a_1^\dagger a_2 + a_2^\dagger a_1\right)$	$-\frac{1}{2}\begin{pmatrix} 0 & I \\ I & 0 \end{pmatrix}\sigma_2$
J_2	$\frac{1}{2i}\left(a_1^\dagger a_2 - a_2^\dagger a_1\right)$	$\frac{i}{2}\begin{pmatrix} 0 & -I \\ I & 0 \end{pmatrix} I$
J_3	$\frac{1}{2}\left(a_1^\dagger a_1 - a_2^\dagger a_2\right),$	$\frac{1}{2}\begin{pmatrix} -I & 0 \\ 0 & I \end{pmatrix}\sigma_2$
S_0	$\frac{1}{2}\left(a_1^\dagger a_1 + a_2 a_2^\dagger\right),$	$\frac{1}{2}\begin{pmatrix} I & 0 \\ 0 & I \end{pmatrix}\sigma_2$
K_1	$-\frac{1}{4}\left(a_1^\dagger a_1^\dagger + a_1 a_1 - a_2^\dagger a_2^\dagger - a_2 a_2\right)$	$\frac{i}{2}\begin{pmatrix} I & 0 \\ 0 & -I \end{pmatrix}\sigma_1$
K_2	$+\frac{i}{4}\left(a_1^\dagger a_1^\dagger - a_1 a_1 + a_2^\dagger a_2^\dagger - a_2 a_2\right)$	$\frac{i}{2}\begin{pmatrix} I & 0 \\ 0 & I \end{pmatrix}\sigma_3$
K_3	$\frac{1}{2}\left(a_1^\dagger a_2^\dagger + a_1 a_2\right)$	$-\frac{i}{2}\begin{pmatrix} 0 & I \\ I & 0 \end{pmatrix}\sigma_1$
Q_1	$-\frac{i}{4}\left(a_1^\dagger a_1^\dagger - a_1 a_1 - a_2^\dagger a_2^\dagger + a_2 a_2\right)$	$-\frac{i}{2}\begin{pmatrix} I & 0 \\ 0 & -I \end{pmatrix}\sigma_3$
Q_2	$-\frac{1}{4}\left(a_1^\dagger a_1^\dagger + a_1 a_1 + a_2^\dagger a_2^\dagger + a_2 a_2\right)$	$\frac{i}{2}\begin{pmatrix} I & 0 \\ 0 & I \end{pmatrix}\sigma_1$
Q_3	$\frac{i}{2}\left(a_1^\dagger a_2^\dagger - a_1 a_2\right)$	$\frac{1}{2}\begin{pmatrix} I & 0 \\ 0 & I \end{pmatrix}\sigma_2$

The off-diagonal matrix J_2 couples the first and second oscillators without changing the overall volume of the four-dimensional phase space. However, in order to construct the closed set of commutation relations, we need the three additional generators: J_1, K_3, and Q_3. The commutation relations given in Equations (34) are clearly consequences of Heisenberg's uncertainty relations.

As for the $O(3,2)$ group, the generators are five-by-five matrices, applicable to (x, y, z, t, s), where t and s are time-like variables. These matrices can be written as

$$J_1 = \begin{pmatrix} 0 & 0 & 0 & 0 & 0 \\ 0 & 0 & -i & 0 & 0 \\ 0 & i & 0 & 0 & 0 \\ 0 & 0 & 0 & 0 & 0 \\ 0 & 0 & 0 & 0 & 0 \end{pmatrix}, \quad J_2 = \begin{pmatrix} 0 & 0 & i & 0 & 0 \\ 0 & 0 & 0 & 0 & 0 \\ -i & 0 & 0 & 0 & 0 \\ 0 & 0 & 0 & 0 & 0 \\ 0 & 0 & 0 & 0 & 0 \end{pmatrix}, \quad J_3 = \begin{pmatrix} 0 & -i & 0 & 0 & 0 \\ i & 0 & 0 & 0 & 0 \\ 0 & 0 & 0 & 0 & 0 \\ 0 & 0 & 0 & 0 & 0 \\ 0 & 0 & 0 & 0 & 0 \end{pmatrix},$$

$$K_1 = \begin{pmatrix} 0 & 0 & 0 & i & 0 \\ 0 & 0 & 0 & 0 & 0 \\ 0 & 0 & 0 & 0 & 0 \\ i & 0 & 0 & 0 & 0 \\ 0 & 0 & 0 & 0 & 0 \end{pmatrix}, \quad K_2 = \begin{pmatrix} 0 & 0 & 0 & 0 & 0 \\ 0 & 0 & 0 & i & 0 \\ 0 & 0 & 0 & 0 & 0 \\ 0 & i & 0 & 0 & 0 \\ 0 & 0 & 0 & 0 & 0 \end{pmatrix}, \quad K_3 = \begin{pmatrix} 0 & 0 & 0 & 0 & 0 \\ 0 & 0 & 0 & 0 & 0 \\ 0 & 0 & 0 & i & 0 \\ 0 & 0 & i & 0 & 0 \\ 0 & 0 & 0 & 0 & 0 \end{pmatrix},$$

$$Q_1 = \begin{pmatrix} 0 & 0 & 0 & 0 & i \\ 0 & 0 & 0 & 0 & 0 \\ 0 & 0 & 0 & 0 & 0 \\ 0 & 0 & 0 & 0 & 0 \\ i & 0 & 0 & 0 & 0 \end{pmatrix}, \quad Q_2 = \begin{pmatrix} 0 & 0 & 0 & 0 & 0 \\ 0 & 0 & 0 & 0 & i \\ 0 & 0 & 0 & 0 & 0 \\ 0 & 0 & 0 & 0 & 0 \\ 0 & i & 0 & 0 & 0 \end{pmatrix}, \quad Q_3 = \begin{pmatrix} 0 & 0 & 0 & 0 & 0 \\ 0 & 0 & 0 & 0 & 0 \\ 0 & 0 & 0 & 0 & i \\ 0 & 0 & 0 & 0 & 0 \\ 0 & 0 & i & 0 & 0 \end{pmatrix},$$

$$S_0 = \begin{pmatrix} 0 & 0 & 0 & 0 & 0 \\ 0 & 0 & 0 & 0 & 0 \\ 0 & 0 & 0 & 0 & 0 \\ 0 & 0 & 0 & 0 & -i \\ 0 & 0 & 0 & i & 0 \end{pmatrix}. \tag{41}$$

Next, we are interested in eliminating all the elements in the fifth row. The six generators J_i and K_i are not affected by this operation, but Q_1, Q_2, Q_3, and S_0 become

$$P_1 = \begin{pmatrix} 0 & 0 & 0 & 0 & i \\ 0 & 0 & 0 & 0 & 0 \\ 0 & 0 & 0 & 0 & 0 \\ 0 & 0 & 0 & 0 & 0 \\ 0 & 0 & 0 & 0 & 0 \end{pmatrix}, \quad P_2 = \begin{pmatrix} 0 & 0 & 0 & 0 & 0 \\ 0 & 0 & 0 & 0 & i \\ 0 & 0 & 0 & 0 & 0 \\ 0 & 0 & 0 & 0 & 0 \\ 0 & 0 & 0 & 0 & 0 \end{pmatrix}, \quad P_3 = \begin{pmatrix} 0 & 0 & 0 & 0 & 0 \\ 0 & 0 & 0 & 0 & 0 \\ 0 & 0 & 0 & 0 & i \\ 0 & 0 & 0 & 0 & 0 \\ 0 & 0 & 0 & 0 & 0 \end{pmatrix},$$

$$P_0 = \begin{pmatrix} 0 & 0 & 0 & 0 & 0 \\ 0 & 0 & 0 & 0 & 0 \\ 0 & 0 & 0 & 0 & 0 \\ 0 & 0 & 0 & 0 & -i \\ 0 & 0 & 0 & 0 & 0 \end{pmatrix}, \tag{42}$$

respectively. While J_i and K_i generate Lorentz transformations on the four dimensional Minkowski space, these Q_i and S_0 in the form of the P_i, P_0 matrices generate translations along the x, y, z, and t directions respectively. We shall study this aspect in detail in Section 4.

4. Contraction of $O(3,2)$ to the Inhomogeneous Lorentz Group

We can contract $O(3,2)$ according to the procedure introduced by Inönü and Wigner [3]. They introduced the procedure for transforming the four-dimensional Lorentz group into the three-dimensional Galilei group. Here, we shall contract the boost generators belonging to the time-like s variable, Q_i , along with the rotation generator between the two time-like variables, S_0.

Here, we illustrate the Inönü-Wigner procedure using the concept of squeeze transformations. For this purpose, let us introduce the squeeze matrix

$$C(\epsilon) = \begin{pmatrix} 1/\epsilon & 0 & 0 & 0 & 0 \\ 0 & 1/\epsilon & 0 & 0 & 0 \\ 0 & 0 & 1/\epsilon & 0 & 0 \\ 0 & 0 & 0 & 1/\epsilon & 0 \\ 0 & 0 & 0 & 0 & \epsilon \end{pmatrix}. \tag{43}$$

This matrix commutes with J_i and K_i. The story is different for Q_i and S_0.

For Q_1,

$$C\, Q_1\, C^{-1} = \begin{pmatrix} 0 & 0 & 0 & 0 & i/\epsilon^2 \\ 0 & 0 & 0 & 0 & 0 \\ 0 & 0 & 0 & 0 & 0 \\ 0 & 0 & 0 & 0 & 0 \\ i\epsilon^2 & 0 & 0 & 0 & 0 \end{pmatrix}, \tag{44}$$

which, in the limit of small ϵ, becomes

$$Q_1' = \begin{pmatrix} 0 & 0 & 0 & 0 & i/\epsilon^2 \\ 0 & 0 & 0 & 0 & 0 \\ 0 & 0 & 0 & 0 & 0 \\ 0 & 0 & 0 & 0 & 0 \\ 0 & 0 & 0 & 0 & 0 \end{pmatrix}. \tag{45}$$

We then make the inverse squeeze transformation:

$$C^{-1}\, Q_1'\, C = \begin{pmatrix} 0 & 0 & 0 & 0 & i \\ 0 & 0 & 0 & 0 & 0 \\ 0 & 0 & 0 & 0 & 0 \\ 0 & 0 & 0 & 0 & 0 \\ 0 & 0 & 0 & 0 & 0 \end{pmatrix}. \tag{46}$$

Thus, we can write this contraction procedure as

$$P_1 = \lim_{\epsilon \to 0} \left(\epsilon^2\, C\, Q_1\, C^{-1} \right), \tag{47}$$

where the explicit five-by-five matrix is given in Equation (42). Likewise

$$P_2 = \lim_{\epsilon \to 0} \left(\epsilon^2\, C\, Q_2\, C^{-1} \right), \quad P_3 = \lim_{\epsilon \to 0} \left(\epsilon^2\, C\, Q_3\, C^{-1} \right), \quad P_0 = \lim_{\epsilon \to 0} \left(\epsilon^2\, C\, S_0\, C^{-1} \right). \tag{48}$$

These four contracted generators lead to the five-by-five transformation matrix, as can be seen from

$$\exp \left\{ -i \left(aP_1 + bP_2 + cP_3 + dP_0 \right) \right\} \tag{49}$$

performing translations in the four-dimensional Minkowski space:

$$\begin{pmatrix} 1 & 0 & 0 & 0 & a \\ 0 & 1 & 0 & 0 & b \\ 0 & 0 & 1 & 0 & c \\ 0 & 0 & 0 & 1 & -d \\ 0 & 0 & 0 & 0 & 1 \end{pmatrix} \begin{pmatrix} x \\ y \\ z \\ t \\ 1 \end{pmatrix} = \begin{pmatrix} x+a \\ y+b \\ z+c \\ t-d \\ 1 \end{pmatrix}. \tag{50}$$

In this way, the space-like directions are translated and the time-like t component is shortened by an amount d. This means the group $O(3,2)$ derivable from the Heisenberg's uncertainty relations becomes the inhomogeneous Lorentz group governing the Poincaré symmetry for quantum mechanics and quantum field theory. These matrices correspond to the differential operators

$$P_x = -i\frac{\partial}{\partial x}, \quad P_y = -i\frac{\partial}{\partial y}, \quad P_z = -i\frac{\partial}{\partial z}, \quad P_0 = i\frac{\partial}{\partial t}, \tag{51}$$

respectively. These translation generators correspond to the Lorentz-covariant four-momentum variable with

$$p_1^2 + p_2^2 + p_3^2 - p_0^2 = \text{constant}. \tag{52}$$

This energy-momentum relation is widely known as Einstein's $E = mc^2$.

5. Concluding Remarks

According to Dirac [1], the problem of finding a Lorentz-covariant quantum mechanics reduces to the problem of finding a representation of the inhomogeneous Lorentz group. Again, according to Dirac [2], it is possible to construct the Lie algebra of the group $O(3,2)$ starting from two oscillators. We have shown in our earlier paper [12] that this $O(3,2)$ group can be contracted to the inhomogeneous Lorentz group according to the group contraction procedure introduced by Inönü and Wigner [3].

In this paper, we noted first that the symmetry of a single oscillator is generated by three generators. Two independent oscillators thus have six generators. We have shown that there are four additional generators needed for the coupling of the two oscillators. Thus there are ten generators. These ten generators can then be linearly combined to produce ten generators which were spelled out in Dirac's 1963 paper.

For the two-oscillator system, there are four step-up and step-down operators. There are therefore sixteen quadratic forms [9]. Among those, only ten of them are in Dirac's 1963 paper [2]. Why ten? Dirac needed those ten to construct the Lie algebra for the $O(3,2)$ group. At the end of the same paper, he stated that this Lie algebra is the same as that for the $Sp(4)$ group, which preserves the minimum uncertainty for each oscillator.

In this paper, we started with the block-diagonal matrix given in Equation (28) for two totally independent oscillators with six independent generators. We then added one two-by-two Hermitian matrix of Equation (29) with four generators for the off-diagonal blocks. The result is the four-by-four Hermitian matrix given in Equation (30). This four-by-four Hermitian matrix has ten independent operators which can be linearly combined to the ten operators chosen by Dirac. Thus, in this paper, we have shown how the two-oscillators are coupled, and how this coupling introduces additional symmetries.

Paul A. M. Dirac made life-long efforts to make quantum mechanics consistent with special relativity, starting from 1927 [4]. While we exploited the contents of his paper published in 1963 [2], it is of interest to review his earlier efforts. In his earlier papers, Dirac started with quantum mechanics and special relativity as two different branches of science based on two different mathematical bases.

In this paper, based on Dirac's two papers [1,2], we concluded that both quantum mechanics and special relativity can be derived from the same mathematical base. A brief review of Dirac's earlier efforts is given in Appendix A.

Author Contributions: Each author contributed equally to this article.

Funding: This research received no external funding.

Conflicts of Interest: The authors declare no conflict of interest.

Appendix A

As we all know, quantum mechanics and special relativity were developed along two separate routes. As early as 1927, Dirac was interested in understanding whether these two scientific disciplines are compatible with each other. In his paper of 1927 [4], Dirac noted the the existence of the time-energy uncertainty relation without excitations. He called this the "c-number" time–energy uncertainty relation. Dirac pointed out that the space-time asymmetry makes it difficult to construct the uncertainty relation in the Lorentz-covariant world.

In 1945, Dirac considered the four-dimensional harmonic oscillator wave functions applicable to the four-dimensional space and time. In so doing, Dirac was considering localized bound states. The space and time variables in his case are the separations between two constituents, like the proton and electron in the hydrogen atom.

It was shown later that Dirac's concern about the c-number time–energy uncertainty is not necessary in view of the fact that a massive particle at rest has only three space-like dimensions [13]. According to Wigner [14], the little group for the massive particle is isomorphic to $O(3)$ [14]. With this understanding, we can use a circle in the z t plane as shown in Figure A1, where z and t are longitudinal and time separations respectively.

In his 1949 paper [15], Dirac introduced the light-cone coordinate system which tells us that the Lorentz boost is a squeeze transformation. This aspect is also illustrated in Figure A1. It is then not difficult to see how the circle looks to a moving observer.

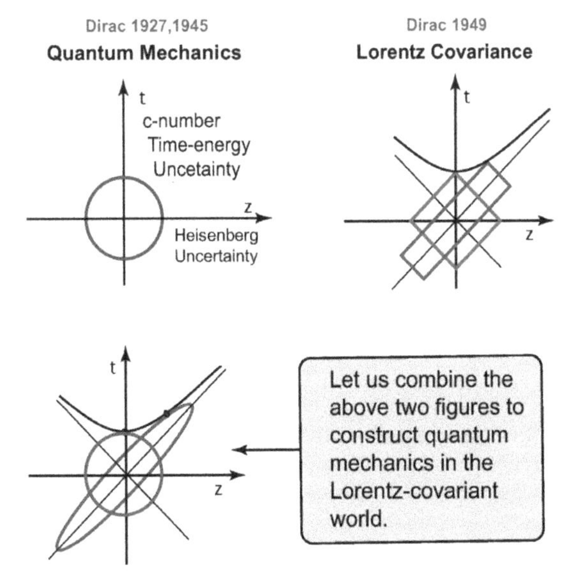

Figure A1. Dirac's three papers. His 1927 and 1945 papers can be described by a circle in the longitudinal space-like and time-like coordinate. Dirac introduced the light-cone coordinate system in 1949. In this system, the Lorentz boost is a squeeze transformation. It is then natural to synthesize these two figures to a squeezed circle or an ellipse. Figure A2 will illustrate how this elliptic squeeze manifests itself in the real world.

The next question is whether this elliptic squeeze has anything to do with the real world. One hundred years ago, Niels Bohr and Albert Einstein met occasionally to discuss physics. Their interests were different. Bohr was worrying about the electron orbit in the hydrogen atom. Einstein was interested in how things look to moving observers. Then the question arises: How would the hydrogen atom look to a moving observer? This was a metaphysical issue during the period of Bohr and Einstein, as there were no hydrogen atoms moving fast enough to exhibit this Einstein effect.

Fifty years later, the physics world was able to produce many protons from particle accelerators. In 1964 [16], Gell-Mann observed that the proton is a bound state of the more fundamental particles called "quarks" according to the quantum mechanics applicable also to the hydrogen atom.

However, according to Feynman [17,18], when the proton moves very fast, it appears as a collection of a large number of free-moving light-like partons with a wide-spread momentum distribution, as described in Figure A2. Feynman's parton picture was entirely based on what we observe in laboratories.

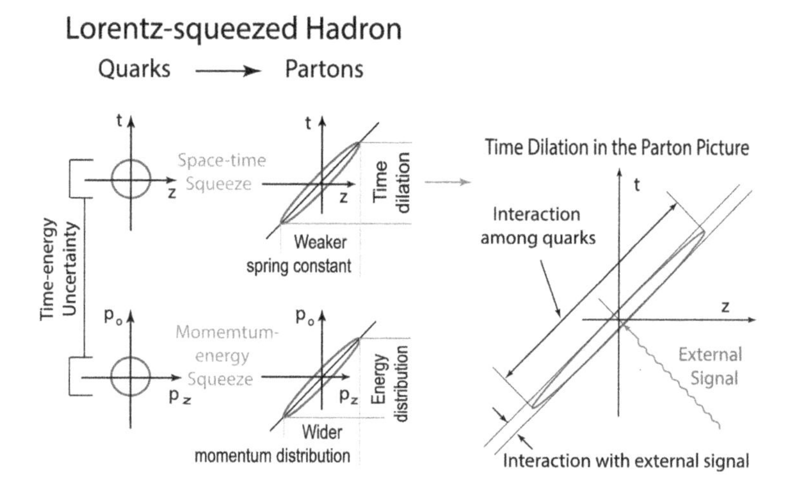

Figure A2. In the harmonic-oscillator regime, the momentum–energy wave function takes the same mathematical form as that of the space-time wave functions. This figure shows that the quark model and the parton model are two different aspects of one Lorentz-covariant entity. In 1969 [17], Feynman observed that the fast-moving proton appears as a collection of a large number of light-like partons with a wide-spread momentum distribution, and short interaction time with the external signal. This figure is a graphical illustration of the 1977 paper by Kim and Noz [19]. This figure is from a recent book by the present authors [20].

Unlike the hydrogen atom, the proton can become accelerated, and its speed could be very close to that of light. Thus the Bohr–Einstein issue became the Gell-Mann–Feynman issue, as illustrated in Figure A3. The question is whether Gell-Mann's quark model and Feynman's parton picture are two different aspects of one Lorentz-covariant entity. This question was addressed by Kim and Noz 1977 [19] and was explained in detail by the present authors with a graphical illustration given in Figure A2.

Figure A3. The Bohr–Einstein issue is 100 years old. Fifty years later, it became the quark–parton puzzle, based on observations made in high-energy laboratories.

References

1. Dirac, P.A.M. Forms of Relativistic Dynamics. *Rev. Mod. Phys.* **1949**, *21*, 392–399. [CrossRef]
2. Dirac, P.A.M. A Remarkable Representation of the 3 + 2 de Sitter Group. *J. Math. Phys.* **1963**, *4*, 901–909. [CrossRef]
3. Inönü, E.; Wigner, E.P. On the Contraction of Groups and their Representations. *Proc. Natl. Acad. Sci. USA* **1953**, *39*, 510–524. [CrossRef] [PubMed]
4. Dirac, P.A.M. The Quantum Theory of the Emission and Absorption of Radiation. *Proc. R. Soc. (London)* **1927**, *A114*, 243–265. [CrossRef]
5. Han, D.; Kim, Y.S.; Noz, M.E. Linear canonical transformations of coherent and squeezed states in the Wigner phase space. *Phys. Rev. A* **1988**, *37*, 807–814. [CrossRef] [PubMed]
6. Kim, Y.S.; Wigner, E.P. Canonical transformation in quantum mechanics. *Am. J. Phys.* **1990**, *58*, 439–447. [CrossRef]
7. Kim, Y.S.; Noz, M.E. *Phase Space Picture of Quantum Mechanics*; World Scientific Publishing Company: Singapore, 1991.
8. Dodonov, V.V.; Man'ko, V.I. *Theory of Nonclassical States of Light*; Taylor & Francis: London, UK; New York, NY, USA, 2003.
9. Han, D.; Kim, Y.S.; Noz, M.E. $O(3,3)$-like Symmetries of Coupled Harmonic Oscillators. *J. Math. Phys.* **1995**, *36*, 3940–3954. [CrossRef]
10. Abraham, R.; Marsden, J.E. *Foundations of Mechanics*, 2nd ed.; Benjamin Cummings: Reading, MA, USA, 1978.
11. Goldstein, H. *Classical Mechanics*, 2nd ed.; Addison-Wesley: Reading, MA, USA, 1980.
12. Başkal, S.; Kim, Y.S.; Noz, M.E. Poincaré Symmetry from Heisenberg's Uncertainty Relations. *Symmetry* **2019**, *11*, 409. [CrossRef]
13. Kim, Y.S.; Noz, M.E.; Oh, S.H. Representations of the Poincaré group for relativistic extended hadrons. *J. Math. Phys.* **1979**, *20*, 1341–1344. [CrossRef]
14. Wigner, E. On unitary representations of the inhomogeneous Lorentz group. *Ann. Math.* **1939**, *40*, 149–204. [CrossRef]
15. Dirac, P.A.M. Unitary Representations of the Lorentz Group. *Proc. R Soc. (London)* **1945**, *A183*, 284–295.
16. Gell-Mann, M. A Schematic Model of Baryons and Mesons. *Phys. Lett.* **1964**, *8*, 214–215. [CrossRef]
17. Feynman, R.P. Very High-Energy Collisions of Hadrons. *Phys. Rev. Lett.* **1969**, *23*, 1415–1417. [CrossRef]

18. Bjorken, J.D.; Paschos, E.A. Electron-Proton and γ-Proton Scattering and the Structure of the Nucleon. *Phys. Rev.* **1969**, *185*, 1975–1982. [CrossRef]
19. Kim, Y.S.; Noz, M.E. Covariant harmonic oscillators and the parton picture. *Phys. Rev. D* **1977**, *15*, 335–338. [CrossRef]
20. Başkal, S; Kim. Y.S.; Noz, M.E. *Physics of the Lorentz Group, IOP Concise Physics*; Morgan & Claypool Publisher: San Rafael, CA, USA; IOP Publishing: Bristol, UK, 2015.

 *quantum reports*

Article

Mutually Unbiased Bases and Their Symmetries

Gernot Alber * and Christopher Charnes

Institut für Angewandte Physik, Technische Universität Darmstadt, D-64289 Darmstadt, Germany; ccharnes@gmail.com

* Correspondence: gernot.alber@physik.tu-darmstadt.de; Tel.: +49-6151-16-20400

Received: 30 September 2019; Accepted: 6 November 2019; Published: 8 November 2019

Abstract: We present and generalize the basic ideas underlying recent work aimed at the construction of mutually unbiased bases in finite dimensional Hilbert spaces with the help of group and graph theoretical concepts. In this approach finite groups are used to construct maximal sets of mutually unbiased bases. Thus the prime number restrictions of previous approaches are circumvented and this construction principle sheds new light onto the intricate relation between mutually unbiased bases and characteristic geometrical structures of Hilbert spaces.

Keywords: mutually unbiased bases; group representations; graphs; quantum information

---

1. Introduction

Mutually unbiased bases of Hilbert spaces, as originally pioneered by Schwinger [1], are not only of mathematical interest by exhibiting characteristic geometric properties of Hilbert spaces, but they have also interesting practical applications in quantum technology. Current applications range from quantum state discrimination [2] and quantum state reconstruction [3,4], to quantum error correction [5,6] and quantum key distribution [7]. They also have been used as signature schemes for CDMA systems in various radio communication technologies [8].

Since the early work of Schwinger [1] the influential work of Wootters and Fields [3] has exhibited intriguing relations between mutually unbiased bases and discrete mathematics. A major result of these authors established that in a Hilbert space of d dimensions the maximum possible number of mutually unbiased bases is $(d+1)$ provided such bases exist. Mutually unbiased bases that saturate this bound are called complete. Previously many investigations have constructed complete sets of mutually unbiased bases (see e.g., [9–16]). Typically, these investigations exploit variants of the two constructions proposed by Wootters and Fields [3] and rely on the properties of Galois fields in odd and even characteristics. Within this framework it is possible to construct systematically maximal sets of mutually unbiased bases in Hilbert spaces whose dimensions are prime powers. Although these developments have exhibited numerous interesting structural properties of complete sets of mutually unbiased bases in prime-power dimensional Hilbert spaces, many questions remain open. Especially interesting is the question of the construction of complete sets of mutually unbiased bases in Hilbert spaces whose dimensions are not prime powers. The lowest dimensional example is dimension $d = 6$ for which it is still unknown whether there are mutually unbiased bases saturating the upper bound of $d + 1 = 7$ originally established by Wootters and Fields.

Here we discuss and generalize a recently developed group and graph theoretical method aimed at the systematic construction of large sets of mutually unbiased bases. This approach stems from the early ideas of Charnes and Beth [17] which were recently developed in [18,19]. The underlying idea in this approach is the systematic use of groups as the setting for constructing large sets of mutually unbiased bases. An important new feature of this framework is the formulation of the construction of systems of mutually unbiased bases as a clique finding problem in Cayley graphs of groups which are naturally associated with sets of mutually unbiased bases. Besides the possible practical advantages, this method is independent of prime power restrictions of previous techniques and thus may offer interesting novel conceptual advantages and links to other areas of mathematics.

The purpose of this manuscript is to present the central ideas of this group and graph theoretical method in a self contained way, and to exhibit new connections between mutually unbiased bases and the symmetries encoded in the related basis groups and basis graphs. Thus, we will explore the theme for which Hilbert space dimensions the Cayley graphs of basis groups are the 1-skeletons of polyhedra in Euclidean 3-space cf. polytopal graphs [20]. The examples of polytopal graphs presented are restricted to low dimensional Hilbert spaces, i.e., $d = 2, 3, 4$, and thus do not address the still open questions concerning dimension $d = 6$. However, these examples demonstrate interesting new links between mutually unbiased bases and the symmetries of graphs which are not apparent with the more orthodox constructions based on Galois fields in prime power dimensions.

2. Mutually Unbiased Bases and Their Construction by Finite Groups

Based on the early work of Charnes and Beth [17] we summarize in this section the basic definitions encompassing the relations between mutually unbiases bases, their basis groups and associated Cayley graphs which are capable of encoding characteristic features of mutually unbiased bases of Hilbert spaces [18,19]. In particular, based on a recent theorem of Charnes [19], which establishes a structural link between complete multipartite Cayley graphs of finite groups and complete sets of mutually unbiased bases, all polytopal basis graphs in Euclidean 3-space are determined. The cliques of these graphs yield complete sets of mutually unbiased bases.

2.1. *Mutually Unbiased Bases—Basic Concepts*

Two orthonormal bases, say $B := \{|B_i\rangle; i = 1, \cdots, d\}$ and $C := \{|C_i\rangle; i = 1, \cdots, d\}$, of a d-dimensional Hilbert space $\mathcal{H}^d$ with scalar product $\langle .|.\rangle$ are called mutually unbiased if and only if the relation

$$| \langle B_i|C_j\rangle |^2 = \frac{1}{d} \tag{1}$$

is independent of the chosen pair (B_i, C_j). Simple well known examples are the eigenstates of any pair of Pauli spin operators in the case of $d = 2$ or the eigenstates of the quantum mechanical position and momentuma operators for $d = \infty$. In the following, however, we shall restrict our considerations to finite dimensional Hilbert spaces.

Subsequent non-selective quantum measurements of two observables associated with mutually unbiased states completely erase any quantum information contained in an arbitrarily prepared quantum state. This becomes apparent if for example we consider two such observables, namely

$$\begin{aligned} \hat{O}_B &= \sum_{i=1}^{d} b_i |B_i\rangle\langle B_i|, \ b_i \in \mathbb{R}, \\ \hat{O}_C &= \sum_{j=1}^{d} c_j |C_j\rangle\langle C_j|, \ c_j \in \mathbb{R}, \end{aligned} \tag{2}$$

and an arbitrary quantum state with density operator $\hat{\rho}$. The subsequent non-selective measurement [21] of observables $\hat{O}_B$ and $\hat{O}_C$ yields the chaotic quantum state $\hat{I}/d$, i.e.,

$$\hat{\rho}' = \sum_{j=1}^{d} |C_j\rangle\langle C_j| \left(\sum_{i=1}^{d} |B_i\rangle\langle B_i|\hat{\rho}|B_i\rangle\langle B_i| \right) |C_j\rangle\langle C_j| = \frac{\mathrm{Tr}(\hat{\rho})}{d} \sum_{j=1}^{d} |C_j\rangle\langle C_j| = \frac{\hat{I}}{d}, \tag{3}$$

thus erasing all previous quantum information contained in the quantum state $\hat{\rho}$. In Equation (3) we have used the completeness relation $\hat{I} = \sum_{j=1}^{d} |C_j\rangle\langle C_j|$ in the Hilbert space $\mathcal{H}^d$.

2.2. Mutually Unbiased Bases and Their Encoding by Unitary Matrices

It should be noted that within quantum theory the ordering of an orthonormal basis is physically relevant. This is apparent from Equation (2), for example, because each basis vector $|B_i\rangle$ can be associated with a different physically measurable eigenvalue b_i of the associated observable $\hat{O}_B$. Therefore, the different elements of an orthonormal basis B are distinguishable physically.

Hence in our subsequent discussion we select an arbitrarily chosen orthonormal ordered basis $(|\alpha\rangle; \alpha = 1, \cdots, d)$ of a finite dimensional Hilbert space $\mathcal{H}^d$. Based on this choice any other ordered orthonormal basis, say $B := (|B_i\rangle; i = 1, \cdots, d)$, can be mapped onto a unitary matrix $M_B \in U(d)$ by

$$(M_B)_{i\alpha} := \langle B_i|\alpha\rangle^* \tag{4}$$

with $*$ denoting complex conjugation. In this mapping row i of the matrix M_B contains the components of the basis vector $|B_i\rangle$ in the canonical basis $(|\alpha\rangle; \alpha = 1, \cdots, d)$. The group of d-dimensional unitary matrices $U(d)$ acts transitively on all ordered orthonormal bases of the Hilbert space $\mathcal{H}^d$ by right multiplication. So to each pair of ordered orthonormal bases, say B and C, there corresponds a unique unitary matrix U satisfying the relation

$$M_B U = M_C. \tag{5}$$

Consequently the defining property (1) of mutually unbiases bases can be reformulated in terms of the matrices associated with different ordered orthonormal bases. Thus, two ordered orthonormal bases B and C are mutually unbiased if and only if for all $i, j \in \{1, \cdots, d\}$

$$|\langle B_i|C_j\rangle|^2 = |\left(M_C M_B^\dagger\right)_{ji}|^2 = \frac{1}{d}. \tag{6}$$

Note that the map $\hat{O}_B \longrightarrow B \longrightarrow M_B$ takes into account the distinguishability of the orthonormal basis vectors associated with different eigenvalues of the observable $\hat{O}_B$, contrary to previous approaches [12].

2.3. Mutually Unbiased Bases and Their Basis Groups

With a set of $n+1$ pairwise mutually unbiased ordered orthonormal bases $\{B^{(0)}, B^{(1)}, \cdots, B^{(n)}\}$ of a Hilbert space $\mathcal{H}^d$ one can associate a basis group G, which is generated by the corresponding unitary matrices, i.e.,

$$G = \left\langle M_{B^{(0)}}, M_{B^{(1)}}, \cdots, M_{B^{(n)}} \right\rangle \subset U(d). \tag{7}$$

This subgroup of the unitary group in d dimensions $U(d)$ has the following properties:

- One of the matrices, e.g., $M_{B^{(0)}}$, is the unit matrix E_d. So it can be removed from the generating set, i.e.,

$$G = \left\langle M_{B^{(1)}}, \cdots, M_{B^{(n)}} \right\rangle. \tag{8}$$

- G is a subgroup of $U(d)$ which has finite or infinite order.
- Not all pairs of elements of G correspond to mutually unbiased bases.
- The structure of the mutually unbiased bases contained in G can be captured by an associated Cayley graph.

2.4. Basis Groups of Mutually Unbiased Bases and Their Cayley Graphs

To each (finite) basis group G there is an associated Cayley graph $\Gamma(G, S)$ defined by the following properties:

- The vertices of $\Gamma(G,S)$ are the group elements of G.
- A generating set $S \subset G$ is defined as all the elements of G which are mutually unbiased to the canonical basis, i.e., mutually unbiased to E_d in the case of a d dimensional Hilbert space. (S does not contain the identity matrix E_d.) Therefore, $z \in S$ implies $z^{-1} \in S$, i.e., $S = S^{-1}$.
- The edge set of $\Gamma(G,S)$ is defined as follows. Two vertices, say x and y, of the graph $\Gamma(G,S)$ are connected by an edge, if and only if $yx^{-1} \in S$, or equivalently if and only if there is an $s \in S$ with $y = sx$. The totality of edges obtained in this way comprises the edge set of $\Gamma(G,S)$.

These Cayley graphs $\Gamma(G,S)$ have the following basic properties:

- As $S^{-1} = S$, the graphs $\Gamma(G,S)$ are simple undirected graphs, i.e., they do not have multiple edges or vertex loops.
- The graphs are represented by symmetric $N \times N$ adjacency matrices with $N = |G|$. Their rows and columns are indexed by the group elements. These adjacency matrices have 0 on the diagonal positions and 0 or 1 elsewhere. Their entries are calculated using Equation (6).
- If two elements of the set S, say $M_{B(i)}, M_{B(j)} \in S$, are mutually unbiased not only with respect to the canonical basis but also among themselves, the set S also contains the matrix $M_{B(i)} M^{\dagger}_{B(j)} \in S$.
- Right multiplication by group elements preserves the adjacency relation of $\Gamma(G,S)$, so G is a subgroup of the automorphism group of $\Gamma(G,S)$.
- Since Cayley graphs are connected, there is an edge connected path between every pair of vertices of $\Gamma(G,S)$.
- As Cayley graphs are regular, each vertex of $\Gamma(G,S)$ is connected to the same number of neighbouring vertices, i.e., it has constant valency. The valency k of a graph is the number of non-zero entries in any row or column of its adjacency matrix.

It should now be apparent that the cliques of a Cayley graph $\Gamma(G,S)$, i.e., the complete subgraphs in which any two vertices are joined by an edge, correspond to mutually unbiased bases. In view of this correspondence, the clique number $\omega(\Gamma(G,S))$ of the Cayley graph $\Gamma(G,S)$, i.e., the size of its largest clique, is not only a mathematically interesting characteristic property of $\Gamma(G,S)$ but it also determines the maximal number of mutually unbiased bases characterized by this graph.

2.5. Maximal Sets of Mutually Unbiased Bases and the Structure of Their Associated Cayley Graphs

The physical relevance of the clique number $\omega(\Gamma(G,S))$ raises the interesting question whether there is a relationship between the maximal possible number of mutually unbiased bases in a d-dimensional Hilbert space, i.e., $d+1$, and the structure of the corresponding Cayley graphs $\Gamma(G,S)$. The following recent theorem [19] demonstrates that for finite basis groups G there is such a relationship.

Theorem 1. *(Charnes Theorem 2 [19]) Let G be a finite basis group of order N with S a generating set of mutually unbiased bases in a Hilbert space $\mathcal{H}^d$. The corresponding Cayley graph $\Gamma(G,S)$ of valency k has a clique of maximum size $d+1$ whenever the condition*

$$\frac{N}{N-k} = \omega(\Gamma(G,S)) = d+1 \tag{9}$$

is fulfilled. In such a case $\Gamma(G,S)$ is a k-regular and complete multipartite graph.

A detailed proof of this theorem is presented in [19]. Here we just outline its basic idea. For this purpose let us consider a Cayley graph $\Gamma(G,S)$ with N vertices and with constant valency k. It is known [22,23] that the clique number $\omega(\Gamma(G,S))$ of such a Cayley graph, i.e., the largest clique size, is lower bounded by the relation

$$\frac{N}{N-k} \leq \omega(\Gamma(G,S)). \tag{10}$$

In addition, Yildirim [23] has shown that this inequality is saturated for complete multipartite graphs, i.e.,

$$\frac{N}{N-k} = \omega(\Gamma(G,S)). \tag{11}$$

Therefore, a sufficient condition that a Cayley graph $\Gamma(G,S)$ yields maximal sets of $d+1$ mutually unbiased bases in a d-dimensional Hilbert space is given by the nested inequalities

$$d+1 = \frac{N}{N-k} \leq \omega(\Gamma(G,S)) \leq d+1 \tag{12}$$

and the associated Cayley graphs $\Gamma(G,S)$ are complete multipartite, as stated in the theorem.

2.6. *Maximal Sets of Mutually Unbiased Bases and Associated Polyhedra in Euclidean 3-Space*

According to the previous section k-regular complete multipartite graphs satisfying Equation (9) play an important role in the group theoretical construction of maximal sets of mutually unbiased bases. We will now explore their relation to polyhedra in Euclidean 3-space [20].

Let us start our discussion with the definitions of k-regularity and complete multipartiteness of graphs. A graph is k-regular if every vertex has exactly k edges. Furthermore, a graph is complete multipartite if its vertices can be partitioned into independent sets, also called colour classes, in such a way that

- vertices within an independent set are not connected by any edge and
- there is an edge between every pair of vertices from different independent sets.

In Figure 1 the complete multipartite graph $K_{2,2,2}$ is an example of a k-regular complete multipartite graph. Its valency is $k = 4$ and the vertices belong to three independent sets each containing two vertices. However, this graph is not only complete multipartite with constant valency, it is also the 1-skeleton of a regular polyhedron in Euclidean 3-space, namely an octahedron. Therefore, the interesting question arises whether there are other polyhedra in Euclidean 3-space, whose 1-skeletons are k-regular complete multipartite graphs and are relevant in determining maximal sets of mutually unbiased bases. Interestingly, all such polyhedra can be determined by combining the condition of Equation (9) with the Steinitz criterion [24] for polytopal graphs in Euclidean 3-space and by making use of the four-color theorem [25].

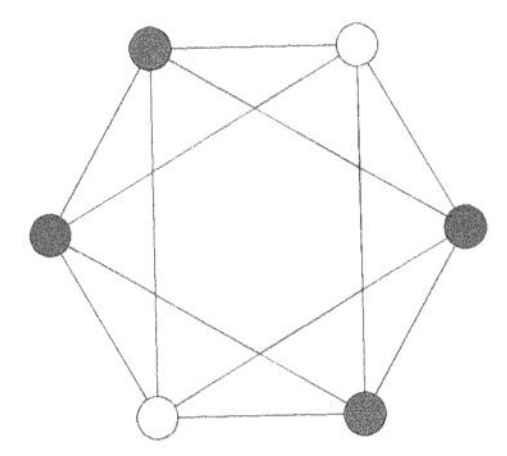

Figure 1. The graph $K_{2,2,2}$ as an example of a k-regular complete multipartite graph which is also the 1-skeleton of an octahedron in Euclidean 3-space: Each vertex has exactly $k = 4$ edges. The vertices can be partitioned into 3 independent sets (color classes) each containing 2 vertices. Vertices within the same color class are not connected by an edge and there is an edge between every pair of vertices within different color classes.

According to the Steinitz criterion a graph is polytopal in Euclidean 3-space iff the graph is planar and 3-connected [24]. A graph is planar if it can be drawn in the plane so that its edges do not intersect, and a graph is 3-connected if there is an at most 3-connected path between every two vertices of the graph.

According to the four-color theorem [25] all planar graphs can be colored using at most 4 colors. Therefore, by Equation (9) and the four-color theorem a maximal set of $d+1$ mutually unbiased bases can be constructed in a d-dimensional Hilbert space using a finite group G of order N with a generating set S and associated Cayley graph $\Gamma(G,S)$ of valency k, if the following relation

$$\frac{N}{N-k} = d+1 \leq 4 \tag{13}$$

is statisfied. Therefore, the necessary condition for the existence of complete multipartite polytopal graphs $\Gamma(G,S)$ which saturate the bound $d+1$ for complete sets of mutually unbiased bases is that the Hilbert space has dimension $d=2$ or 3.

In order to determine the possible orders N of the groups and the possible valencies k a further relation is needed. The Descartes-Euler relation [26], involving the number of vertices f_0, edges f_1 and facets f_2 of a finite convex polyhedron, establishes the equation $f_0 - f_1 + f_2 = 2$. This relation gives the additional constraint. It places an upper bound on the possible values of the valencies of the form

$$k \leq 5, \tag{14}$$

because every 3-polytopal graph has a vertex of valency at most 5 [24]. Consequently the dimensions d of the Hilbert spaces, the orders N and valencies k of all Cayley graphs $\Gamma(G,S)$ can be determined. Such triples (N,k,d) are the feasibility parameters of polytopal graphs in Euclidean 3-space which yield maximal sets of mutually unbiased bases. They are summarized in Table 1.

Table 1. Orders of basis groups $\mid G \mid = N$, valencies k of basis Cayley graphs $\Gamma(G,S)$, dimensions of the Hilbert spaces d and polytopal Cayley graphs $\Gamma(G,S)$ in Euclidean 3-space for which maximal sets of mutually unbiased bases can be constructed.

N	$k = Nd/(d+1)$	d	**Polyhedron in 3-Space**
3	2	2	triangle (degenerate)
6	4	2	octahedron
4	3	3	tetrahedron

3. Examples of Maximal Sets of Mutually Unbiased Bases, Their Basis Groups and Cayley Graphs

In this section examples are presented which exemplify the theoretical developments of the previous section in Hilbert spaces of low dimensions, i.e., $d = 2, 3$ and 4. These examples include complete multipartite polytopal Cayley graphs in Euclidean 3-space as well as more general scenarios.

3.1. A Cyclic Basis Group for $d = 2$ with an Octahedral Cayley Graph

In two dimensional Hilbert spaces a one-parameter family of cyclic basis groups $G_\varphi =< M_\varphi >$ of order $N = 6$, i.e., $M_\varphi^6 = E_2$, yielding maximal sets of mutually unbiased bases is generated by the matrix

$$M_\varphi = \frac{1}{\sqrt{2}} \begin{pmatrix} e^{-i\pi/4} & e^{i\varphi} \\ -e^{-i\varphi} & e^{i\pi/4} \end{pmatrix} \tag{15}$$

with $\varphi \in [0, 2\pi)$. The group generators defining the Cayley graph $\Gamma(G_\varphi, S_\varphi) = K_{2,2,2}$ are

$$S_\varphi = \{M_\varphi, M_\varphi^2, M_\varphi^\dagger, M_\varphi^{2\dagger}\}. \tag{16}$$

This Cayley graph is the 1-skeleton of an octahedron. It is 4-regular, complete multipartite and its vertices are partitioned into $N/(N-k) = 3 = d+1$ independent sets $I_i = \{M_\varphi^i, M_\varphi^{i+3}\}$ ($i \in \{1,2,3\}$) each containing $N - k = 2$ elements. It is apparent that this Cayley graph satisfies the feasibility constraints of Table 1. The number of maximal mutually unbiased bases, i.e., complete subgraphs K_3, is $(N-k)^{d+1} = 2^3 = 8$. Since the basis group is cylic the defining representation of G_φ splits into the direct sum of 2 one dimensional representations [27].

3.2. A Non-Abelian Basis Group for $d = 2$ with an Octahedral Cayley Graph

In two dimensional Hilbert spaces a non-Abelian basis group $G =< M_1, M_2 >$ is generated by the following matrices

$$M_1 = \frac{1}{\sqrt{2}} \begin{pmatrix} -1 & i \\ -i & 1 \end{pmatrix}, \quad M_2 = \begin{pmatrix} 0 & -e^{i\pi/4} \\ e^{3i\pi/4} & 0 \end{pmatrix} \tag{17}$$

satisfying the defining relations $M_1^2 = M_2^2 = (M_1M_2)^3 = E_2$. This representation of the symmetric group S_3 is irreducible. A generating set of S_3 used to define the Cayley graph $\Gamma(G, S)$ is

$$S = \{M_2M_1M_2, M_1, M_1M_2, M_2M_1\}. \tag{18}$$

Once again this Cayley graph is $K_{2,2,2}$ and it is the 1-skeleton of an octahedron thus satisfying the feasibility parameters of Table 1. It is 4-regular and complete multipartite with $N/(N-k) = 3 = d+1$ independent sets $I_1 = \{M_2, M_2^2\}$, $I_2 = \{M_1M_2, M_2M_1M_2\}$ and $I_3 = \{M_1, (M_1M_2)^2\}$ each containing $N-k = 2$ elements. Furthermore, the number of maximal mutually unbiased bases, i.e., of complete subgraphs K_3, is $(N-k)^{d+1} = 2^3 = 8$. Comparing this graph with the previous example demonstrates that isomorphic Cayley graphs can be associated with different basis groups and generating sets, i.e., $K_{2,2,2} \cong \Gamma(G_\varphi, S_\varphi) \cong \Gamma(G, S)$.

3.3. A Non-Abelian Basis Group for $d = 3$ with a Non Polytopal Cayley Graph

In three dimensional Hilbert spaces the following matrices R_1 and R_2, where $\omega := exp(\frac{2i\pi}{3})$, i.e.,

$$R_1 = \frac{1}{3} \begin{pmatrix} \omega-\omega^2 & -2\omega-\omega^2 & -2\omega-\omega^2 \\ \omega+2\omega^2 & -2\omega-\omega^2 & \omega+2\omega^2 \\ \omega+2\omega^2 & \omega+2\omega^2 & -2\omega-\omega^2 \end{pmatrix}, \quad R_2 = \frac{1}{3} \begin{pmatrix} \omega-\omega^2 & \omega-\omega^2 & \omega-\omega^2 \\ \omega-\omega^2 & -2\omega-\omega^2 & \omega+2\omega^2 \\ \omega-\omega^2 & \omega+2\omega^2 & -2\omega-\omega^2 \end{pmatrix}, \tag{19}$$

satisfy the defining relations $R_1^4 = E_3$, $R_1^2 = R_2^2$; $R_2^{-1}R_1R_2 = R_1^{-1}$. Thus they generate the non-Abelian basis group $Q_8 =< R_1, R_2 >$, which is isomorphic to the quaternion group of order $| Q_8 |= N = 8$. The defining representation of Q_8 is reducible and splits into irreducible representations as $1 \oplus 2$. The entries of the matrices $3R_1$ and $3R_2$ are the Eisenstein integers. The associated Cayley graph is defined by the following set S of generators of Q_8

$$S = \{R_1, R_2, R_2R_1, R_1^\dagger, R_2^\dagger, R_1^\dagger R_2^\dagger\}. \tag{20}$$

The resulting Cayley graph $\Gamma(Q_8, S)$ is $k = 6$-regular and complete multipartite. Each of its $N/(N-k) = 4 = d+1$ independent sets has size $N-k = 2$ (compare with Figure 2). In contrast to the two previous examples $\Gamma(Q_8, S)$ is not a polytopal graph in Euclidean 3-space.

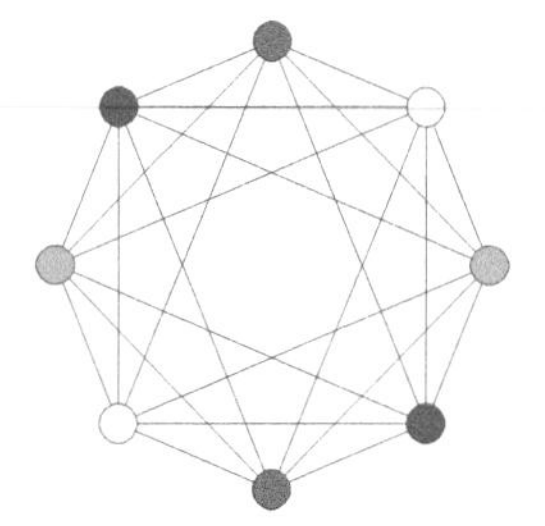

Figure 2. The Cayley graph $\Gamma(Q_8, S) = K_{2,2,2,2}$ as defined by the set S of Equation (20): This complete multipartite Cayley graph is $k = 6$-regular with 4 independent sets each containing 2 vertices. It is not a polytopal graph in Euclidean 3-space as it does not fulfill the conditions of Table 1.

The number of maximal mutually unbiased bases, i.e., complete subgraphs K_4 of $\Gamma(Q_8, S)$, is $(N-k)^{d+1} = 2^4 = 16$. A sample of four representative mutually unbiased bases, corresponding to the K_4 subgraphs of $\Gamma(Q_8, S)$, is: $\{R_1^2, R_2R_1, R_2^3, R_1^3\}, \{R_1^2, R_1, R_2, R_2^3R_1\}, \{E_3, R_2R_1, R_2^3, R_1^3\}$, $\{E_3, R_1, R_2, R_2^3R_1\}$. For the complete set of 16 mutually unbiased bases see [19]. In order to determine the number of physically distinguishable mutually unbiased bases one has to take into account the projective structure of quantum theory implying that a pure quantum state is represented by a ray in Hilbert space. Therefore, orthonormal bases which differ by a global phase have to be identified because they are indistinguishable physically. As the basis group Q_8 has non trivial centers not all 2^4 cliques of the associated Cayley graph $K_{2,2,2,2}$ yield physically distinguishable complete sets of mutually unbiased bases.

3.4. *An Icosahedral Basis Group for $d = 4$ with a Non Polytopal Cayley Graph*

In four dimensional Hilbert spaces the matrices T_1 and T_2, i.e.,

$$T_1 = \frac{1}{2}\begin{pmatrix} 1 & i & i & -1 \\ -i & -1 & 1 & -i \\ -i & 1 & -1 & -i \\ -1 & i & i & 1 \end{pmatrix}, \quad T_2 = \begin{pmatrix} 0 & 0 & 0 & -i \\ 0 & 1 & 0 & 0 \\ i & 0 & 0 & 0 \\ 0 & 0 & 1 & 0 \end{pmatrix}, \tag{21}$$

satisfying the defining relations $T_1^2 = T_2^3 = (T_1T_2)^5 = E_4$ generate the non-Abelian basis group $I_{60} =< T_1, T_2 >$, which is isomorphic to the icosahedral group of order $\mid I_{60} \mid= N = 60$. This 4-dimensional representation of I_{60} is irreducible. This basis group is a simple group, i.e., it has no proper normal subgroups [27].

The adjacency matrix of the associated Cayley graph graph $\Gamma(I_{60}, S)$ is determined by this representation of the group I_{60} and the defining relation of Equation (6) for mutually unbiased bases. The generators S of the Cayley graph $\Gamma(I_{60}, S)$ are defined by this adjacency matrix. This Cayley graph is 48-regular so that the set S contains 48 elements. Furthermore, it is complete multipartite with $N/(N-k) = 5 = d+1$ independent sets each containing $N - k = 12$ elements.

Although I_{60} is the group of proper three dimensional rotations of the icosahedron, the graph $\Gamma(I_{60}, S)$ is not polytopal in Euclidean 3-space. But the basis group I_{60} is a subgroup of the automorphism group of $\Gamma(I_{60}, S)$, as required by the general properties of basis groups and their associated Cayley graphs. The number of maximal mutually unbiased bases, i.e., the number of complete subgraphs K_5 of $\Gamma(I_{60}, S)$, is $(N-k)^{d+1} = 12^5 = 248,832$. As the basis group I_{60} is a simple group all these 12^5 cliques yield physically distinguishable complete sets of mutually unbiased bases.

4. Conclusions

We have discussed and generalized a recently developed group and graph theoretial approach aiming at the construction of large sets of mutually unbiased bases in finite dimensional Hilbert spaces.

In this approach the construction of mutually unbiased bases in a Hilbert space of given dimension is reformulated as a clique finding problem of a Cayley graph associated with a finite basis group. This approach offers the possibility to enlarge and possibly also to complete already known systems of mutually unbiased basis systems. As this approach is independent of prime number restrictions of previous formulations, such as the ones in [9–16], it sheds new light onto the connections between the structure of mutually unbiased bases of Hilbert spaces and other areas of mathematics.

In this manuscript we have explored a connection to geometry by classifying all the polytopal graphs in Euclidean 3-space which are the possible Cayley graphs of basis groups supporting maximal sets of mutually unbiased bases. It has been shown that apart from the degenerate case of a two dimensional triangle such polyhedral constructions can only occur in Hilbert space dimensions $d = 2$ and $d = 3$ either by octahedra in the case $d = 2$ or by tetrahedra in the case $d = 3$. The Cayley graphs of the two dimensional examples presented in Sections 3.1 and 3.2 are isomorphic octahedra exemplifying these polytopal constructions in Euclidean 3-space. In particular, these examples demonstrate that different basis groups may lead to isomorphic Cayley graphs. The Cayley graphs of the three and four dimensional examples discussed in Sections 3.3 and 3.4, however, are of a more general nature. They do not belong to the set of polytopal constructions in Euclidean 3-space and introduce new complete sets of mutually unbiased bases. In particular, these two latter examples demonstrate the general property discussed in Section 2.4 that a basis group, such as Q_8 (I_{60}), is always a subgroup of the automorphism group of the associated Cayley graph, such as $K_{2,2,2,2}$ ($K_{12,12,12,12,12}$). This property establishes an interesting general relation between the symmetry encoded in a basis group and the symmetry encoded in its associated Cayley graph which is expected to be useful for the construction of complete sets of mutually unbiased bases in higher dimensional Hilbert spaces.

Our investigations and the low dimensional examples presented here constitute the first steps in a systematic exploration of this group and graph-theoretical approach. They hint at interesting connections between structures of mutually unbiased bases of finite dimensional Hilbert spaces and symmetries of Cayley graphs which will be explored in future work.

Author Contributions: Conceptualization, G.A. and C.C.; methodology, G.A. and C.C.; validation, G.A. and C.C.; investigation, G.A. and C.C.

Funding: This research was funded by the German Academic Exchange Service (DAAD).

Conflicts of Interest: The authors declare no conflict of interest.

References

1. Schwinger, J. Unitary operator bases. *Proc. Natl. Acad. Sci. USA* **1960**, *46*, 570–579. [CrossRef] [PubMed]
2. Ivanovic, I.D. Geometrical description of quantal state determination. *J. Phys. A: Math. Theor.* **1981**, *14*, 3241–3245. [CrossRef]
3. Wootters, W.K.; Fields, B.D. Optimal state-determination by mutually unbiased measurements. *Ann. Phys.* **1989**, *191*, 363–381. [CrossRef]
4. Yuan, H.; Zhon, Z.-W.; Guo, G.-S. Quantum state tomography via mutually unbiased measurements in driven cavity QED systems. *New J. Phys.* **2016**, *18*, 043013. [CrossRef]
5. Gottesman, D. Class of quantum error-correcting codes saturating the quantum Hamming bound. *Phys. Rev. A* **1996**, *54*, 1862–1868. [CrossRef] [PubMed]
6. Calderbank, A.R.; Rains, E.M.; Shor, P.W.; Sloane, N.J.A. Quantum error correction and orthogonal geometry. *Phys. Rev. Lett.* **1997**, *78*, 405–408. [CrossRef]
7. Bruß, D. Optimal eavesdropping in quantum cryptography with six states. *Phys. Rev. Lett.* **1998**, *81*, 3018–3021. [CrossRef]
8. Heath, R.W.; Strohner, T.; Paulraj, A.J. On quasi-orthogonal signatures for CDMA systems. *IEEE Trans. Inf. Theory* **2006**, *52*, 1217–1225. [CrossRef]
9. Lawrence, J.; Brukner, C.; Zeilinger, A. Mutually unbiased binary observable sets on N qubits. *Phys. Rev. A* **2002**, *65*, 032320. [CrossRef]

10. Šulc, P.; Tolar, J. Group theoretical construction of mutually unbiased bases in Hilbert spaces of prime dimensions. *J. Phys. A: Math. Theor.* **2007**, *40*, 15099–15111. [CrossRef]
11. Garcia, A.; Romero, J.L.; Klimov, A.B. Generation of bases with definite factorization for an n-qubit system and mutually unbiased sets construction. *J. Phys. A: Math. Theor.* **2010**, *43* 385301. [CrossRef]
12. Brierley, S.; Weigert, S.; Bengtsson, I. All mutually unbiased bases in dimensions two to five. *Quantum Inf. Comp.* **2010**, *10*, 803–820.
13. van Dam, W.; Howard, M. Bipartite entangled stabilizer mutually unbiased bases as maximum cliques of Cayley graphs. *Phys. Rev. A* **2011**, *84*, 012117. [CrossRef]
14. Klimov, A.B.; Björk, G.; Sánchez-Soto, L.L. Optimal quantum tomography of permutationally invariant qubits. *Phys. Rev. A* **2013**, *87*, 012109. [CrossRef]
15. Spengler, C.; Kraus, B. Graph-state formalism for mutually unbiased bases. *Phys. Rev. A* **2013**, *88*, 052323. [CrossRef]
16. García, A.; Klimov, A.B. Complete sets of mutually unbiased operators in n-qudit systems. *Phys. Scr. T* **2014**, *160*, 014012. [CrossRef]
17. Charnes, C.; Beth, T. Groups, graphs and mutually unbiased bases. In Proceedings of ERATO conference on Quantum Information Science 2005, Tokyo, Japan, 26–30 April 2005; pp. 73–74.
18. Alber, G. Charnes, C. Mutually unbiased bases: a group and graph theoretical approach. *Phys. Scr.* **2018**, *94*, 1–8. [CrossRef]
19. Charnes, C. Group representations, graphs and mutually unbiased bases. *Linear Algebra Appl.* **2019**, in preparation.
20. Coxeter, H.S.M. *Regular Polytopes*; Dover Publications: New York, NY, USA, 1973.
21. Holevo, A.S. *Statistical Structure of Quantum Theory*; Springer: Berlin, Germany, 2001.
22. Wilf, H.S. Spectral bounds for the clique and independence numbers in graphs. *J. Comb. Theory Ser. B* **1986**, *40*, 113–117. [CrossRef]
23. Yildirim, E.A. A simpler characterization of a spectral lower bound on the clique number. *Math. Methods Oper. Res.* **2010**, *71*, 267–281. [CrossRef]
24. Kalai, G. Polytope skeletons and paths. In *Handbook of Discrete and Computational Geometry-Third Edition*; Goodman, J.E., O'Rourke, J., Tóth, C.D., Eds.; CRC Press: Boca Raton, FL, USA, 2017; pp. 505–532.
25. Gonthier, G. Formal proof—The four-color theorem. *Not. Am. Math. Soc.* **2008**, *55*, 1382–1393.
26. Ziegler, G.M.; Blatter, C. Euler's polyhedron formula—A starting point of today's polytope theory. *Elem. Math.* **2007**, *62*, 184–192. [CrossRef] [PubMed]
27. Murnaghan, F.D. *The Theory of Group Representations*; Dover Publications: New York, NY, USA, 2005.

Article

A Quantum Charged Particle under Sudden Jumps of the Magnetic Field and Shape of Non-Circular Solenoids

Viktor V. Dodonov [1,2,*] **and Matheus B. Horovits** [1,3]

[1] Institute of Physics, University of Brasilia, P.O. Box 04455, Brasília 70919-970, DF, Brazil; matheus.horovits@gmail.com
[2] International Center for Physics, University of Brasilia, Brasilia 70910-900, DF, Brazil
[3] Instituto Federal de Brasília, Campus Estrutural, Brasília 71255-200, DF, Brazil
* Correspondence: vdodonov@fis.unb.br

Received: 12 September 2019; Accepted: 22 October 2019; Published: 28 October 2019

Abstract: We consider a quantum charged particle moving in the xy plane under the action of a time-dependent magnetic field described by means of the linear vector potential of the form $\mathbf{A} = B(t)\left[-y(1+\beta), x(1-\beta)\right]/2$. Such potentials with $\beta \neq 0$ exist inside infinite solenoids with non-circular cross sections. The systems with different values of β are not equivalent for nonstationary magnetic fields or time-dependent parameters $\beta(t)$, due to different structures of induced electric fields. Using the approximation of the stepwise variations of parameters, we obtain explicit formulas describing the change of the mean energy and magnetic moment. The generation of squeezing with respect to the relative and guiding center coordinates is also studied. The change of magnetic moment can be twice bigger for the Landau gauge than for the circular gauge, and this change can happen without any change of the angular momentum. A strong amplification of the magnetic moment can happen even for rapidly decreasing magnetic fields.

Keywords: circular gauge; Landau gauge; arbitrary linear gauge; stepwise variation; center-of-orbit coordinates; relative coordinates; elliptic and hyperbolic solenoids; angular momentum; magnetic moment; squeezing

---

1. Introduction

The motion of a quantum non-relativistic charged particle in a uniform stationary magnetic field, $\mathbf{B} = (0,0,B) = \text{rot}\,\mathbf{A}$, has been studied since the dawn of quantum mechanics, beginning with the papers by Kennard, Darwin, and Fock [1–3]. These authors used the "circular" gauge of the vector potential $A_c = [\boldsymbol{B} \times \boldsymbol{r}]/2 = B(-y,x,0)/2$. Another gauge, $\boldsymbol{A}_L = B(-y,0,0)$, was considered by Landau [4]. Although the form of solutions to the Schrödinger equation is rather different for these two gauges, the physical results, such as the magnetization of the free electron gas, were identical.

The Schrödinger equation in the case of time-dependent uniform magnetic field was solved by Malkin, Man'ko, and Trifonov [5] for the "circular" gauge of the vector potential and by Dodonov, Malkin, and Man'ko for the Landau gauge [6]. However, in these two cases, the physical results turned out quite different, e.g., comparing the transition probabilities between the energy levels. This can be explained by different geometries of the induced electric field $\boldsymbol{E}(\boldsymbol{r},t) = -\partial \boldsymbol{A}(\boldsymbol{r},t)/\partial(ct)$ (we use the Gaussian system of units). Other manifestations of the nonequivalence between the Landau and circular gauges in the case of time-dependent magnetic fields were observed in studies [7,8] devoted to the problem of generation of squeezed states of charged particles in magnetic fields.

In this paper, we suppose that the homogeneous magnetic field B, directed along z-axis, is described by means of a general linear vector potential,

$$A(t) = B(t)\left[-y(1+\beta), x(1-\beta), 0\right]/2, \tag{1}$$

with arbitrary values of real parameter β. The circular gauge corresponds to $\beta = 0$, whereas the Landau gauge corresponds to $\beta = \pm 1$. Our main goal is to see how the mean values of energy, angular momentum, and other quantities can change in the case of time-dependent magnetic field with arbitrary values of parameter β. In particular, interesting questions are whether it is possible to "cool" the particle by changing the magnetic field or the shape of solenoid, or how strong is the change of the magnetic moment due to the change of the magnetic field? Unfortunately, the general treatment of the time-dependent problem with arbitrary functions $B(t)$ and/or $\beta(t)$ is very difficult. For this reason, we confine ourselves in this paper with the simplified model of instantaneous changes of parameters. Although abrupt changes of parameters are idealizations of real processes, they are frequently used for the analysis of various physical processes [9–17].

Our plan is as follows. In Section 2, we show how the general linear vector potential can be created inside solenoids with non-circular cross sections. In Section 3, we present the basic equations and explicit expressions for the transformation matrix, relating the relative and guiding center coordinates before and after the jump. Section 4 is devoted to concrete formulas describing the changes of the mean energy, magnetic moment, and mean positions of the guiding center, as well as the squeezing with respect to the relative and guiding center coordinates. Section 5 contains a discussion of the results, including a justification of the sudden jump model.

2. Vector Potential Inside an Infinite Solenoid with an Arbitrary Cross Section

Our starting point is the well-known formula for the vector potential created by an arbitrary distribution of the electrical current density $\mathbf{j}$ (in the Gauss system of units) [18]:

$$\mathbf{A}(\mathbf{r}) = \frac{1}{c}\int \frac{\mathbf{j}(\mathbf{r}')}{|\mathbf{r}-\mathbf{r}'|}dV. \tag{2}$$

Consider the cylindrical solenoid of length $2Z$ and an arbitrary cross section. Two examples are shown in Figure 1.

Suppose that the current density differs from zero in a thin slab of thickness τ, being directed perpendicular to the cylinder axis z and having the constant absolute value j_0. The element of the integration volume dV can be chosen as $dV = \tau dl dz$, where dl is the infinitesimal length element along the cylinder boundary in the horizontal plane xy. We are interested in the vector potential in the plane $z = 0$ (assuming that $-Z < z < Z$ for the points of the solenoid surface), keeping in mind to take the limit $Z \to \infty$. We can write $\mathbf{j}dV = \mathcal{I}\mathbf{dl}dz$, where $\mathcal{I} = j_0\tau$ is the surface current density and $\mathbf{dl}$ is the infinitesimal vector along the contour of the cross-section. Then, (2) takes the form

$$\mathbf{A}(x,y) = \frac{\mathcal{I}}{c}\oint \mathbf{dl}\int_{-Z}^{Z}\frac{dz}{(\rho^2+z^2)^{1/2}}, \qquad \rho^2 = (x-\xi)^2 + (y-\eta)^2, \tag{3}$$

where (ξ, η) is the point on the cylinder surface in the plane $z = 0$, so that $\mathbf{dl} = (d\xi, d\eta)$. The integral over dz can be easily calculated with the aid of the substitution $z = \rho \sinh(\chi)$:

$$\int_{-Z}^{Z} \frac{dz}{(\rho^2 + z^2)^{1/2}} = 2\chi_Z = 2 \ln \left[\left(Z + \sqrt{Z^2 + \rho^2} \right) / \rho \right].$$

For $Z \gg \rho$, the right-hand side can be written as $2\ln(2Z) + \rho^2/(2Z^2) - 2\ln\rho$. As $\oint \mathbf{dl} \ln(2Z) = 0$, we obtain the following limit of (3) for $Z \to \infty$:

$$\mathbf{A}(x, y) = -\frac{\mathcal{I}}{c} \oint \mathbf{dl} \ln \left(\rho^2 \right). \tag{4}$$

This general formula holds for any infinite cylindrical solenoid, with an arbitrary cross-section (provided the surface current density does not depend on z). (One can be unhappy to see the dimensional quantity ρ^2 in the argument of the logarithmic function. However, one can make this argument dimensionless, adding to the right-hand side of (4) the expression $\oint \mathbf{dl} \ln(S) = 0$, where S can be an arbitrary constant with the dimension of square of length. In any case, no error with the dimensionality will appear in the final results).

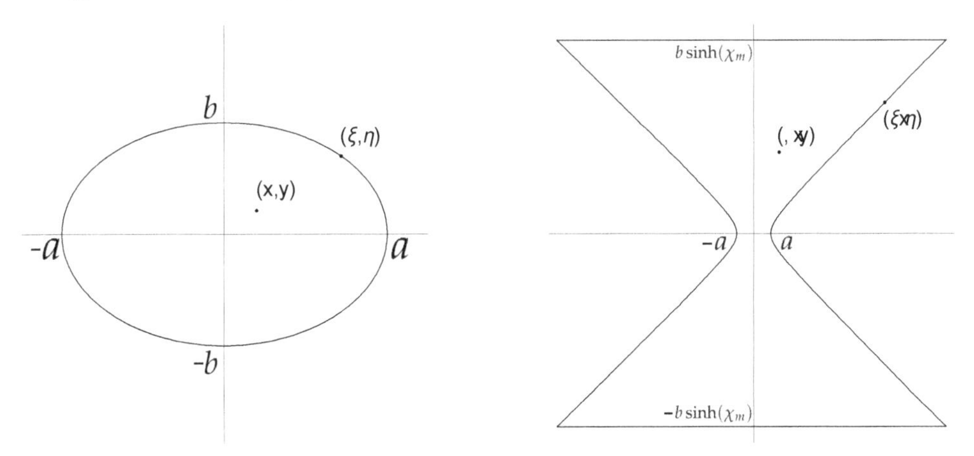

Figure 1. (**Left**) The cross section of the elliptic solenoid with the axes ratio $a/b = 2$, yielding the vector potential (1) with $\beta = 1/3$. (**Right**) The cross section of the hyperbolic solenoid with $\beta \approx -3/2$.

For the elliptical solenoid with the axes $2a$ and $2b$, one can use the parametrization $\xi = a\cos\varphi$ and $\eta = b\sin\varphi$. Then, the two components of the vector potential can be written as

$$\begin{pmatrix} A_x \\ A_y \end{pmatrix} = -\frac{\mathcal{I}}{c} \int_0^{2\pi} d\varphi \begin{pmatrix} -a\sin\varphi \\ b\cos\varphi \end{pmatrix} \ln \left[(x - a\cos\varphi)^2 + (y - b\sin\varphi)^2 \right]. \tag{5}$$

One can see that the change of the integration variable $\varphi = \psi - \pi/2$, together with the substitutions $x \to y$, $y \to -x$ and $a \leftrightarrow b$, transforms the integral for A_x into the integral for A_y. Therefore, it is sufficient to calculate A_x. However, the calculation of integral (5) for arbitrary four parameters, a, b, x, y, is rather complicated. Therefore, we assume that $|x| \ll a$ and $|y| \ll b$. This means that we consider the motion in some relatively small region inside the solenoid with big enough perpendicular sizes. One can see that integral (5) equals zero for $x = y = 0$, as the integrand is an odd function in this case. Taking into account only linear terms with respect to x and y, we arrive at the formula

$$A_x = -\frac{2aby\mathcal{I}}{c}\int_0^{2\pi}\frac{d\varphi\sin^2\varphi}{a^2\cos^2\varphi + b^2\sin^2\varphi}. \tag{6}$$

The integral proportional to x equals zero, as the related integrand is an odd function of φ. The integral in (6) can be written as

$$\int_0^{2\pi}\frac{d\psi(1-\cos\psi)}{a^2+b^2+(a^2-b^2)\cos\psi} = -\oint_{|z|=1}\frac{dz}{iz}\frac{(z-1)^2}{(a^2-b^2)(z^2+1)+2(a^2+b^2)z},$$

where $\psi = 2\varphi$ and $z = \exp(i\psi)$. The integrand of the contour integral has two poles inside the circle $|z| = 1$: $z_1 = 0$ and $z_2 = (b-a)/(b+a)$. Using the residue method, we arrive at the formula

$$\int_0^{2\pi}\frac{d\varphi\,\sin^2\varphi}{a^2\cos^2\varphi + b^2\sin^2\varphi} = \frac{2\pi}{b(a+b)}. \tag{7}$$

Consequently, the vector potential components near the center of the elliptic solenoid have the form

$$A_x = -\frac{Bay}{a+b}, \qquad A_y = \frac{Bbx}{a+b}, \qquad B = 4\pi\mathcal{I}/c, \tag{8}$$

where B is the strength of the uniform magnetic field inside the solenoid. Comparing (1) with (8), we obtain the relation between parameter β and the semi-axes of the elliptic solenoid:

$$\beta = (a-b)/(a+b). \tag{9}$$

Actually, the linear vector potential of the form (1) exists for an arbitrary shape of the solenoid cross section, possessing a symmetry with respect to reflections from the x and y axes, within a small (compared with the transverse solenoid sizes) region near the center. For example, let us consider the rectangular cross section $-a \le \xi \le a$ and $-b \le \eta \le b$. Then, Formula (4) yields

$$A_y = \frac{\mathcal{I}}{c}\int_{-b}^{b} d\eta\,\ln\left[\frac{(\eta-y)^2+(x+a)^2}{(\eta-y)^2+(x-a)^2}\right]. \tag{10}$$

The corresponding indefinite integrals can be calculated exactly:

$$\int dz\,\ln\left(z^2+g^2\right) = z\ln\left(z^2+g^2\right) - 2z + 2g\tan^{-1}(z/g).$$

However, the resulting exact formula for A_y is very cumbersome. On the other hand, expanding the integrand in (10) with respect to x and y, we obtain in the linear approximation:

$$A_y = \frac{\mathcal{I}}{c}\int_{-b}^{b}\frac{4axd\eta}{a^2+\eta^2} = (2/\pi)Bx\tan^{-1}(b/a). \tag{11}$$

In this case,

$$\beta = 1 - (4/\pi)\tan^{-1}(b/a). \tag{12}$$

To show how the vector potential (1) with $|\beta| > 1$ can be created, let us consider the concave solenoid with two hyperbolic boundaries, $(\xi/a)^2 - (\eta/b)^2 = 1$, parametrized as $\xi = \pm a\cosh(\chi)$ and

$\eta = b\sinh(\chi)$. As the current must flow along the closed circuit, we suppose that $-\chi_m < \chi < \chi_m$, so that the circuit is closed by two straight lines $\eta = \pm b\sinh(\chi_m)$. In this geometry, the component A_y is determined only by the current in the hyperbolic boundaries. It can be written as (if the current flows in the counterclockwise direction)

$$A_y = \frac{\mathcal{I}}{c}\int_{-\chi_m}^{\chi_m} b\cosh\chi d\chi \ln\left[\frac{(b\sinh\chi - y)^2 + (x + a\cosh\chi)^2}{(b\sinh\chi - y)^2 + (x - a\cosh\chi)^2}\right]. \tag{13}$$

In the linear approximation, with respect to variables x and y, we arrive at the integral

$$A_y = \frac{4abx\mathcal{I}}{c}\int_{-\chi_m}^{\chi_m} \frac{d\chi\cosh^2\chi}{a^2\cosh^2\chi + b^2\sinh^2\chi}. \tag{14}$$

This integral can be calculated exactly [19]. The result is

$$A_y = \frac{2Babx}{\pi\left(a^2+b^2\right)}\left[\chi_m + \frac{b}{a}\tan^{-1}\left(\frac{b}{a}\tanh\chi_m\right)\right]. \tag{15}$$

Consequently,

$$\beta = 1 - \frac{4ab}{\pi\left(a^2+b^2\right)}\left[\chi_m + \frac{b}{a}\tan^{-1}\left(\frac{b}{a}\tanh\chi_m\right)\right]. \tag{16}$$

If $\chi_m \ll 1$, then β is close to the unity. But β can be less than -1 for sufficiently large values of parameter χ_m. In particular, if $\chi_m \gg 1$ and $a = b$, then $\beta \approx 1/2 - 2\chi_m/\pi$.

The linear approximation for the vector potential can be justified if the transverse solenoid size is much larger than the Larmor radius of the charged particle $R_L = |v/\omega|$. For the quantum particle with the minimum energy $\hbar\omega_0/2$, one has $R_{L0} = \sqrt{\hbar/(m\omega_0)} = \sqrt{c\hbar/(eB_0)}$. To have an idea on the orders of magnitude of parameters, we consider an electron moving in a magnetic field of order of $B_0 = 10^3$ G. Then, $\omega_0 \sim 2\times 10^{10}$ s^{-1} and $R_{L0} \sim 10^{-5}$ cm. Therefore, the linear approximation is quite good for solenoids whose transverse dimensions are much bigger than 1 mm and magnetic fields are stronger than 1 G.

3. Main Equations

We consider a quantum nonrelativistic spinless particle of mass m and charge e, whose motion in the xy plane is governed by the Hamiltonian

$$\hat{H} = \hat{\pi}^2/(2m), \qquad \pi = p - eA/c. \tag{17}$$

If the magnetic field $B = B_0$ does not depend on time, then Hamiltonian (17) admits two linear integrals of motion,

$$\hat{X} = \hat{x} + \hat{\pi}_y/(m\omega_0), \qquad \hat{Y} = \hat{y} - \hat{\pi}_x/(m\omega_0), \qquad \omega_0 = eB_0/mc, \tag{18}$$

which are merely the coordinates of the center of a circle, which the particle rotates around with the cyclotron frequency ω_0. The importance of these integrals of motion was emphasized by many authors during decades [4,17,20–25].

The second pair of physical observables consists of two relative coordinates:

$$\hat{\xi} = \hat{x} - \hat{X} = -\hat{\pi}_y/(m\omega_0), \qquad \hat{\eta} = \hat{y} - \hat{Y} = \hat{\pi}_x/(m\omega_0). \tag{19}$$

Then Hamiltonian (17) with $B = B_0 = const$ can be written as

$$\hat{H} = m\omega_0^2 \left(\hat{\xi}^2 + \hat{\eta}^2\right)/2. \tag{20}$$

Due to the commutation relations,

$$[\hat{\xi}, \hat{\eta}] = [\hat{Y}, \hat{X}] = i\hbar/(m\omega_0), \qquad [\hat{\xi}, \hat{X}] = [\hat{\xi}, \hat{Y}] = [\hat{\eta}, \hat{X}] = [\hat{\eta}, \hat{Y}] = 0, \tag{21}$$

the eigenvalues of operator (20) assume discrete values $\hbar\omega_0(n + 1/2)$. Moreover, these eigenvalues have infinite degeneracy [4], because they do not depend on the mean values of operators $\hat{X}$ and $\hat{Y}$ (or their functions). In addition to the energy, there exists another quadratic integral of motion, which can be considered as the generalized angular momentum (the formulas below are identical for the classical variables and quantum operators; therefore, we do not put here the symbol ^ over operators):

$$L = x\pi_y - y\pi_x + m\omega_0 \left(x^2 + y^2\right)/2 = xp_y - yp_x + m\omega_0\beta \left(x^2 - y^2\right)/2. \tag{22}$$

It coincides formally with the canonical angular momentum $L_{can} = xp_y - yp_x$ in the case of "circular" gauge of the vector potential. It follows from (22) that the "kinetic" angular momentum, defined as

$$L_{kin} = x\pi_y - y\pi_x, \tag{23}$$

is not a conserved quantity, and it can vary with time in the generic case [20,26,27]. On the other hand, the "intrinsic" angular momentum

$$J = \xi\pi_y - \eta\pi_x = -m\omega_0 \left(\xi^2 + \eta^2\right) = -2H/\omega_0 \tag{24}$$

is obviously conserved for the constant magnetic field. Formulas (22) and (23) can be written in terms of "geometric" variables (different from the phase plane coordinates) as

$$L = m\omega_0 \left(X^2 + Y^2 - \xi^2 - \eta^2\right)/2, \qquad L_{kin} = -m\omega_0 \left(\xi^2 + \eta^2 + X\xi + Y\eta\right). \tag{25}$$

It is convenient to use the vector $\mathbf{q} = (\xi, \eta, X, Y)$, whose components can be either classical variables, quantum operators, or mean values of these operators. The initial values, before the variation of parameters, are contained in vector $\mathbf{q}_0$. Vector $\mathbf{q}_f$ contains the information about the geometrical coordinates after the variations of parameters. Due to the linearity of the equations of motion, these two vectors are connected by means of some transformation matrix $\mathbf{\Lambda}$:

$$\mathbf{q}_f = \mathbf{\Lambda}\mathbf{q}_0, \qquad \mathbf{\Lambda} = \left\| \begin{array}{cc} \lambda_1 & \lambda_2 \\ \lambda_3 & \lambda_4 \end{array} \right\|. \tag{26}$$

Here, λ_j are four 2×2 blocks of the 4×4 matrix $\mathbf{\Lambda}$.

The mean energy in the stationary magnetic field equals $\mathcal{E} = m\omega_0^2\langle\hat{\xi}^2 + \hat{\eta}^2\rangle/2$. It can be written as the sum of two independent parts:

$$\mathcal{E} = \mathcal{E}_c + \mathcal{E}_q, \qquad \mathcal{E}_c = m\omega_0^2 \left(\langle\xi\rangle^2 + \langle\eta\rangle^2\right)/2, \qquad \mathcal{E}_q = m\omega_0^2 \left(\sigma_{\xi\xi} + \sigma_{\eta\eta}\right)/2, \tag{27}$$

where $\sigma_{\alpha\beta} = \langle\hat{\alpha}\hat{\beta} + \hat{\beta}\hat{\alpha}\rangle/2 - \langle\hat{\alpha}\rangle\langle\hat{\beta}\rangle$ with $\alpha, \beta = \xi, \eta, X, Y$. The quantity $\mathcal{E}_c$ coincides with the energy of the classical particle moving along the mean trajectory $(\langle\xi\rangle, \langle\eta\rangle)$, whereas the quantum correction

$\mathcal{E}_q$ in (27) arises due to quantum fluctuations, described by means of the variances $\sigma_{\xi\xi}$ and $\sigma_{\eta\eta}$ of the relative coordinates.

3.1. Change of the Classical Part of Energy

The change of the classical part of the mean energy can be written as some quadratic form with respect to the initial quantum-mechanical mean values of the relative and guiding center coordinates: $\Delta\mathcal{E}_c = (m\omega_0^2/2)\sum W_{\alpha\beta}\langle\alpha\rangle_0\langle\beta\rangle_0$, where coefficients $W_{\alpha\beta}$ are certain bilinear combinations of the elements of matrix $\mathbf{\Lambda}$. Therefore, $\Delta\mathcal{E}_c$ depends on four initial parameters. In particular, these parameters can be almost always chosen in such a way that the final classical part of energy will be equal to zero. For example, it is sufficient to solve the equations $\langle\xi\rangle_f = \langle\eta\rangle_f = 0$ with respect to the initial values $\langle X\rangle_0$ and $\langle Y\rangle_0$ with fixed values of $\langle\xi\rangle_0$ and $\langle\eta\rangle_0$. The answer in terms of elements of matrix Λ is as follows,

$$\langle X\rangle_0 = \left[(\Lambda_{14}\Lambda_{21} - \Lambda_{24}\Lambda_{11})\langle\xi\rangle_0 + (\Lambda_{14}\Lambda_{22} - \Lambda_{24}\Lambda_{12})\langle\eta\rangle_0\right] / \left[\Lambda_{13}\Lambda_{24} - \Lambda_{23}\Lambda_{14}\right], \tag{28}$$

$$\langle Y\rangle_0 = \left[(\Lambda_{23}\Lambda_{11} - \Lambda_{13}\Lambda_{21})\langle\xi\rangle_0 + (\Lambda_{23}\Lambda_{12} - \Lambda_{13}\Lambda_{22})\langle\eta\rangle_0\right] / \left[\Lambda_{13}\Lambda_{24} - \Lambda_{23}\Lambda_{14}\right]. \tag{29}$$

However, this is an extremely peculiar situation.

Therefore, let us consider a rarefied gas of charges in the uniform magnetic field. Neglecting the interaction between charges (i.e., assuming the magnetic field to be strong enough), we can write the total energy as the sum of independent single-particle energies. However, the positions of guiding centers of the gas particles can be quite arbitrary, as well as the concrete values of the relative coordinates of each particle (obeying the restriction $\langle\xi\rangle_0^2 + \langle\eta\rangle_0^2 = \rho_0^2 = 2\mathcal{E}_c^{(0)}/(m\omega_0^2)$. In such a case, the most interesting quantity is the average value of $\Delta\mathcal{E}_c$, where the averaging is performed over the whole ensemble of charges with arbitrary initial quantum-mechanical mean values. Designating such an additional averaging by means of the over-line (to distinguish from the primary quantum-mechanical averaging), it seems natural to assume the absence of initial correlations and the isotropic distribution of non-zero mean values:

$$\overline{\langle\alpha\rangle_0\langle\beta\rangle_0} = 0 \quad \text{if } \alpha \neq \beta; \quad \overline{\langle\xi\rangle_0^2} = \overline{\langle\eta\rangle_0^2} = \mathcal{E}_c^{(0)}/(m\omega_0^2) = \rho_0^2/2, \quad \overline{\langle X\rangle_0^2} = \overline{\langle Y\rangle_0^2} = \mathcal{R}_0^2/2. \tag{30}$$

Here, $\mathcal{R}_0^2$ is the average square of the distance between the classical guiding center position and the center of solenoid, whereas ρ_0^2 is the average square of the classical radius of orbit. Under these assumptions, the average change of the classical part of energy can be written as

$$\overline{\Delta\mathcal{E}_c} = m\omega_0^2\left(F_\rho\rho_0^2 + F_R\mathcal{R}_0^2\right)/2, \tag{31}$$

$$F_\rho = \Theta^2\left(\Lambda_{11}^2 + \Lambda_{21}^2 + \Lambda_{12}^2 + \Lambda_{22}^2\right)/2 - 1 = \Theta^2\mathrm{Tr}\left(\lambda_1\tilde{\lambda}_1\right)/2 - 1, \tag{32}$$

$$F_R = \Theta^2\left(\Lambda_{13}^2 + \Lambda_{23}^2 + \Lambda_{14}^2 + \Lambda_{24}^2\right)/2 = \Theta^2\mathrm{Tr}\left(\lambda_2\tilde{\lambda}_2\right)/2, \tag{33}$$

where $\Theta = \omega_f/\omega_0$ and $\tilde{\lambda}_j$ means the transposed matrix. Coefficient F_R is always non-negative. However, coefficient F_ρ can be negative, thus indicating a possibility to cool the gas by means of fast variations of the magnetic field.

3.2. Change of the Quantum Part of Energy

It is convenient to combine all covariances in the single 4×4 symmetrical covariance matrix $\sigma = \|\sigma_{\alpha\beta}\|$. Then, the linear transformation (26) implies the following relation between the final (σ_f) and initial (σ_0) covariance matrices [28],

$$\sigma_f = \mathbf{\Lambda}\sigma_0\tilde{\mathbf{\Lambda}}, \tag{34}$$

where $\tilde{\Lambda}$ is the transposed matrix. The change of the quantum part of the mean energy,

$$\Delta\mathcal{E}_q = \left(m\omega_f^2/2\right)\left(\sigma_{\xi\xi} + \sigma_{\eta\eta}\right)_f - \left(m\omega_0^2/2\right)\left(\sigma_{\xi\xi} + \sigma_{\eta\eta}\right)_0, \tag{35}$$

depends on the evolution matrix Λ and the initial covariance matrix σ_0. In general, σ_0 can have 10 independent elements (obeying some restrictions due to the uncertainty relations). Therefore, it is difficult to analyze the problem for the most general initial states. We consider the simplest situation, when the initial state possesses some rotational symmetry inside the pairs (ξ, η) and (X, Y). Therefore, we assume that non-zero initial elements of matrix σ are $\sigma_{\xi\xi} = \sigma_{\eta\eta} = G$ and $\sigma_{XX} = \sigma_{YY} = \chi$, i.e.,

$$\sigma(0) = \left\| \begin{array}{cc} GI_2 & 0 \\ 0 & \chi I_2 \end{array} \right\|, \tag{36}$$

where I_2 is the 2×2 unity matrix. Then,

$$\Delta\mathcal{E}_q = \left(m\omega_f^2/2\right)\mathrm{Tr}\left(G\Lambda_1\tilde{\Lambda}_1 + \chi\Lambda_2\tilde{\Lambda}_2\right) - m\omega_0^2 G. \tag{37}$$

In view of the commutation relations (21), two positive parameters in matrix (36) must obey the restrictions $G \geq \hbar/|2m\omega_0|$ and $\chi \geq \hbar/|2m\omega_0|$. In particular, the variance matrix (36) can describe the thermal quantum state with different temperatures of the $(\xi\eta)$ and (XY) subsystems. However, this is by no means the only possible state, because the covariance matrix describes uniquely only the Gaussian states (unless $G = \chi = \hbar/|2m\omega_0|$).

3.3. Change of the Average Guiding Center Position

Calculations similar to that of the preceding subsections give the following formulas for the "classical" and "quantum" parts of the change of the average square of the radius of the guiding center position.

$$\overline{\Delta\mathcal{R}_c^2} \equiv \overline{\langle X\rangle_f^2 + \langle Y\rangle_f^2} - \mathcal{R}_0^2 = \frac{1}{2}\mathrm{Tr}\left(\Lambda_3\tilde{\Lambda}_3\rho_0^2 + \Lambda_4\tilde{\Lambda}_4\mathcal{R}_0^2\right) - \mathcal{R}_0^2, \tag{38}$$

$$\Delta\mathcal{R}_q^2 \equiv \left(\sigma_{XX} + \sigma_{YY}\right)_f - \left(\sigma_{XX} + \sigma_{YY}\right)_0 = \mathrm{Tr}\left(G\Lambda_3\tilde{\Lambda}_3 + \chi\Lambda_4\tilde{\Lambda}_4\right) - 2\chi. \tag{39}$$

One can see that, under conditions (30) and (36), the formulas for the "classical" and "quantum" parts of the quantities $\Delta\mathcal{E}$ and $\Delta\mathcal{R}^2$ are, in fact, identical, if one makes the formal substitutions $\chi \to \mathcal{R}_0^2/2$ and $G \to \rho_0^2/2$ (remembering that parameters χ and G are limited from below, whereas $\mathcal{R}_0^2$ and ρ_0^2 can be arbitrary non-negative numbers). Therefore, in the subsequent sections, we give explicit formulas for the "quantum" parts of the quantities under study only.

3.4. Change of the Angular Momentum

Due to the choice of the initial variance matrix in the form (36), the initial mean value of the quantum part of the angular momentum operator (25) equals $\langle\hat{L}\rangle_q = m\omega_0(\chi - G)$. Equations (25), (34) and (36) yield the following formula for the change of this quantity,

$$\Delta L_q = m\omega_0\left\{2(G - \chi) + \Theta\mathrm{Tr}\left[G\left(\Lambda_3\tilde{\Lambda}_3 - \Lambda_1\tilde{\Lambda}_1\right) + \chi\left(\Lambda_4\tilde{\Lambda}_4 - \Lambda_2\tilde{\Lambda}_2\right)\right]\right\}, \qquad \Theta = \omega_f/\omega_0. \tag{40}$$

3.5. Evolution of the Magnetic Moment

The operator of magnetic moment $\hat{\mathcal{M}}$ can be introduced in several ways. One can start from the thermodynamic relation $\langle \mathcal{M} \rangle = -\partial \langle \mathcal{E} \rangle / \partial B$ and define $\hat{\mathcal{M}} = -\partial \hat{H} / \partial B$. This definition was used, e.g., in [20] for the circular gauge ($\beta = 0$). For the general linear gauge, such a definition results in the formula

$$\hat{\mathcal{M}}_1 = \frac{e}{2mc} \left[\hat{x}\hat{\pi}_y - \hat{y}\hat{\pi}_x - \beta \left(\hat{x}\hat{\pi}_y + \hat{y}\hat{\pi}_x \right) \right] \equiv \frac{e}{2mc} \left(\hat{L}_{kin} - \beta \hat{\Lambda} \right). \tag{41}$$

However, one can use as the starting point the standard definition of the classical magnetic moment

$$\mathbf{M} = \frac{1}{2c} \int dV \left[\mathbf{r} \times \mathbf{j} \right] \tag{42}$$

Then, using the expression for the quantum probability current density

$$\mathbf{j} = \frac{ie\hbar}{2m} \left(\psi \nabla \psi^* - \psi^* \nabla \psi \right) - \frac{e^2}{mc} \mathbf{A} \psi^* \psi, \tag{43}$$

one can write the right-hand side of (42) as the mean value of operator

$$\hat{\mathcal{M}}_2 = \frac{e}{2mc} \hat{L}_{kin}. \tag{44}$$

Therefore, the relation $\mathcal{M} = eL/(2mc)$ does not hold for the time independent angular momentum (22).

Formulas (41) and (44) coincide only for $\beta = 0$ (note that namely this special gauge was used in the overwhelming number of papers devoted to the motion of quantum particles in the magnetic field). However, writing the operator $\hat{\Lambda}$ in (41) as $\hat{\Lambda} = m\omega_0 \left(\hat{\eta}^2 - \hat{\xi}^2 + \hat{Y}\hat{\eta} - \hat{X}\hat{\xi} \right)$, one can see that the average value of $\hat{\Lambda}$ over the period $2\pi/\omega_0$ equals zero (as $\hat{X}$ and $\hat{Y}$ do not depend on time, whereas $\hat{\xi}(t)$ and $\hat{\eta}(t)$ oscillate with frequency ω_0 in the Heisenberg picture; in addition, the difference $\hat{\eta}^2 - \hat{\xi}^2$ oscillates with frequency $2\omega_0$). Using the symbol $\langle\langle \cdots \rangle\rangle$ for the double averaging (over the quantum state and over the period of rotation in the magnetic field), we have $\langle\langle \hat{\mathcal{M}}_1 \rangle\rangle = \langle\langle \hat{\mathcal{M}}_2 \rangle\rangle$. Moreover, in view of (24) and (25), we have $\langle\langle \hat{L}_{kin} \rangle\rangle = \langle\langle \hat{J} \rangle\rangle$. Therefore, $\langle\langle \hat{M} \rangle\rangle = -e\mathcal{E}/(mc\omega_0)$, for any definition of the magnetic moment operator. (Actually, some authors used this formula as the definition of the magnetic moment of a charged particle moving in the magnetic field [29].) This relation explains why different choices of the vector potential gauge do not influence the final results for the magnetization of the free electron gas in the time-independent magnetic field. Designating $\Delta\mathcal{M} = \langle\langle \hat{M} \rangle\rangle_f - \langle\langle \hat{M} \rangle\rangle_0$, we obtain the following expression for the quantum part of the change of the magnetic moment in the nonstationary case,

$$\Delta\mathcal{M}_q = (e\omega_0/c) \left[G - (\Theta/2) \mathrm{Tr} \left(G\lambda_1\tilde{\lambda}_1 + \chi\lambda_2\tilde{\lambda}_2 \right) \right]. \tag{45}$$

3.6. The Transformation Matrix for the Sharp Jump of Parameters

For the constant magnetic field with the cyclotron frequency ω_0, the canonical coordinates (x, y, p_x, p_y) are related to the "geometrical" coordinates (ξ, η, X, Y) as follows,

$$\hat{x} = \hat{\xi} + \hat{X}, \qquad \hat{y} = \hat{\eta} + \hat{Y}, \tag{46}$$

$$\hat{p}_x = m\omega_0 \left[\hat{\eta}(1-\beta) - \hat{Y}(1+\beta) \right]/2, \quad \hat{p}_y = m\omega_0 \left[\hat{X}(1-\beta) - \hat{\xi}(1+\beta) \right]/2. \tag{47}$$

The canonical coordinates (or mean values of the corresponding operators in the quantum case) do not change during the instantaneous finite jump of the parameters (as the wave function cannot change instantaneously in the nonrelativistic quantum mechanics; the same result can be obtained by integrating the equations of motion for the canonical variables during the infinitesimally short time of the jump). However, the relations between the canonical and geometrical variables are different before and after the jump. Therefore, one can easily find the elements of matrix $\mathbf{\Lambda}$, solving the equations

$$\hat{\xi}_0 + \hat{X}_0 = \hat{\xi}_f + \hat{X}_f, \qquad \omega_0 \left[\hat{X}_0(1-\beta_0) - \hat{\xi}_0(1+\beta_0)\right] = \omega_f \left[\hat{X}_f(1-\beta_f) - \hat{\xi}_f(1+\beta_f)\right]$$

and similar equations for the pair (η, Y). If $\omega_f = \Theta\omega_0$ and $\Delta\beta = \beta_f - \beta_0$, then

$$\mathbf{\Lambda} = \frac{1}{2\Theta} \left\| \begin{array}{cccc} L_- & 0 & K_- & 0 \\ 0 & L_+ & 0 & K_+ \\ K_+ & 0 & L_+ & 0 \\ 0 & K_- & 0 & L_- \end{array} \right\|, \tag{48}$$

where

$$K_\pm = \kappa\left(1 \pm \beta_f\right) \pm \Delta\beta, \quad L_\pm = 2 + K_\pm, \quad \kappa \equiv \Theta - 1. \tag{49}$$

Matrix (48) equals the unity matrix if $\kappa = \Delta\beta = 0$. For this reason, many formulas in the subsequent section have the most compact form if they are written in terms of parameter κ, instead of Θ.

4. Results

4.1. Energy Evolution

Equations (37), (48) and (49) yield the following expression for the quantum part of the energy change,

$$\Delta\mathcal{E}_q = \left(m\omega_0^2/4\right)\left\{\left[\left(\kappa\beta_f + \Delta\beta\right)^2 + \kappa^2\right](G+\chi) + 4\kappa G\right\}. \tag{50}$$

The energy always increases after the jump of the gauge parameter β with the fixed value of the magnetic field ($\kappa = 0$): $\Delta\mathcal{E}_q = m\omega_0^2(\Delta\beta)^2(G+\chi)/4$. On the other hand, $\Delta\mathcal{E}_q$ can be negative for the fixed β and negative values of parameter $\kappa = \Theta - 1$ inside the interval

$$-\frac{4G}{(1+\beta^2)(G+\chi)} < \kappa < 0. \tag{51}$$

The minimum is achieved in the middle of this interval:

$$\Delta\mathcal{E}_q^{(min)} = -\frac{\mathcal{E}_q^{(0)} G}{(1+\beta^2)(G+\chi)}, \qquad \mathcal{E}_q^{(min)} = \frac{\mathcal{E}_q^{(0)}\left[G\beta^2 + \chi\left(1+\beta^2\right)\right]}{(1+\beta^2)(G+\chi)}. \tag{52}$$

In the special case of $G = \chi$, we have

$$\Delta\mathcal{E}_q = \frac{1}{2}\mathcal{E}_q^{(0)}\left[2\kappa + \kappa^2\left(1+\beta^2\right)\right], \qquad \kappa_{min} = -\frac{1}{1+\beta^2}, \quad \mathcal{E}_q^{(min)} = \mathcal{E}_q^{(0)}\frac{1+2\beta^2}{2+2\beta^2}. \tag{53}$$

In this case, the quantum part of the energy can be reduced to the half of the initial value for $\beta = 0$ (the circular gauge) and $\Theta = 0$ (the total switch off the magnetic field), but it can be reduced by only 25% for $\beta = 1$ (the Landau gauge) and $\Theta = 1/2$. For $\kappa = -1$ (or $\Theta = 0$), we have

$$\Delta\mathcal{E}_q = \frac{1}{4}m\omega_0^2\left[\left(1+\beta^2\right)(G+\chi) - 4G\right], \qquad \Delta\mathcal{E}_q\big|_{G=\chi} = \frac{1}{2}\mathcal{E}_q^{(0)}\left(\beta^2 - 1\right). \tag{54}$$

Another interesting special case is $\kappa = -2$ (when $\Theta = -1$: the instantaneous inversion of the magnetic field). Then,

$$\Delta\mathcal{E}_q = m\omega_0^2 \left[\left(1+\beta^2\right)\chi - \left(1-\beta^2\right)G \right], \qquad \Delta\mathcal{E}_q\big|_{G=\chi} = 2\mathcal{E}_q^{(0)}\beta^2. \tag{55}$$

For $|\beta| = 1$ (the Landau gauge), the energy always increases, and the increase depends on the initial fluctuations of the guiding center position only. On the other hand, if $\beta = 0$, then $\Delta\mathcal{E}_q = m\omega_0^2(\chi - G)$, so that the *final* quantum part of energy is determined by the guiding center fluctuations: $\mathcal{E}_f = m\omega_0^2\chi$.

4.2. Shift of the Guiding Center and Variation of the Angular Momentum

The "classical" and "quantum" parts of the guiding center mean position shift are described by Equations (38) and (39). The explicit formula for the quantum part is as follows,

$$\Delta\mathcal{R}_q^2 = \frac{1}{2\Theta^2}\left\{ \left[\left(\kappa\beta_f + \Delta\beta\right)^2 + \kappa^2 \right](G+\chi) - 4\kappa\Theta\chi \right\} \tag{56}$$

For small changes of the magnetic field, when $\Theta \approx 1$ and $|\kappa| \ll 1$, we have $\Delta\mathcal{R}_q^2 \approx -2\kappa\chi$. According to (56), $\Delta\mathcal{R}_q^2 \to \infty$ for $\Theta \to 0$. However, this is a consequence of Equation (18), because in the absence of the field the particle moves along a straight line, whose curvature radius is obviously infinite.

If $\kappa = 0$ but $\Delta\beta \neq 0$, then $\Delta\mathcal{R}_q^2$ is always positive. A more interesting case is $\Delta\beta = 0$. For example, if $G = \chi$, then

$$\Delta\mathcal{R}_q^2 = \frac{G\kappa^2}{\Theta^2}\left(\beta^2 - \frac{\Theta+1}{\Theta-1}\right). \tag{57}$$

In this case, if $\Theta < -1$, then $\Delta\mathcal{R}_q^2 < 0$ for $\beta = 0$, whereas $\Delta\mathcal{R}_q^2 > 0$ for $\beta^2 = 1$, with the critical value $\beta_c^2 = (|\Theta| - 1)/(|\Theta| + 1)$, when $\Delta\mathcal{R}_q^2$ changes its sign.

The angular momentum does not change its value for any values of parameters β and Θ, as Equation (40) leads to the formula $\Delta L_q \equiv 0$. Actually, this is the direct consequence of Equations (22), (36) and the instant jump approximation, because the canonical coordinates do not change during the instant jump, whereas $\langle \hat{x}^2 - \hat{y}^2 \rangle = 0$, due to the diagonal form of the initial covariance matrix. Situations with $\Delta L_q \neq 0$ can happen for asymmetric initial conditions.

4.3. Change of the Mean Magnetic Moment

Using Equations (45) and (48), we obtain the following formula for the relative change of quantum part of the mean magnetic moment $\Delta\mu \equiv \Delta\mathcal{M}_q/\mathcal{M}_0$ (where $\mathcal{M}_0 = -(e/c)\omega_0 G$ is the initial magnetic moment):

$$\Delta\mu = (4\Theta)^{-1}\left[\left(\kappa\beta_f + \Delta\beta\right)^2 + \kappa^2 \right](1+g), \qquad g = \chi/G. \tag{58}$$

We see that the sign of change is determined completely by the sign of ratio $\Theta = \omega_f/\omega_0$. If $\kappa = 0$, then any change of the solenoid shape increases the magnetization: $\Delta\mu = (1+g)(\Delta\beta)^2/4$.

If $\Delta\beta = 0$, then $\Delta\mu = \kappa^2\left(1+\beta^2\right)(1+g)/(4\Theta)$, so that the Landau gauge yields a twice bigger relative change of the magnetic moment than the circular gauge. Bigger changes can be achieved in hyperbolic solenoids with $|\beta| > 1$. Note that function $\kappa^2/(4\Theta) = (\Theta - 2 + 1/\Theta)/4$ is symmetrical with respect to the transformation $\Theta \to 1/\Theta$. Consequently, the increase and decrease of the magnetic field result in the same relative change of the magnetic moment (by the absolute value). If $\Theta = -1$ (the instant inversion of the magnetic field), then $\Delta\mu = -\left(1+\beta^2\right)(1+g)$, so that $\langle\langle\hat{M}\rangle\rangle_f = -\langle\langle\hat{M}\rangle\rangle_0$ for $g = 1$ and $\beta = 0$, but $\langle\langle\hat{M}\rangle\rangle_f = -2\langle\langle\hat{M}\rangle\rangle_0$ for the Landau gauge.

4.4. Generation of Squeezing

The squeezed states of a free charged particle moving under the homogeneous magnetic field (with the circular gauge of the vector potential) were considered by several authors [16,30,31], but they calculated the squeezing coefficients with respect to the canonical pairs of variables, such as x, p_x and y, p_y, whose physical meaning is not quite clear. Therefore, it was suggested in [32–34] to analyze the variances in the pairs (X, Y) and (ξ, η). The states possessing variances of any element of the pairs (X, Y) or (ξ, η) less than $\hbar/2m\omega_0$ were named "geometrical squeezed states" (GSS) in [7], to emphasize that all the observables (X, Y, ξ, η) have the meaning of coordinates in the usual ("geometrical") space, and not in the phase space. Only the circular gauge was considered in [32–34]. Two gauges—the circular and Landau ones—were compared in [7].

An interesting problem raised in [7] is how one could create GSS, starting from coherent states (having all variances equal to $\hbar/2m\omega_0$). For the single-mode systems, such a problem can be solved effectively by using quadratic Hamiltonians with suitably chosen time-dependent coefficients [28]. However, can this can be done using time-dependent magnetic fields in two dimensions? It appeared that the answer depends on the choice of time-dependent gauge. Namely, it was shown in [7] that no squeezing can be obtained for any geometrical observable (X, Y, ξ, η) for an arbitrary time-dependent magnetic field described by means of the circular gauge of the vector potential. On the other hand, some degree of squeezing can be obtained in the case of Landau gauge. These observations explain our interest to the general vector potential (1).

Let us suppose, for the sake of simplicity, that the initial state was coherent with respect to all geometrical observables, i.e., $G = \chi = \hbar/|2m\omega_0|$. Then, we have the following form of matrix $\sigma_f = G\Lambda\tilde{\Lambda}$ immediately after the jump, $\sigma_f = \left(\hbar/|2m\omega_f|\right)\sigma_*$, where

$$\sigma_* = \frac{1}{4|\Theta|}\left\| \begin{array}{cccc} L_-^2 + K_-^2 & 0 & L_+K_- + L_-K_+ & 0 \\ 0 & L_+^2 + K_+^2 & 0 & L_+K_- + L_-K_+ \\ L_+K_- + L_-K_+ & 0 & L_+^2 + K_+^2 & 0 \\ 0 & L_+K_- + L_-K_+ & 0 & L_-^2 + K_-^2 \end{array} \right\|. \tag{59}$$

The relative squeezing with respect to the new ground state value $\hbar/|2m\omega_f|$ immediately after the jump is characterized by two coefficients:

$$\sigma_- = \sigma_{*\xi\xi} = \sigma_{*YY} = \frac{L_-^2 + K_-^2}{4|\Theta|} = \frac{1 + \left[(1 - \Delta\beta) + \kappa\left(1 - \beta_f\right)\right]^2}{2|\Theta|}, \tag{60}$$

$$\sigma_+ = \sigma_{\eta\eta} = \sigma_{XX} = \frac{L_+^2 + K_+^2}{4|\Theta|} = \frac{1 + \left[(1 + \Delta\beta) + \kappa\left(1 + \beta_f\right)\right]^2}{2|\Theta|}. \tag{61}$$

If $\kappa = 0$, then squeezing is possible provided $|1 \pm \Delta\beta| < 1$. The maximum squeezing of 50% can be achieved for $\Delta\beta = \pm 1$. If $\Delta\beta = 0$, then

$$\sigma_\pm = \frac{1 + (\Theta \pm \kappa\beta)^2}{2|\Theta|}, \tag{62}$$

so that no squeezing is possible for the circular gauge ($\beta = 0$), in accordance with the authors of [7]. On the other hand, $\sigma_- = 1/|\Theta|$ for $\beta = 1$. In any case, squeezing is possible only for $2|\Theta| > 1$. The minimal value $\sigma_{min} = (2|\Theta|)^{-1}$ can be achieved for $\beta = \pm\Theta/(\Theta - 1)$. If $|\kappa| \ll 1$, then $\sigma_\pm \approx 1 \pm \beta\kappa$, so that squeezing is possible for any $\beta \neq 0$. Note the symmetry relations $\sigma_+(\Theta, \beta) = \sigma_-(\Theta, -\beta)$ and $\sigma_\pm(\Theta, \beta) = \sigma_\pm(-\Theta, -\beta)$. Figure 2 shows typical plots of $\sigma_\pm(\Theta, \beta)$ for some fixed positive values of β.

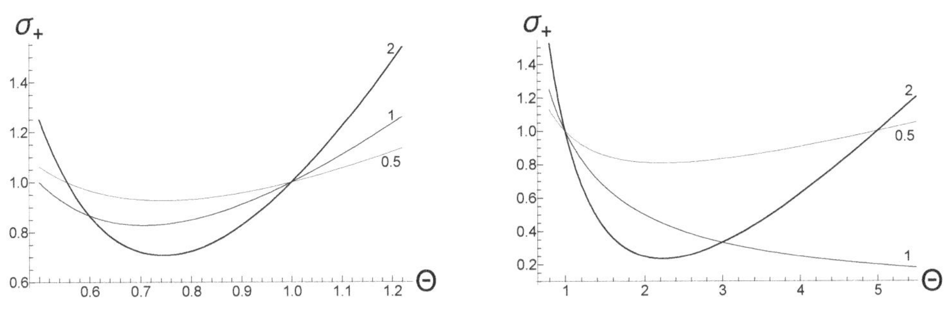

Figure 2. The functions $\sigma_+(\Theta)$ (**left**) and $\sigma_-(\Theta)$ (**right**) for some fixed positive values of the gauge parameter β (shown near the respective curves).

5. Discussion

The results of the preceding section show that, indeed, the choice of the gauge parameter β (i.e., the choice of the solenoid shape) can influence significantly on changes of many physical quantities after the stepwise variation of the magnetic field. Moreover, it appears that the directions of these changes can be quite different for different quantities. For example, the change of the mean value of the angular momentum does not depend on β (for the chosen symmetric initial conditions). The energy change can be stronger for the circular gauge ($\beta = 0$) than for the Landau gauge with $|\beta| = 1$. On the contrary, no squeezing can be created from the initial coherent state for $\beta = 0$, whereas it can be quite significant for $|\beta| = 1$. However, the most interesting results, probably, are related to the change of the average magnetic moment.

This change is twice bigger for the Landau gauge than for the circular one. The magnetic moment increases with the increase of the magnetic field, and this behavior seems natural. An unexpected result is the increase of the magnetic moment when the magnetic field decreases. An especially strange consequence is the infinite magnetic moment after the instant switching off the magnetic field. Of course, the sharp change of the magnetic field in time is an idealization, which can be justified if the magnetic field varies rapidly during a small time interval $\delta t \ll 2\pi/\omega_0$. This means that the particle practically does not change its position, and its wave function does not change its value, during the interval δt. On the other hand, the time-dependent magnetic field in the empty space inside the solenoid cannot be strictly homogeneous, due to the time dependence of the induced electric field. However, this inhomogeneity can be neglected if the Larmor radius is much smaller than the scale of spatial variations of the time-dependent electromagnetic field, which is of the order of the characteristic wave length $\lambda \sim c\delta t$. Thus, we arrive at the restrictions $\sqrt{\hbar/(mc^2\omega_0)} \ll \delta t \ll 2\pi/\omega_0$, which are self-consistent if the cyclotron frequency satisfies the condition $\omega_0 \ll (2\pi)^2 mc^2/\hbar$, which is equivalent to $B_0 \ll (2\pi)^2 m^2c^3/(e\hbar)$. For the electron, one obtains $B_0 \ll 10^{15}\,$G. Consequently, the model considered in the paper seems to be well justified for typical magnetic fields used in laboratories, which do not exceed $10^6\,$G (see also [16]). For $\omega_0 \sim 10^{10}\,\text{s}^{-1}$ ($B_0 \sim 500\,$G), the restrictions on the parameter δt takes the form $3 \times 10^{-16}\,\text{s} \ll \delta t \ll 6 \times 10^{-10}\,\text{s}$. These restrictions are much softer for atomic ions, whose masses are five orders of magnitudes higher than the electron mass.

Strictly speaking, we cannot exclude a possibility that the model of homogeneous and rapidly changing in time magnetic field can be invalid for $\Theta \to 0$, when the quantum Larmor radius

$R_L = \sqrt{\hbar c/|eB_f|}$ becomes too large. This special case needs a separate detailed investigation. In this connection, it would be interesting to find solutions to the problem (approximate analytical or numerical), considering more realistic time dependencies of the magnetic field $B(t)$ and the gauge parameter $\beta(t)$. Note, however, that $R_L \sim 2\,\mu\text{m}$ for $B_f \sim 1\,\text{G}$, which can be considered as practically zero field if $B_0 \sim 1\,\text{T}$.

Among other results, it is worth mentioning a possibility of cooling the system of charged particles by means of fast variations of the magnetic field. Actually, this is a subtle problem, because one has to take into account many other effects, such as the interaction between the charged particles in the gas, to make realistic estimations. On the other hand, the problem of interaction does not exist, e.g., for single ions in traps. In any case, it would be interesting to verify our results experimentally, using solenoids with different shapes of the cross sections.

Author Contributions: The authors made equal contributions to the paper.

Funding: This research received no external funding.

Acknowledgments: The authors thank the referees for valuable comments and suggestions. VVD acknowledges the partial support from the Brazilian funding agency CNPq.

Conflicts of Interest: The authors declare no conflicts of interest.

References

1. Kennard, E.H. Zur Quantenmechanik einfacher Bewegungstypen. *Z. Phys.* **1927**, *44*, 326–352. [CrossRef]
2. Darwin, C.G. Free motion in wave mechanics. *Proc. R. Soc. Lond. A* **1927**, *117*, 258–293. [CrossRef]
3. Fock, V. Bemerkung zur Quantelung des harmonischen Oszillators im Magnetfeld. *Z. Phys.* **1928**, *47*, 446–448. [CrossRef]
4. Landau, L. Diamagnetismus der Metalle. *Z. Phys.* **1930**, *64*, 629–637. [CrossRef]
5. Malkin, I.A.; Man'ko, V.I.; Trifonov, D.A. Coherent states and transition probabilities in a time-dependent electromagnetic field. *Phys. Rev. D* **1970**, *2*, 1371–1385. [CrossRef]
6. Dodonov, V.V.; Malkin, I.A.; Man'ko, V.I. Coherent states of a charged particle in a time-dependent uniform electromagnetic field of a plane current. *Physica* **1972**, *59*, 241–256. [CrossRef]
7. Dodonov, V.V.; Man'ko, V.I.; Polynkin, P.G. Geometrical squeezed states of a charged particle in a time-dependent magnetic field. *Phys. Lett. A* **1994**, *188*, 232–238. [CrossRef]
8. Dodonov, V.V.; Horovits, M.B. Squeezing of relative and center of orbit coordinates of a charged particle by stepwise variations of a uniform magnetic field with an arbitrary linear vector potential. *J. Rus. Laser Res.* **2018**, *39*, 389–400. [CrossRef]
9. Janszky, J.; Yushin, Y.Y. Squeezing via frequency jump. *Opt. Commun.* **1986**, *59*, 151–154. [CrossRef]
10. Graham, R. Squeezing and frequency changes in harmonic oscillations. *J. Mod. Opt.* **1987**, *34*, 873–879. [CrossRef]
11. Bechler, A. Generation of squeezed states in a homogeneous magnetic field. *Phys. Lett. A* **1988**, *130*, 481–482. [CrossRef]
12. Ma, X.; Rhodes, W. Squeezing in harmonic oscillators with time-dependent frequencies. *Phys. Rev. A* **1989**, *39*, 1941–1947. [CrossRef] [PubMed]
13. Kiss, T.; Janszky, J.; Adam, P. Time evolution of harmonic oscillators with time-dependent parameters: A step-function approximation. *Phys. Rev. A* **1994**, *49*, 4935–4942. [CrossRef] [PubMed]
14. Kira, M.; Tittonen, I.; Lai, W.K.; Stenholm, S. Semiclassical computations of time-dependent tunneling. *Phys. Rev. A* **1995**, *51*, 2826–2837. [CrossRef] [PubMed]
15. Tittonen, I.; Stenholm, S.; Jex, J. Effect of a phase step on two-level atoms in a cavity. *Opt. Commun.* **1996**, *124*, 271–276. [CrossRef]
16. Delgado, F.C.; Mielnik, B. Magnetic control of squeezing effects. *J. Phys. A: Math. Gen.* **1998**, *31*, 309–320.
17. Mielnik, B.; Ramírez, A. Magnetic operations: a little fuzzy mechanics? *Phys. Scr.* **2011**, *84*, 045008. [CrossRef]
18. Landau, L.D.; Lifshitz, E.M. *The Classical Theory of Fields*, 3rd ed.; Pergamon: Oxford, UK, 1971; pp. 101–103.

19. Gradshteyn, I.S.; Ryzhik, I.M. *Table of Integrals, Series, and Products*, 7th ed.; Academic: Amsterdam, The Netherlands, 2007.
20. Johnson, M.H.; Lippmann, B.A. Motion in a constant magnetic field. *Phys. Rev.* **1949**, *76*, 828–832. [CrossRef]
21. Avron, J.E.; Herbst, I.W.; Simon, B. Separaltion of center of mass in homogeneous magnetic fields. *Ann. Phys.* **1978**, *114*, 431–451. [CrossRef]
22. Johnson, B.R.; Hirschfelder, J.O.; Yang, K.-H. Interaction of atoms, molecules, and ions with constant electric and magnetic fields. *Rev. Mod. Phys.* **1983**, *55*, 109–153. [CrossRef]
23. Von Baltz, R. Guiding center motion of two interacting $n = 0$ Landau electrons in two dimensions. *Phys. Lett. A* **1984**, *105*, 371–373. [CrossRef]
24. Kowalski, K.; Rembieliński, J. Coherent states of a charged particle in a uniform magnetic field. *J. Phys. A Math. Gen.* **2005**, *38*, 8247–8258. [CrossRef]
25. Dodonov, V.V. Coherent states and their generalizations for a charged particle in a magnetic field. In *Coherent States and Their Applications: A Contemporary Panorama*; Antoine, J.-P., Bagarello, F., Gazeau, J.-P., Eds.; Springer: Berlin, Germany, 2018; pp. 311–338.
26. Li, C.-F.; Wang, Q. The quantum behavior of an electron in a uniform magnetic field. *Physica B* **1999**, *269*, 22–27. [CrossRef]
27. Greenshields, C.R.; Stamps, R.L.; Franke-Arnold, S; Barnett, S.M. Is the angular momentum of an electron conserved in a uniform magnetic field? *Phys. Rev. Lett.* **2014**, *113*, 240404. [CrossRef]
28. Dodonov, V.V. Parametric excitation and generation of nonclassical states in linear media. In *Theory of Nonclassical States of Light*; Dodonov, V.V., Man'ko, V.I., Eds.; Taylor & Francis: London, UK, 2003; pp. 153–218.
29. Parker, L. On the magnetic moment of a charged particle in a changing magnetic field. *Nuovo Cim. B* **1965**, *XL*, 99–108. [CrossRef]
30. Jannussis, A.; Vlahos, E.; Skaltsas, D.; Kliros, G.; Bartzis, V. Squeezed states in the presence of a time-dependent magnetic field. *Nuovo Cim. B* **1989**, *104*, 53–66. [CrossRef]
31. Baseia, B.; Mizrahi, S.S.; Moussa, M.H.Y. Generation of squeezing for a charged oscillator and a charged particle in a time-dependent electromagnetic field. *Phys. Rev. A* **1992**, *46*, 5885–5889. [CrossRef]
32. Dodonov, V.V.; Kurmyshev, E.V.; Man'ko, V.I. Correlated coherent states. In *Classical and Quantum Effects in Electrodynamics*; Nova Science: Commack, NY, USA, 1988; pp. 169–199.
33. Aragone, C. New squeezed Landau states. *Phys. Lett. A* **1993**, *175*, 377–381. [CrossRef]
34. Ozana, M.; Shelankov, A.L. Squeezed states of a particle in magnetic field. *Phys. Solid State* **1998**, *40*, 1276–1282. [CrossRef]

Article

Descriptions of Relativistic Dynamics with World Line Condition

Florio Maria Ciaglia [1], Fabio Di Cosmo [2,3,*], Alberto Ibort [2,3] and Giuseppe Marmo [4,5]

1 Max-Planck-Institut für Mathematik in den Naturwissenschaften, Inselstraße 22, 04103 Leipzig, Germany; ciaglia@mis.mpg.de
2 Dep.to de Matematica, Univ. Carlos III de Madrid. Av. da de la Universidad, 30, 28911 Leganes, Madrid, Spain; albertoi@math.uc3m.es
3 ICMAT, Instituto de Ciencias Matematicas (CSIC-UAM-UC3M-UCM), Nicolás Cabrera, 13-15, Campus de Cantoblanco, UAM, 28049 Madrid, Spain
4 INFN-Sezione di Napoli, Complesso Universitario di Monte S. Angelo. Edificio 6, via Cintia, 80126 Napoli, Italy; marmo@na.infn.it
5 Dipartimento di Fisica "E. Pancini", Università di Napoli Federico II, Complesso Universitario di Monte S. Angelo Edificio 6, via Cintia, 80126 Napoli, Italy
* Correspondence: fcosmo@math.uc3m.es

Received: 14 September 2019; Accepted: 15 October 2019; Published: 19 October 2019

Abstract: In this paper, a generalized form of relativistic dynamics is presented. A realization of the Poincaré algebra is provided in terms of vector fields on the tangent bundle of a simultaneity surface in $\mathbb{R}^4$. The construction of this realization is explicitly shown to clarify the role of the commutation relations of the Poincaré algebra versus their description in terms of Poisson brackets in the no-interaction theorem. Moreover, a geometrical analysis of the "eleventh generator" formalism introduced by Sudarshan and Mukunda is outlined, this formalism being at the basis of many proposals which evaded the no-interaction theorem.

Keywords: relativistic dynamics; no-interaction theorem; world line condition

---

In memory of E.C.G. Sudarshan, who was interested, for almost three decades, in problems of covariant description of relativistic interacting particles.

1. Introduction

Seventy years ago, Dirac [1] argued that a covariant description of relativistic dynamics means to find a realization of the Lie algebra of the Poincaré group in terms of observables and a Poisson bracket. In the so-called instant form, Dirac argued that the role of the single Hamiltonian in non relativistic dynamics should be replaced by four Hamiltonians, one of them being the total energy of the system, the other three being the boosts. Quoting Dirac, "the equations of motion should be expressible in the Hamiltonian form. This is necessary for a transition to the quantum theory to be possible."

All subsequent papers, therefore, always assumed the realization of the Poincaré algebra to be given in terms of Poisson Brackets on dynamical variables on a carrier space which would attribute three positions and three momenta (or velocities) to each particle. The additional requirement that every particle evolution would be associated with an invariant world-line on space-time, gave rise to the world-line condition and, in the hands of Sudarshan and collaborators, ended up with the no-interaction theorem [2]. For the sake of clarity, let us recall what are the assumptions at the base of this theorem: The physical system is described in the canonical formalism by a phase space with an associated Poisson bracket; the three-dimensional position coordinates of each particle at a common physical time (instant form of dynamics) represent half of a system of canonical variables for this phase space; under the Euclidean subgroup of the Poincaré group (characteristic of the instant form) the canonical and geometrical transformation laws for these coordinates coincide; if in any state of motion, as seen from a given reference frame, the world-lines of the particles are drawn in space-time, then the canonical transformations which relates this description with the one seen from a different frame preserve the objective reality of these lines. These conditions express the two aspects of relativistic invariance in the description of a physical system: From one point of view physical laws have to be invariant under changes of reference frames (relativistic invariance); on the other hand some physical quantities transform in a specific way under changes of reference frames due to the action of the Lorentz group on space-time events (manifest invariance). Then the no-interaction theorem states that, when considering a system of particles, the only dynamical evolution which is compatible with the above assumptions is the free dynamical evolution. This conclusion is in agreement with the relativistic principle of constancy of the speed of light, which forbids instantaneous interactions propagating faster than the speed of light. These results have supported the development of field theories: Particles locally exchange energy and momenta with fields which have their own degrees of freedom and their own dynamics.

Later on, in the 1980s, many relativists and particle physicists took up the problem of providing a covariant description of relativistic interacting particles without the intervention of fields, that is, a kind of "action-at-a-distance" covariant under the Poincaré group (For a detailed discussion about the meaning of an action-at-a-distance in the context of classical and quantum mechanics we refer to Reference [3], where a comparison with local field theories is also presented). In these attempts (see, for instance, References [4–6]) interactions were described via constraints, using Dirac description of Hamiltonian constrained systems [7]. In the beginning these models were supposed to violate the world line condition (WLC). However, Sudarshan and Mukunda proposed an interpretation in terms of an "eleventh generator" formalism [8]. The set of constraints used to define the interaction among particles depend on an additional variable, parameterising the curves which constitute a state of the motion for the system. The generator of this "evolution" is independent of the other ten generators of the Poincaré group, obtaining the so-called "eleventh generator" formalism. They also noticed that the notion of WLC can be meaningfully implemented in this framework, providing examples of interacting particles satisfying the WLC. The difference with the assumptions of the no-interaction theorem, indeed, consists in the "dynamical" choice of the time parameter: In the constrained formalism the time depends on the state of the motion of the system and there is not a neat separation between kinematics and dynamics, as in Dirac's form of relativistic dynamics. In other words, the variables interpreted as the world line positions after implementing all the constraints do not need to coincide with the canonical positions of the phase space initially associated with the physical system. Later on [9], it was proven that all these constrained descriptions can be derived from a common "covering phase space" via suitable reduction procedures.

However, the additional requirement of separability, that is, the requirement that in the instant form clusters of particles very far apart should behave as non-interacting, gave rise to a novel version of the no-interaction theorem [10]. All these results pointed out again that a suitable covariant description of relativistic interacting particles cannot be formulated without the intervention of fields.

With the hope to clarify the role of the commutation relations of the Poincaré algebra versus their description in terms of Poisson Brackets, in this paper we propose a solution of the problem posed by Dirac in terms of vector fields, we call it a Newtonian realization, the dynamical vector field being a second order one, and then we require this realization to be compatible with a Lagrangian description (a covariant inverse problem for the full Poincaré algebra, not just the dynamcs). We also provide a geometrical description of the "eleventh generator" formalism by means of Jacobi Brackets. We find that in this approach "canonical positions" do not coincide with "geometrical positions", that is, positions in space-time. We shall present these various aspects in the particular case of one particle relativistic systems with possible "external forces", a situation which would arise after implementing the "separability condition".

2. A Geometrical Formulation of Dirac's Problem

In modern geometrical terms, the problem addressed by Dirac may be formulated as follows. The Lie agebra of any Lie group G, say l_G, canonically defines a Poisson structure on the dual space $l_G^* = Lin(l_G, \mathbb{R})$ by means of the following construction. With every element $u \in l_G$, we associate a linear function $\hat{u} \in \mathcal{F}(l_G^*)$ and we define a Poisson Bracket on $\mathcal{F}(l_G^*)$ by setting

$$\{\hat{u}, \hat{v}\} = \widehat{[u, v]}, \tag{1}$$

or, if $\alpha \in l_G^*$

$$\{\hat{u}, \hat{v}\}(\alpha) = \alpha([u, v]). \tag{2}$$

This bracket leads to a well defined tensor field even in infinite dimensions provided that l_G may be identified with its double dual $(l_G^*)^*$, that is, whenever l_G is reflexive. Thus, if P denotes the Poincaré group, we introduce the following notations for the elements of l_P:

$$P_\mu, \qquad \mu = 0, 1, 2, 3 \tag{3}$$

form the Abelian part of the Poincaré algebra and

$$M_{\mu\nu}, \qquad \mu, \nu = 0, 1, 2, 3, \tag{4}$$

represent rotations and boosts. Then, the commutation relations read

$$[P_\mu, P_\nu] = 0 \tag{5}$$

$$[M_{\mu\nu}, P_\rho] = g_{\nu\rho} P_\mu - g_{\mu\rho} P_\nu \tag{6}$$

$$[M_{\mu\nu}, M_{\rho\sigma}] = g_{\nu\rho} M_{\mu\sigma} - g_{\mu\rho} M_{\nu\sigma} + g_{\nu\sigma} M_{\rho\mu} - g_{\mu\sigma} M_{\rho\nu}, \tag{7}$$

where $g_{\mu\nu}$ is the Minkowski metric on $\mathbb{R}^4$ and they can be used to explicitly characterize also the Poisson algebra on the dual space l_P^*.

Now, the problem as stated by Dirac amounts to find a "symplectic realization" or a "Lagrangian realization" [11,12] of the Poisson manifold $(l_P^*, \{\cdot, \cdot\})$, associated with the Poincaré group. This means we need to find a Poisson map:

$$\mu : T^*Q \to l_P^* \tag{8}$$

in the phase-space case or a Poisson map

$$\mu_{\mathcal{L}} : TQ \to l_P^* \tag{9}$$

in the Lagrangian one. In the latter situation, we have to consider a Lagrangian-dependent Poisson bracket, since the tangent bundle is not equipped with a canonical symplectic structure.

Remark 1. *This formulation of the problem would be consistent with Wigner's approach to the classification of elementary particles as irreducible unitary representations of the Poincaré group. The classical counterpart would correspond to the coadjoint orbits of the Poincaré group acting on the dual of the Lie algebra.*

Our "Newtonian realization" would correspond, instead, to a map:

$$\nu : l_P \to \chi(\mathcal{M}) \tag{10}$$

such that the Lie bracket in the Lie algebra of the Poincaré group is realized in terms of the commutator between the vector fields on some carrier space $\mathcal{M}$. Note that a similar idea has been developed in Reference [13].

3. Newtonian Realization in the Instant Form

Let us start with the vector space $\mathbb{R}^4$ with globally defined coordinate functions x^μ, $\mu = 0,1,2,3$ and equipped with the Minkowski metric tensor

$$g = -dx^0 \otimes dx^0 + \sum_{j=1}^{3} dx^j \otimes dx^j .$$

In this space, a simultaneity surface for an inertial reference frame is diffeomorphic with $\mathbb{R}^3$, that is, it represents the space of positions at a given instant. The coordinate functions for this copy of $\mathbb{R}^3$ are given by the x_j's with $j = 1,2,3$. Positions and velocities are points of the tangent bundle $T\mathbb{R}^3$, which will be our carrier space $\mathcal{M}$. Thus, we have to find a realization of the Poincaré algebra in terms of vector fields on $T\mathbb{R}^3$.

It is clear that, since space rotations and space translations form the Euclidean group which preseves every simultaneity leaf associated with a given inertial frame, we can exploit the tangent lift of their standard realization on $\mathbb{R}^3$. On the other hand, in Dirac parlance, the complicated generators will be those corresponding to the "Hamiltonians", that is, the dynamics and the three boosts. It is at this point that the world-line condition plays a preminent role. Let Γ denote the second order dynamical vector field and K_1, K_2, K_3 the vector fields representing the boosts. Then, we require that the following conditions hold:

$$L_{K_j} x_l = x_j L_\Gamma x_l = x_j \dot{x}_l \tag{11}$$

$$L_{K_j} \dot{x}_l = \dot{x}_j \dot{x}_l + x_j L_\Gamma \dot{x}_l - \delta_{jl} . \tag{12}$$

These vectors fields express the world-line condition as stated by Sudarshan and collaborators [14]. Moreover, we notice that K_j is a non-linear vector field, it does not respect the tangent bundle structure of $T\mathbb{R}^3$ and it is the sum of a Newtonoid [15] vector field and a vertical lift of a translation for all $j = 1,2,3$. Indeed, we have

$$K_j = x_j \dot{x}_l \frac{\partial}{\partial x_l} + \dot{x}_j \dot{x}_l \frac{\partial}{\partial \dot{x}_l} + x_j L_\Gamma \dot{x}_l \frac{\partial}{\partial \dot{x}_l} - \delta_{jl} \frac{\partial}{\partial \dot{x}_l} = x_j \Gamma + \dot{x}_j \Delta - \delta_{jl} \left(\frac{\partial}{\partial x_l} \right)^V , \tag{13}$$

where $\Delta = \dot{x}_j \frac{\partial}{\partial \dot{x}_j}$ is the dilation vector field along the fibers of the tangent bundle $T\mathbb{R}^3$ [15,16] and $\left(\frac{\partial}{\partial x_j} \right)^V$ denotes the vertical lift of the vector field $\frac{\partial}{\partial x_j}$ [17]. In summary, the Newtonian realization of the Poincaré algebra is given by

$$P_j = -\frac{\partial}{\partial x_j}, \qquad P_0 = \Gamma \tag{14}$$

$$J_l = \epsilon_{ljk} \left(x_j \frac{\partial}{\partial x_k} + \dot{x}_j \frac{\partial}{\partial \dot{x}_k} \right), \quad K_j = x_j \Gamma + \dot{x}_j \Delta + (P_j)^V . \tag{15}$$

We eventually notice that this realization of the Poincaré algebra is a dynamical one, that is, it depends on the second order dynamics Γ.

By requiring these vector fields to satisfy the commutation relations of the Poincaré algebra, we get a system of partial differential equations for the accelerations, that is, we derive a system of PDE for the functions a_1, a_2, a_3 which appear in the second order dynamical vector field

$$\Gamma = \dot{x}_j \frac{\partial}{\partial x_j} + a_j \frac{\partial}{\partial \dot{x}_j} .$$

In particular, the commutation relations with the spatial translations impose that the accelerations cannot depend on the positions x_j. The commutation relations with the rotations, instead, imply the following expression:

$$a_j = \dot{x}_j f(\dot{x}^2)\,. \tag{16}$$

Finally, let us consider the commutation relation $[K_1, K_2] = J_3$. This relation is verified if and only if the following set of PDE holds true:

$$2(x_1\dot{x}_2 - x_2\dot{x}_1)\dot{x}_3\left[(1-\dot{x}^2)\tfrac{\partial f}{\partial \dot{x}^2} + f\right] = 0 \tag{17}$$

$$-x_2 f + 2(x_1\dot{x}_2 - x_2\dot{x}_1)\dot{x}_1\left[(1-\dot{x}^2)\tfrac{\partial f}{\partial \dot{x}^2} + f\right] = 0 \tag{18}$$

$$x_1 f + 2(x_1\dot{x}_2 - x_2\dot{x}_1)\dot{x}_2\left[(1-\dot{x}^2)\tfrac{\partial f}{\partial \dot{x}^2} + f\right] = 0\,. \tag{19}$$

The only solution to this system is $f = 0$, which means that the only dynamics, as intended by Dirac, compatible with the world-line condition must admit a generator Γ which has vanishing accelerations.

An additional result can be obtained if one requires that the realization by means of vector fields allows for a compatible non-trivial Lagrangian function [14,18], that is, for any of the vector fields in the Newtonian realization the following condition must be satisfied:

$$L_X \omega_{\mathcal{L}} = 0\,, \tag{20}$$

or, equivalently,

$$L_X \theta_{\mathcal{L}} = dF_X\,, \tag{21}$$

due to the contractability of $T\mathbb{R}^3$. By using $L_{K_m}\theta_{\mathcal{L}} = dF_m$, $L_\Gamma \theta_{\mathcal{L}} = d\mathcal{L}$, $L_{P_m}\theta_{\mathcal{L}} = L_{J_m}\theta_{\mathcal{L}} = 0$ we find that

$$L_{K_m}\mathcal{L} = L_\Gamma F_m\,, \quad m = 1,2,3\,. \tag{22}$$

Since Γ is the vector field describing the dynamics of a non-interacting particle, we can find a solution for F_m having the form

$$F_m = x_m h(\dot{x}^2)\,, \tag{23}$$

where $\dot{x}^2 = \sum_j \dot{x}_j \dot{x}_j$ and we can add to it any constant of the motion. Then Equation (22) becomes

$$2\left(\dot{x}^2 - 1\right)\frac{\partial \mathcal{L}}{\partial \dot{x}^2} = h\left(\dot{x}^2\right)\,. \tag{24}$$

The commutation relations of the Poincaré algebra, however, impose some additional conditions. Indeed, since

$$[K_m, P_n] = \delta_{mn}\Gamma\,, \tag{25}$$

we have that

$$L_{P_m} L_{K_n}\theta_{\mathcal{L}} = -\delta_{mn} L_\Gamma \theta_{\mathcal{L}}\,, \tag{26}$$

which implies the following equation for F_n

$$L_{P_m} F_m = -\mathcal{L}\,. \tag{27}$$

This is an equation involving only the Lagrangian function

$$2(\dot{x}^2 - 1)\frac{\partial \mathcal{L}}{\partial \dot{x}^2} = \mathcal{L}\,, \tag{28}$$

which integrates to

$$\mathcal{L} = c\sqrt{1 - \sum_j (\dot{x}_j)^2}\,, \tag{29}$$

with $c \in \mathbb{R}$.

This result is quite remarkable because, by requiring the Newtonian realization of the Poincaré algebra to allow for a Lagrangian description, we obtain that the Lagrangian is unique. This uniqueness property is very startling when we recall that, if we only require the dynamical vector field

$$\Gamma = \dot{x}_j \frac{\partial}{\partial x_j} \tag{30}$$

to allow for a Lagrangian description, we find an infinite family of solutions provided not only by any function $\mathcal{L} = \mathcal{L}(\dot{x}^2)$ but also $\mathcal{L} = \mathcal{L}(\dot{x}_1, \dot{x}_2, \dot{x}_3)$. In other words, among the infinitely many Lagrangian functions providing a description of the free dynamics, there is only one which admits the above Newtonian realization of the Poincaré algebra as generalized Noether symmetries.

4. The Eleventh-Generator Formalism

In References [8], Mukunda and Sudarshan introduced an eleventh-generator formalism. We would like to unveil the geometric content of their formalism because it forms the prototype for the proposals made in the eighties to evade the no-interaction theorem by means of the Dirac-Bergmann constraint formalism [4–6].

We start with what Dirac calls an elementary solution of the symplectic realization of the Poisson manifold defined by l_P^*, that is, the phase-space $T^*\mathbb{R}^4$ with coordinates (x_μ, p_μ) and with symplectic structure given by the natural one,

$$\omega = dp_\mu \wedge dx^\mu = g^{\mu\nu} dp_\mu \wedge dx_\nu\,, \tag{31}$$

which is the exterior differential of the one form

$$\theta_0 = p_\mu dx^\mu = g^{\mu\nu} p_\mu dx_\nu\,. \tag{32}$$

By setting $P_\mu = p_\mu$, $M_{\mu\nu} = x_\mu p_\nu - x_\nu p_\mu$ we obtain Dirac's elementary solution of the problem in terms of Poisson Brackets and generating functions.

Starting from this solution, we can construct another solution by adding an element to the Poincaré algebra. Specifically, on $T^*\mathbb{R}^4$ we consider the submanifold

$$\Sigma_m = \left\{ (x_\mu, p_\mu) | p_\mu p^\mu = m^2 \right\}\,. \tag{33}$$

If we consider the natural immersion of Σ_m into $T^*\mathbb{R}^4$, it is possible to write the pull-back

$$i^*_{\Sigma_m} \theta_0 = \theta_m\,. \tag{34}$$

On Σ_m, the form θ_m defines a contact structure [19,20], since $\theta_m \wedge (d\theta_m)^3 \neq 0$ represents a volume form. Then, it is possible to define a Lie algebra structure [21] on $\mathcal{F}(\Sigma_m)$ by setting

$$[f, g]_m\, \theta_m \wedge (d\theta_m)^3 = (f dg - g df) \wedge (d\theta_m)^3 + 2 df \wedge dg \wedge \theta_m \wedge (d\theta_m)^2\,. \tag{35}$$

The given expression shows that the bracket, being defined only in terms of θ_m which is Poincaré invariant, is indeed preserved by the action of the Poincaré group given above. As a matter of fact, this bracket can be described in the more common setting of a bivector field and a vector field, say (Λ_m, Γ_m), defined by

$$i_{\Gamma_m}\left(\theta_m \wedge (d\theta_m)^3\right) = (d\theta_m)^3 \tag{36}$$

$$i_{\Lambda_m}\left(\theta_m \wedge (d\theta_m)^3\right) = 3\theta_m \wedge (d\theta_m)^2 \,. \tag{37}$$

Then, the bracket in Equation (35) can be defined as follows:

$$[f,g]_m = \Lambda_m(df,dg) + fL_{\Gamma_m}g - gL_{\Gamma_m}f\,, \tag{38}$$

and the Jacobi identity holds true because of the following properties:

$$[\Lambda_m, \Lambda_m] = 2\Gamma_m \wedge \Lambda_m \tag{39}$$

$$[\Gamma_m, \Lambda_m] = 0\,, \tag{40}$$

where the bracket between multivector fields is the Schouten bracket [17].

By using the pair (Λ_m, Γ_m), it is possible to associate with any function f a first order differential operator $\tilde{X}_f$ and a vector field X_f, respectively given by

$$\tilde{X}_f = \Lambda_m(df,\cdot) + f\Gamma_m - L_{\Gamma_m}f \tag{41}$$

$$X_f = \Lambda_m(df,\cdot) + f\Gamma_m\,. \tag{42}$$

It should be noticed that constant functions are not mapped into the null vector field, but in both cases we have $X_c = c\Gamma_m$, for any $c \in \mathbb{R}$. Moreover, the Lie bracket does not satisfy the Leibniz rule but we have

$$[f,gh]_m = [f,g]_m\,h + g\,[f,h]_m - [f,1]_m\,gh\,, \tag{43}$$

showing an important difference between $[f,g]_m$ and $\{f,g\}$, where the first is called Jacobi bracket and the second Poisson Bracket. It is common to call X_f the Hamiltonian vector field associated with f. However, on the subalgebra of functions which are constants of the motion for Γ_m, that is, $L_{\Gamma_m}f = 0$, the Jacobi bracket reduces to a Poisson Bracket.

It turns out that the usual generating functions of the Poincaré algebra given in $T^*\mathbb{R}^4$, when pulled back to Σ_m, provide a solution of the Dirac problem because they are constants of the motion for Γ_m and therefore generate a realization of the Poincaré algebra in terms of Poisson Brackets. To compare the Jacobi algebra with the Poisson Bracket on $T^*\mathbb{R}^4$, we consider the following bivector field

$$\Lambda = \left(g^{\mu\nu} - \frac{p^\mu p^\nu}{p_\mu p^\mu}\right)\frac{\partial}{\partial p^\mu} \wedge \frac{\partial}{\partial x^\mu} = g^{\mu\nu}\frac{\partial}{\partial p^\mu} \wedge \frac{\partial}{\partial x^\mu} - \Delta \wedge \Gamma\,, \tag{44}$$

where

$$\Delta = p^\mu \frac{\partial}{\partial p^\mu}\,, \quad \Gamma = \frac{p^\mu}{p_\nu p^\nu}\frac{\partial}{\partial x^\mu}\,. \tag{45}$$

All these contravariant tensor fields are actually tangent to the leaves of the foliation defined by $p_\mu p^\mu = m^2$ when m changes and we remove the manifold defined by $p_\mu p^\mu = 0$. Then, on each mass-shell Σ_m, we have the following pair

$$\Lambda_m = \left(g^{\mu\nu} - \tfrac{p^\mu p^\nu}{m^2}\right)\tfrac{\partial}{\partial p^\mu} \wedge \tfrac{\partial}{\partial x^\nu} \tag{46}$$

$$\Gamma_m = \tfrac{p^\mu}{m^2}\tfrac{\partial}{\partial x^\mu}\,, \tag{47}$$

and the associated Jacobi bracket is given by the following relations:

$$[x^\rho, x^\sigma]_m = \frac{x^\rho p^\sigma - x^\sigma p^\rho}{m^2} \tag{48}$$

$$[p^\rho, x^\sigma]_m = g^{\rho\sigma} \tag{49}$$

$$[p^\rho, p^\sigma]_m = 0\,. \tag{50}$$

By using the generating functions which are constants of the motion for Γ, that is, $M_{\mu\nu}$ and P_μ, we find the associated vector fields:

$$X_{\mu\nu} = x_\mu \frac{\partial}{\partial x^\nu} - x_\nu \frac{\partial}{\partial x^\mu} + p_\mu \frac{\partial}{\partial p^\nu} - p_\nu \frac{\partial}{\partial p^\mu} \tag{51}$$

$$Y_\mu = \frac{\partial}{\partial x^\mu}\,. \tag{52}$$

Therefore, in this realization, the Poincaré algebra, represented in terms of vector fields, contains a central element given by Γ_m which plays the role of the eleventh generator.

5. A "Frozen Phase-Space" Realization

In the same spirit of the previous section, we start with the phase-space $T^*\mathbb{R}^4$, we remove the manifold defined by $p_\mu p^\mu = 0$ and we consider the one-form

$$\theta = \frac{\theta_0}{\sqrt{p_\mu p^\mu}}, \tag{53}$$

which, on a given mass-shell Σ_m, would be

$$\tilde{\theta}_m = \frac{\theta_m}{m}\,. \tag{54}$$

Then, the two-form $d\theta = \frac{d\theta_0}{\sqrt{p_\mu p^\mu}} - \frac{p_\mu dp^\mu}{\sqrt{p_\mu p^\mu}} \wedge \theta_0$ is a degenerate two-form whose kernel is generated by

$$\Delta = p^\mu \frac{\partial}{\partial p^\mu} \quad \text{and} \quad \Gamma = \frac{p^\mu}{p_\nu p^\nu}\frac{\partial}{\partial x^\mu}\,. \tag{55}$$

Indeed, θ is invariant under dilation because it is homogeneous of degree zero in the momenta. The fact that Γ is in the kernel of $d\theta$ follows by direct computation.

As the infinitesimal generators which realize the Poincaré algebra commute with Δ and Γ, they descend to the quotient manifold which is symplectic and six dimensional. We obtain in this way another solution of Dirac's problem in terms of Poisson brackets. However, there is no evolution on points of this quotient manifold because we quotiented out the dynamics represented by Γ and, therefore, this realization is on a "frozen phase-space" as it was called by P. Bergmann and A. Komar [22].

6. A Lagrangian Solution to the Dirac Problem

In this last section, it is useful to show how it is possible to provide a realization of the Poincaré algebra in the Lagrangian formalism on $T\mathbb{R}^4$ [23]. By means of natural geometric coordinates in $T\mathbb{R}^4$, we can consider the Lagrangian function

$$\mathcal{L} = m\sqrt{g_{\mu\nu}\dot{x}^\mu \dot{x}^\nu}\,. \tag{56}$$

For simplicity, in the following computations we set $m = 1$. The associate one-form $\theta_\mathcal{L}$ will be

$$\theta_\mathcal{L} = \frac{g_{\mu\nu}\dot{x}^\mu dx^\nu}{\mathcal{L}}\,. \tag{57}$$

Let $v^2 = g_{\mu\nu}\dot{x}^\mu\dot{x}^\nu$, then

$$d\theta_{\mathcal{L}} = \omega_{\mathcal{L}} = \frac{1}{v^3}\left(g_{\mu\nu}v^2 - \dot{x}_\mu\dot{x}_\nu\right) d\dot{x}^\nu \wedge dx^\mu\,, \tag{58}$$

and we have to remove from $T\mathbb{R}^4$ the submanifold defined by $\mathcal{L} = 0$. We notice that the vector fields

$$\Delta = \dot{x}^\mu \frac{\partial}{\partial \dot{x}^\mu} \quad \text{and} \quad \Gamma = \dot{x}^\mu \frac{\partial}{\partial x^\mu}\,,$$

are both in the kernel of $\omega_{\mathcal{L}}$. By passing to the quotient with respect to Δ and Γ, we get a symplectic six-dimensional manifold. To define a Poisson Bracket on functions on (the open submanifold of) $T\mathbb{R}^4$, we need to define a lift from vector fields on the quotient manifold to vector fields on $T\mathbb{R}^4$. At this purpose, it is possible to use a flat connection whose horizontal leaves are defined by the level sets of the functions $f_1 = g_{\mu\nu}\dot{x}^\mu x^\nu$ and $f_2 = \mathcal{L}$. A connection one-form, a $(1-1)$-tensor field A which satisfies $A^2 = A$ and contains Δ and Γ in its null-space, is given by

$$A = \mathbf{1} - \frac{\dot{x}_\mu d\dot{x}^\mu}{\mathcal{L}^2} \otimes \Delta - \frac{1}{\mathcal{L}} d\left(\frac{\dot{x}_\mu x^\mu}{\mathcal{L}}\right) \otimes \Gamma\,. \tag{59}$$

We observe that

$$\frac{\dot{x}_\mu d\dot{x}^\mu}{\mathcal{L}^2}(\Delta) = 1\,, \quad \frac{1}{\mathcal{L}} d\left(\frac{\dot{x}_\mu x^\mu}{\mathcal{L}}\right)(\Delta) = 0 \tag{60}$$

$$\frac{\dot{x}_\mu d\dot{x}^\mu}{\mathcal{L}^2}(\Gamma) = 0\,, \quad \frac{1}{\mathcal{L}} d\left(\frac{\dot{x}_\mu x^\mu}{\mathcal{L}}\right)(\Gamma) = 1\,, \tag{61}$$

and that

$$[\Delta, \Gamma] = \Gamma\,. \tag{62}$$

By using this connection, we define a bivector field

$$\Lambda = \mathcal{L} g^{\mu\nu}\left[A\left(\frac{\partial}{\partial \dot{x}^\mu}\right) \wedge A\left(\frac{\partial}{\partial x^\nu}\right)\right]\,, \tag{63}$$

whose associated Poisson Brackets will be

$$\{\dot{x}^\rho, x^\sigma\} = \mathcal{L}\left(g^{\rho\sigma} - \frac{\dot{x}^\rho\dot{x}^\sigma}{\mathcal{L}^2}\right) \tag{64}$$

$$\{\dot{x}^\rho, \dot{x}^\sigma\} = 0 \tag{65}$$

$$\{x^\rho, x^\sigma\} = \frac{\dot{x}^\sigma x^\rho - \dot{x}^\rho x^\sigma}{\mathcal{L}}\,. \tag{66}$$

Note that these brackets are very similar to those derived on the phase space with the aid of the Jacobi bracket. The bivector field Λ reads

$$\Lambda = \mathcal{L}\left(g^{\rho\sigma} - \frac{\dot{x}^\rho\dot{x}^\sigma}{\mathcal{L}^2}\right)\frac{\partial}{\partial \dot{x}^\rho} \wedge \frac{\partial}{\partial x^\sigma}\,. \tag{67}$$

This tensor field and the associated brackets are Lorentz invariant. The canonical coordinates on the symplectic quotient are given by the functions:

$$Q^j = \mathcal{L}\frac{K^j}{\dot{x}^0} \tag{68}$$

$$P^j = \frac{\dot{x}^j}{\mathcal{L}}\,, \tag{69}$$

where we have introduced a splitting in space and time so that we may write

$$J_l = \tfrac{1}{\mathcal{L}}\epsilon_{ljk}\dot{x}^j x^k \tag{70}$$

$$K^j = \tfrac{1}{\mathcal{L}}\left(\dot{x}^j x^0 - \dot{x}^0 x^j\right) . \tag{71}$$

We notice that the functions

$$Q^j = \mathcal{L}\frac{K^j}{\dot{x}^0} = -x^j + \dot{x}^j\frac{x^0}{\dot{x}^0} \tag{72}$$

are the so called Newton-Wigner positions. Once more, canonical coordinates and geometrical positions do not coincide.

Remark 2. *It is in order, at this point, to note that the* $(1-1)$*-tensor field*

$$\frac{1}{\mathcal{L}}d\left(\frac{\dot{x}_\mu x^\mu}{\mathcal{L}}\right)\otimes\Gamma \tag{73}$$

behaves like a "dynamical" reference frame [24]. On the tangent bundle, however, the analog of a simultaneity submanifold would be the "horizontal foliation" defined by the level sets of the two functions f_1 *and* f_2*. The dynamics takes from one simultaneity submanifold to another. If we restrict to the dynamically invariant submanifold*

$$k_\mu = \frac{g_{\mu\nu}\dot{x}^\nu}{\mathcal{L}} , \tag{74}$$

we would get $f_1 = k_\nu x^\nu$ *and* $\frac{\Gamma}{\mathcal{L}}$ *would be a reference frame in space-time in the usual meaning when* $k_\mu k^\mu = 1$.

It should be noticed that, in this dynamical approach, the function

$$\tau = \frac{\dot{x}^\mu x_\mu}{\mathcal{L}} \tag{75}$$

behaves like a dynamical time function [25], whereas, the dynamical vector field $\frac{\Gamma}{\mathcal{L}}$ *defines "dynamical clocks" and* $S(\Gamma)$ *would give a dilation vector field along the fibers. In conclusion, the submanifold defined by the level set of the functions*

$$f_1 = g_{\mu\nu}\dot{x}^\mu x^\nu \tag{76}$$

$$f_2 = \mathcal{L} \tag{77}$$

$$k_\mu = \tfrac{g_{\mu\nu}\dot{x}^\nu}{\mathcal{L}} , \quad \mu = 0,1,2,3 , \tag{78}$$

would give a "dynamical simultaneity surface" in space-time.

One could consider the foliation given by the level sets of the two functions f_1 and f_2 as a "generalized instant form" where the instant is now given by the value of the function f_1 which is the dynamical time associated with the vector field $\frac{\Gamma}{\mathcal{L}}$. We close this section providing a Newtonian realization of the Poincaré algebra also in this different Lagrangian framework. Indeed, the vector fields generating boosts and rotation in $T\mathbb{R}^4$ are already tangent to the leaves of the foliation. On the other hand the generators of translations are not. However, a direct computation shows that the following vector fields solve the problem, since they are tangent to the leaves of the foliation and obey the commutation relations of the Poincaré algebra:

$$P_\mu = \frac{\partial}{\partial x^\mu} - \frac{x_\mu}{\mathcal{L}}\Gamma . \tag{79}$$

The various solutions we have presented of the Dirac's problem seem to imply that, if in every description the canonical positions coincide with geometrical positions, an "action-at-distance" compatible with the world-line condition does not seem possible and the intervention of the fields in the final description seems unavoidable.

7. Conclusions

The main idea developed in the eighties uses the constraint formalism to describe interacting multiparticle systems. One starts with a N-particle system by using a redundant set of variables and imposes a sufficient number of constraints (generalized mass-shell relations) in order to ensure $3N$ degrees of freedom. Several proposals were made and a unified geometrical setting was proposed in References [9].

In all these various models, the description of a true physical system of N particles with $3N$ degrees of freedom requires the introduction of constraints, that is, the selection of a Poincaré invariant submanifold in the redundant initial space one has selected. In summary, the motion is generated by the constraints.

The requirement that, when a system breaks up into clusters, each of the clusters will have an evolution parameter independent of the other clusters, implies that no action-at-distance may possibly satisfy the requirement of covariance under the Poincaré group and the world-line condition [10]. Therefore, the conclusion that can be drawn from this presentation is that a relativistically covariant description of interacting particles requires the introduction of fields. Their additional degrees of freedom, indeed, allow for the implementation of interactions which satisfy all the principles of a relativistic theory.

We close with a quotation from Sudarshan and Mukunda [26]: «*This curious result is reminiscent of the EPR "paradox" in quantum theory but the above indicated circle of ideas suggests that correlations between distant objects need not always involve transport of material influences. It may rather depend upon the indecomposable nature of the dynamical system itself. In the present context it is brought about by the imposition of the apparently innocent WLC. As has often been shown by Dirac, there are surprising structural similarities between classical mechanics and quantum mechanics; and often ideas that were identified in quantum mechanics reappear from a deeper study of classical mechanics*».

Author Contributions: All the authors have equally contributed to the elaboration and writing of the paper.

Funding: This research received no external funding.

Acknowledgments: F.D.C. and A.I. would like to thank partial support provided by the MINECO research project MTM2017-84098-P and QUITEMAD++, S2018/TCS-A4342. A.I. and G.M. acknowledge financial support from the Spanish Ministry of Economy and Competitiveness, through the Severo Ochoa Programme for Centres of Excellence in RD(SEV-2015/0554). G.M. would like to thank the support provided by the Santander/UC3M Excellence Chair Programme 2019/2020, and he is also a member of the Gruppo Nazionale di Fisica Matematica (INDAM), Italy.

Conflicts of Interest: The authors declare no conflicts of interest.

References

1. Dirac, P.A.M. Forms of relativistic dynamics. *Rev. Mod. Phys.* **1949**, *21*, 392–399. [CrossRef]
2. Currie, D.G.; Jordan, T.F.; Sudarshan, E.C.G. Relativistic Invariance and Hamiltonian Theories of Interacting Particles. *Rev. Mod. Phys.* **1963**, *35*, 350–375. [CrossRef]
3. Sudarshan, E.C.G. Action-at-a-distance. *Fields Quanta* **1972**, *2*, 175–216.
4. Komar, A. Interacting Relativistic Particles. *Phys. Rev. D* **1978**, *18*, 1887–1893. [CrossRef]
5. Rohrlich, F. Relativistic Hamiltonian dynamics I. Classical mechanics. *Ann. Phys.* **1979**, *117*, 292–322. [CrossRef]
6. Todorov, I.T. Dynamics of relativistic point particles as a problem with constraints. In *JINR-E-2-10125, USSR*; Joint Inst. for Nuclear Research: Dubna, Russia, 1976.
7. Dirac, P.A.M. *Lectures on Quantum Mechanics*; Belfer Graduate School of Science, Yeshiva University: New York, NY, USA, 1964.

8. Mukunda, N.; Sudarshan, E.C.G. Forms of relativistic dynamics with world lines. *Phys. Rev. D* **1981**, *23*, 2210–2217. [CrossRef]
9. Balachandran, A.P.; Marmo, G.; Mukunda, N.; Nilssen, J.S.; Simoni, A.; Sudarshan, E.C.G.; Zaccaria, F. Unified geometrical approach to relativistic particle dynamics. *J. Math. Phys.* **1984**, *25*, 167–176. [CrossRef]
10. Balachandran, A.P.; Dominici, D.; Marmo, G.; Mukunda, N.; Nilssen, J.S.; Samuel, J.; Sudarshan, E.C.G.; Zaccaria, F. Separability in relativistic Hamiltonian particle dynamics. *Phys. Rev. D* **1982**, *26*, 3492–3498. [CrossRef]
11. Abraham, R.; Marsden, J.E. *Foundations of Mechanics*; Benjamin/Cummings Publishing Company: Reading, MA, USA, 1978.
12. Giordano, M.; Marmo, G.; Simoni, A. Symplectic and Lagrangian Realization of Poisson Manifolds. In *Quantization, Coherent States, and Complex Structures*; Antoine, J.-P., Twareque Ali, S., Lisiecki, W., Mladenov, I.M., Odzijewicz, A., Eds.; Springer: Boston MA, USA, 1995; pp. 159–171.
13. Grigore, D.R. On manifest covariance conditions in the Lagrangian formalism. *Int. J. Mod. Phys. A* **1992**, *7*, 4073–4089. [CrossRef]
14. Marmo, G.; Mukunda, N.; Sudarshan, E.C.G. Relativistic particle dynamics-Lagrangian proof of the no-interaction theorem. *Phys. Rev. D* **1984**, *30*, 2110–2116. [CrossRef]
15. Morandi, G.; Ferrario, C.; Lo Vecchio, G.; Marmo, G.; Rubano, C. The inverse problem in the calculus of variations and the geometry of the tangent bundle. *Phys. Rep.* **1990**, *188*, 147–284. [CrossRef]
16. Cariñena, J.F.; Ibort, A.; Marmo, G.; Morandi, G. *Geometry from Dynamics, Classical and Quantum*; Springer: Dordrecht, The Netherlands, 2015.
17. Kolar, I.; Slovak, J.; Michor, P.W. *Natural Operations in Differential Geometry*; Springe: Berlin, Germany, 1993.
18. Balachandran, A.P.; Marmo, G.; Stern, A. A Lagrangian approach to the No-Interaction theorem. *Il Nuovo Cimento* **1982**, *69*, 175–186. [CrossRef]
19. Arnol'd, V.I. *Mathematical Methods of Classical Mechanics*; Springer: Berlin, Germany, 1978.
20. Asorey, M.; Ciaglia, F.M.; Di Cosmo, F.; Ibort, A.; Marmo, G. Covariant Jacobi brackets for test particles. *Mod. Phys. Lett. A* **2017**, *32*, 1750122-15. [CrossRef]
21. Kirillov, A.A. Local Lie algebras. *Russ. Math. Surv.* **1976**, *31*, 55–75. [CrossRef]
22. Bergmann, P.G.; Komar, A.B. Status Report on the quantization of the gravitational field. In *Recent Developments in General Relativity*; Pergamon Press: New York, NY, USA, 1962.
23. Dubrovin, B.A.; Giordano, M.; Marmo, G.; Simoni, A. Poisson brackets on presymplectic manifolds. *Int. J. Mod. Phys. A* **1993**, *8*, 3747–3771. [CrossRef]
24. De Ritis, R.; Marmo, G.; Preziosi, B. A New Look at Relativity Transformations. *Gen. Relat. Gravit.* **1999**, *31*, 1501–1517. [CrossRef]
25. Bhamathi, G.; Sudarshan, E.C.G. Time as dynamical variable. *Phys. Lett. A* **2003**, *317*, 359–362. [CrossRef]
26. Sudarshan, E.C.G.; Mukunda, N. Forms of Relativistic Dynamics with World Line Condition and Separability. *Found. Phys.* **1983**, *13*, 385–393. [CrossRef]

Article

On the Prospects of Multiport Devices for Photon-Number-Resolving Detection

Yong Siah Teo [1,*], Hyunseok Jeong [1], Jaroslav Řeháček [2], Zdeněk Hradil [2], Luis L. Sánchez-Soto [3,4] and Christine Silberhorn [5]

1 Department of Physics and Astronomy, Seoul National University, Seoul 08826, Korea; jeongh@snu.ac.kr
2 Department of Optics, Palacký University, 17. listopadu 12, 77146 Olomouc, Czech Republic; rehacek@optics.upol.cz (J.Ř.); hradil@optics.upol.cz (Z.H.)
3 Max-Planck-Institut für die Physik des Lichts, Staudtstraße 2, 91058 Erlangen, Germany; lsanchez@ucm.es
4 Departamento de Óptica, Facultad de Física, Universidad Complutense, 28040 Madrid, Spain
5 Integrated Quantum Optics Group, Applied Physics, University of Paderborn, 33098 Paderborn, Germany; christine.silberhorn@upb.de
* Correspondence: ys_teo@snu.ac.kr

Received: 10 September 2019; Accepted: 26 September 2019; Published: 29 September 2019

Abstract: Ideal photon-number-resolving detectors form a class of important optical components in quantum optics and quantum information theory. In this article, we theoretically investigate the potential of multiport devices having reconstruction performances approaching that of the Fock-state measurement. By recognizing that all multiport devices are minimally complete, we first provide a general analytical framework to describe the tomographic accuracy (or quality) of these devices. Next, we show that a perfect multiport device with an infinite number of output ports functions as either the Fock-state measurement when photon losses are absent or binomial mixtures of Fock-state measurements when photon losses are present and derive their respective expressions for the tomographic transfer function. This function is the scaled asymptotic mean squared error of the reconstructed photon-number distributions uniformly averaged over all distributions in the probability simplex. We then supply more general analytical formulas for the transfer function for finite numbers of output ports in both the absence and presence of photon losses. The effects of photon losses on the photon-number resolving power of both infinite- and finite-size multiport devices are also investigated.

Keywords: photon-number-resolving detectors; multiport devices; quantum optics; Fock states; quantum tomography; photon losses

---

1. Introduction

Photon-number-resolving (PNR) detection schemes are measurements that play a vital role in quantum information theory. The ability to perform direct photon counting has been shown to fundamentally impact quantum protocols and technologies. These include quantum metrology [1–4], quantum key distribution [5,6], Bell measurements [7] and quantum random number generator [8,9]. In practice, such PNR measurements either do not faithfully resolve photon numbers or do so up to a limited (typically small) number of photons, especially in the emblematic presence of dark counts and photon losses [10,11]. In recent years, there has been significant progress in the quality and type of photon-counting detectors developed through new-generation quantum engineering techniques [12–19].

An alternative class of setups that are widely used to indirectly perform photon counting are the so-called multiport devices [20–22], which are schematically more sophisticated devices that involve multiple beam splitters and several output ports that lead to "on–off" photodetectors for counting the number of split output signal pulses. Such alternative devices are later refashioned using optical-fiber looping [23,24] or multiplexing [25–28] strategies that give exactly the same photon-number-resolving characteristics but with much more efficient and cost-effective architectures.

In this article, we invoke the machinery of quantum tomography to evaluate the performance of general multiport devices. After providing the general descriptions of multiport devices in Section 2 and introducing the concept of informational completeness for such commuting measurements in Section 3, we establish a general framework in Section 4 to certify their tomographic performances using an operational tomographic transfer function that measures the average asymptotic accuracies of reconstructed photon-number distributions (Equation (33)). According to this formalism, we first investigate the performances of multiport devices that have infinitely many output ports with and without photon losses in Section 5. We shall show respectively that these infinitely large devices behave either exactly like a set of Fock-state measurement outcomes or their binomial-noisy mixtures and derive their tomographic transfer functions (Equations (40) and (49)). We will also demonstrate in Section 5.3 that photon losses can severely limit the photon-number resolution of multiport devices and systematically characterize such limitations in terms of informational completeness phase diagrams and the dependence of the maximum photon-loss rate tolerable on the number of photons to be resolved. Finally in Section 6, we shall derive general formulas for the transfer functions for the most general multiport devices with finite output ports (Equations (58) and (60)) and evaluate the effects of photon losses on their photon-number resolving power in Section 6.3.

2. General Physics of Multiport Devices

A multiport device is general laboratory equipment that houses an input port for receiving photonic signals and a fixed number (say s) of output ports. After undergoing multiple splitting of an input photonic pulse inside the device, each output port would then either idle (symbolically labeled as "0") or register a photonic "click" ("1") that originates from the split pulse. As an example, a three-port device would contain $s = 3$ output ports that give a total of $2^3 = 8$ different detection configurations, which are the "000", "001", "010", "100", "011", "101", "110" and "111" detection events. For the purpose of photon-number-distribution reconstruction, we may as well consolidate all the "0-click", "1-click", "2-click" and "3-click" events respectively and describe this multiport device as a measurement of $M = 4$ outcomes. More generally, an s-port device is one that gives 2^s detection configurations that may be organized to yield a total of $M = s + 1$ measurement events.

Any measurement can be described by a *positive operator-valued measure* (POVM), a set of probability operators (outcomes) that is given by

$$\Pi_j \geq 0 \quad \text{such that} \quad \sum_{j=0}^{M-1} \Pi_j = 1\,. \tag{1}$$

A multiport device is no exception, and is therefore mathematically equivalent to a POVM of $M = s + 1$ outcomes, where a "j-click" outcome is some unnormalized mixture $\Pi_j = \sum_n |n\rangle\, \beta_{jn}\, \langle n|$ of Fock states. For a sufficiently large number of data sampling events N, the data obtained from a measurement of such a POVM give probabilities that are linear combinations of the expectation values $\langle |n\rangle \langle n| \rangle$. The photon-number distribution can subsequently be reconstructed. The amplitudes β_{jn} are, in general, complicated functions of all the port efficiencies $\{\eta_j\}$ $\left(\sum_j \eta_j \leq 1\right)$, each of which depends on the physical parameters of the actual device implementation such as beam-splitter ratio, photodetector efficiency, and so on.

In particular, for arbitrary port efficiencies $\sum_j \eta_j \equiv 1 - \epsilon$, the "0-click" outcome Π_0 possesses amplitudes $\beta_{0,n} = \epsilon^n$ that are independent of any other detail of the multiport specifications. In other words, the probability of a "0-click" event for an n-photon input signal is the n-fold product of the loss probability ϵ, which is consistent with the physical fact that photoabsorption and detector losses are the main mechanisms behind all "0-click" events when $n > 0$ in the absence of other kinds of experimental imperfections.

If the light source is effectively described by a quantum state ρ in a Hilbert space of dimension d, so that the probability of detecting $n > d-1$ photons is practically zero, then all "$(j > d-1)$-click" outcomes are correspondingly zero by construction. The outcomes Π_j are hence represented by $d \times d$ positive matrices that sum to the identity matrix. We can define the measurement matrix that concisely and uniquely determine the multiport POVM. To do this, we first emphasize that in this effective Hilbert space, the conditional photon-number probabilities $\rho_n = \langle n|\,\rho\,|n\rangle = \langle |n\rangle\,\langle n|\rangle$ are properly normalized ($\mathrm{tr}\{\rho\} = 1$), so that the total number of independent parameters to be estimated is $d-1$. From Born's rule, we may express the multiport probabilities in terms of $d-1$ independent state parameters inasmuch as

$$p_j = \mathrm{tr}\{\rho\Pi_j\} = \sum_{n=0}^{d-2}(\beta_{jn} - \beta_{j\,d-1})\rho_n + \beta_{j\,d-1}\,. \tag{2}$$

Following the reasonings in quantum-state tomography [29,30], we may define the measurement matrix

$$C = \sum_{j=0}^{d-1}\sum_{n=0}^{d-2} \boldsymbol{e}_j\boldsymbol{e}_n(\beta_{jn} - \beta_{j\,d-1})\,, \tag{3}$$

with the help of the standard computational basis $\boldsymbol{e}_l \cdot \boldsymbol{e}_{l'} = \delta_{l,l'}$, to be the $d \times (d-1)$ rectangular matrix that fully characterizes the multiport POVM for the $d-1$ independent ρ_n parameters. It is clear that this matrix has a zero eigenvalue corresponding to the eigenvector $\boldsymbol{1}_d$ that is represented as a d-dimensional column of ones—$\boldsymbol{1}_d \cdot C = \boldsymbol{0}^{\mathrm{T}}_{d-1}$.

We shall look into an interesting special case where the port efficiencies are all equal to a constant ($\eta_j = \eta$), so that the POVM amplitudes can be shown to take the simple form [20,23,25]

$$\beta_{jn} = (-1)^j\binom{s}{j}\sum_{k=0}^{j}\binom{j}{k}(-1)^k\,[1-\eta(s-k)]^n\,. \tag{4}$$

The self-consistent consequence $\sum_{j=0}^{s}\beta_{jn} = 1$ can be verified straightforwardly. This type of multiport device is commonly used in practice. We mention in passing that the outcomes Π_j may be equivalently expressed as the normal-ordered form

$$\Pi_j = \binom{s}{j} :\left(\mathrm{e}^{-\eta a^\dagger a}\right)^{s-j}\left(1-\mathrm{e}^{-\eta a^\dagger a}\right)^{j}: \tag{5}$$

from which Equation (4) is quickly obtained through the application of the formula

$$:F(a^\dagger a): = F\left(\frac{\mathrm{d}}{\mathrm{d}x}\right)x^{a^\dagger a}\bigg|_{x=1}\,, \tag{6}$$

for any operator function $F(a^\dagger a)$ of the number operator $a^\dagger a$.

Dark counts may be incorporated in a simplistic way by introducing the parameter $\nu > 0$ that defines the average dark-count rate as the transformation $\eta a^\dagger a \to \eta a^\dagger a + \nu$ to Equation (5). Physically, this transformation increases the partially-depleted number operator $\eta a^\dagger a$ due to losses by an additional ν photons on average. In what follows, dark-count rates are assumed to be negligible in the feasible bandwidth of the photodetectors.

3. Informational Completeness of Photon-Number Distribution Measurements

To analyze photon-number-distribution reconstruction with multiport devices, we shall review the tools that are employed in understanding quantum measurements in this context. We recall that an informationally complete (IC) measurement is one that uniquely characterizes a particular set of physically relevant parameters describing a given quantum source of interest. In quantum-state tomography, such a measurement unambiguously reconstructs the quantum state ρ for the source. For our purpose, the set of parameters constitutes the photon-number distribution $\{\rho_n\}$ of a quantum light source, which are the diagonal entries of ρ in the Fock basis as mentioned in Section 2. With respect to the ρ_ns, a POVM is IC when it contains at least d outcomes with a degree of linear independence of d.

The entire machinery for IC quantum-state tomography can be translated for photon-number distribution tomography. The concept of the operator ket is particularly helpful here for notational simplification. For any d-dimensional operator O in the Fock basis, its operator ket $|O\rangle$ is defined as the d-dimensional column vector of its diagonal entries. The photon-number distribution of ρ that is of interest to us is thus summarized by its operator ket $|\rho\rangle$ such that $\mathrm{tr}\{\rho\} = \langle 1|\rho\rangle = 1$. With this, we can define the frame operator

$$\mathcal{F} = \sum_{j=0}^{M-1} \frac{|\Pi_j\rangle\langle\Pi_j|}{\mathrm{tr}\{\Pi_j\}}, \tag{7}$$

for any POVM $\{\Pi_j\}$ comprising M commuting Fock-state mixtures. Hence, an equivalent definition for an IC POVM is the operator invertibility of $\mathcal{F}$. In addition, using the operator-ket notation, the degree of linear independence of the POVM can be checked by inspecting the eigenvalues of the standard Gram matrix

$$\boldsymbol{G} = \sum_{j=0}^{M-1}\sum_{k=0}^{M-1} \boldsymbol{e}_j\boldsymbol{e}_k \langle\Pi_j|\Pi_k\rangle = \boldsymbol{V}\boldsymbol{V}^\dagger, \quad \boldsymbol{V} = \begin{pmatrix} \langle\Pi_1| \\ \vdots \\ \langle\Pi_M| \end{pmatrix}, \tag{8}$$

for vectorial objects.

For multiport devices, Equation (7) is applicable for $M = s+1$. Consequently, it is necessary for the corresponding multiport POVM to have $s \geq d-1$ output ports for it to be IC in a d-dimensional Hilbert space. Moreover, there exists another important feature for these devices. As discussed in Section 2, that the probability of detecting more photons than the number available in the input signal is zero implies that any multiport POVM is necessarily minimally complete when it is IC on the d-dimensional Hilbert space. This means that for such minimal POVMs, there are effectively only $M = d$ nonzero outcomes (each having amplitudes that depend on s) and we can uniquely express the photon-number distribution as

$$|\rho\rangle = \sum_{j=0}^{d-1} |\Theta_j\rangle\, p_j \tag{9}$$

with the help of the d canonical dual operators

$$|\Theta_j\rangle = \mathcal{F}^{-1}\frac{|\Pi_j\rangle}{\mathrm{tr}\{\Pi_j\}}. \tag{10}$$

It can be shown that

$$\langle\Pi_j|\Theta_k\rangle = \delta_{j,k} \tag{11}$$

for any minimal POVM. For this, we use the general property

$$\boldsymbol{W}^\dagger\boldsymbol{V} = 1 = \boldsymbol{V}^\dagger\boldsymbol{W}, \quad \boldsymbol{W} = \begin{pmatrix} \langle\Theta_1| \\ \vdots \\ \langle\Theta_M| \end{pmatrix}, \tag{12}$$

for any set of (canonical) dual operators, so that sandwiching the left equation in (12) with $\boldsymbol{V}$ from the left and $\boldsymbol{V}^\dagger$ from the right gives

$$VW^{\dagger}G = G = GWV^{\dagger}\,. \tag{13}$$

Next, we realize that for any minimal POVM, G is always invertible and we have $VW^{\dagger} = 1 = WV^{\dagger}$.

4. General Framework for the Reconstruction Accuracy of Multiport Devices

4.1. Mean Squared-Error and Its Cramér–Rao Bound

We shall take the mean squared-error (MSE) $\mathcal{D}_{\mathrm{MSE}}$ as the measure of the reconstruction accuracy of the photon-number distribution $|\rho\rangle$. For a given estimator $|\widehat{\rho}\rangle$ of $|\rho\rangle$, since only $\rho_n\big|_{n=0}^{d-2}$ are independent, this measure is defined as

$$\mathcal{D}_{\mathrm{MSE}} = \mathbb{E}_{\mathrm{data}}[(|\widehat{\rho}\rangle - |\rho\rangle)^2]\Big|_{\mathrm{supp}}\,, \tag{14}$$

where the average is taken over all plausible data. The label "supp" means that the inner product is evaluated in the $(d-1)$-dimensional support of the linearly independent parameters of $|\rho\rangle$. The parameter space of $|\rho\rangle$ is the entire d-dimensional probability simplex, since one can always find a quantum state ρ that gives any particular $|\rho\rangle$ (a statistical mixture of Fock states weighted with the ρ_ns, for instance). The boundary of this space is therefore the edges of this simplex.

When $|\rho\rangle$ is off the boundary ($\rho_n \neq 0$), which is the real experimental situation, it is well-known that the scaled MSE with N is bounded from below by the Cramér–Rao bound (CRB) per sampling event,

$$N\mathcal{D}_{\mathrm{MSE}} \geq \mathrm{tr}\{F(\rho)^{-1}\}\,, \tag{15}$$

where $F(\rho)$ is the $(d-1)$-dimensional Fisher information operator (defined per sampling event) for a given $|\rho\rangle$ and POVM. In particular, the unbiased maximum-likelihood (ML) estimator saturates this bound asymptotically in the limit of large N. Boundary $|\rho\rangle$s may be included in the picture by taking appropriate limits. The CRB directly evaluates the reconstruction accuracy of $|\rho\rangle$ where the constraint $\mathrm{tr}\{\rho\} = \langle 1|\rho\rangle = 1$ is obeyed, and supplies the limit of photon-number reconstruction for any $|\rho\rangle$.

For any minimal POVM, the MSE has a simple compact form for single-shot experiments that yield multinomial data statistics, just as for any multiport device. First, we can define the linear estimator of $|\rho\rangle$, in terms of the canonical dual operators and the measured multiport relative frequencies ν_j, as

$$|\widehat{\rho}\rangle = \sum_{j=0}^{d-1} |\Theta_j\rangle\, \nu_j\,, \tag{16}$$

where $\langle \Pi_j|\rho\rangle = \nu_j$ for any minimal POVM. The fact that $\mathbb{E}_{\mathrm{data}}[|\widehat{\rho}\rangle] = |\rho\rangle$ is evident. Second, we recall that this linear estimator is in fact the ML estimator whenever $|\rho\rangle > \mathbf{0}$ for sufficiently large N, so that the linear estimator in Equation (16) saturates the CRB. So, using the identity

$$\mathbb{E}_{\mathrm{data}}[\nu_j\nu_k] = \frac{1}{N}[\delta_{j,k}p_j + (N-1)p_jp_k], \tag{17}$$

for multinomial distributions, we have

$$\begin{aligned}
\mathcal{D}_{\mathrm{MSE}} &= \mathbb{E}_{\mathrm{data}}[\langle\widehat{\rho}|\widehat{\rho}\rangle] - \langle\rho|\rho\rangle\Big|_{\mathrm{supp}} \\
&= \sum_{j=0}^{d-1}\sum_{k=0}^{d-1} \langle\Theta_j|\Theta_k\rangle \left(\mathbb{E}_{\mathrm{data}}[\nu_j\nu_k] - p_jp_k\right)\Big|_{\mathrm{supp}} \\
&= \frac{1}{N}\left(\sum_{j=0}^{d-1} \langle\Theta_j|\Theta_j\rangle\, p_j - \langle\rho|\rho\rangle\right)\Bigg|_{\mathrm{supp}}\,.
\end{aligned} \tag{18}$$

On the other hand for multinomial data statistics, it is known that the Fisher operator takes the form

$$\begin{aligned} F(|\rho\rangle) &= \sum_{l=0}^{d-1} (|\Pi_l\rangle - |1\rangle\, \beta_{l\,d-1}) \frac{1}{p_l} (\langle\Pi_l| - \beta_{l\,d-1}\langle 1|)\Big|_{\text{supp}} \\ &= \boldsymbol{C}^{\text{T}}\boldsymbol{P}^{-1}\boldsymbol{C}, \end{aligned} \tag{19}$$

where $\boldsymbol{P}_j = p_j$. In view of this, we arrive at the identity

$$\begin{aligned} \text{tr}\{F(|\rho\rangle)^{-1}\} &= \text{tr}\{(\boldsymbol{C}^{\text{T}}\boldsymbol{P}^{-1}\boldsymbol{C})^{-1}\} \\ &= \left(\sum_{j=0}^{d-1} \langle\Theta_j|\Theta_j\rangle\, p_j - \langle\rho|\rho\rangle\right)\Bigg|_{\text{supp}} \end{aligned} \tag{20}$$

for any minimal POVM with respect to the photon-number distribution.

4.2. A Measure of Tomographic Performance

The CRB in Equation (20) is a function of $|\rho\rangle$. To obtain an operational performance certifier, one may choose to average over $|\rho\rangle$, which can be carried out in many different ways. We shall follow a similar direction reported in ref. [30] and perform an average over all distributions over the probability simplex. The resulting average CRB

$$\text{TTF} = \mathbb{E}_{|\rho\rangle}[\text{tr}\{F(|\rho\rangle)^{-1}\}] \tag{21}$$

is the tomographic transfer function (TTF) for photon-number distributions, which generalizes previous analytical scopes, such as those in [11,22], that focus on the class of Poissonian distributions to other more exotic yet classically allowed probability distributions $\{p_j\}$ in the $(d-1)$-dimensional simplex. Following through the calculations, using the simplex identities (see Appendix B)

$$\mathbb{E}_{|\rho\rangle}[p_j] = \frac{1}{d} \quad \text{and} \quad \mathbb{E}_{|\rho\rangle}[p_j^2] = \frac{2}{d(d+1)}, \tag{22}$$

we have

$$\mathbb{E}_{|\rho\rangle}[\langle\rho|\rho\rangle]\Big|_{\text{supp}} = \sum_{n=0}^{d-2} \mathbb{E}_{|\rho\rangle}[\rho_n^2] = \frac{2(d-1)}{d(d+1)}. \tag{23}$$

$$\text{TTF}_{\text{multiport}} = \mathbb{E}_{|\rho\rangle}[\text{tr}\{F(\rho)^{-1}\}] = \frac{1}{d}\text{tr}\{\mathcal{F}^{-1}\}\Big|_{\text{supp}} - \frac{2(d-1)}{d(d+1)}. \tag{24}$$

Finally it turns out that Equation (24) may also be obtained from an average of $F(|\rho\rangle)$ uniformly (under the Haar measure) over all pure states ρ), after a reference to Equation (10).

One can proceed to express the first term on the rightmost side of Equation (24) by recognizing that

$$\text{tr}\{\mathcal{F}^{-1}\}\Big|_{\text{supp}} = \text{tr}\{\mathcal{F}^{-1}\} - \langle d-1|\,\mathcal{F}^{-1}\,|d-1\rangle\,, \tag{25}$$

and that the Fock state

$$|d-1\rangle = \frac{|\Pi_{d-1}\rangle}{\text{tr}\{\Pi_{d-1}\}} \tag{26}$$

for any multiport device since in the absence of dark counts, the "j-click" event occurs when there are j photons or more. Then the orthonormality property in Equation (11) dictates that

$$\langle d-1|\,\mathcal{F}^{-1}\,|d-1\rangle = \frac{1}{\text{tr}\{\Pi_{d-1}\}^2}\langle\Pi_{d-1}|\,\mathcal{F}^{-1}\,|\Pi_{d-1}\rangle = \frac{1}{\text{tr}\{\Pi_{d-1}\}}, \tag{27}$$

which brings us to the slightly more explicit expression

$$\mathrm{TTF}_{\mathrm{multiport}}(s,\{\eta_j\}) = \frac{1}{d}\mathrm{tr}\{\mathcal{F}^{-1}\} - \frac{1}{d\,\mathrm{tr}\{\Pi_{d-1}\}} - \frac{2(d-1)}{d(d+1)}\,. \tag{28}$$

The result in Equation (33) assigns a number to the average performance of a multiport device of arbitrary number of output ports s, port efficiencies $\{\eta_j\}$ and loss probability ϵ based on statistical estimation theory.

For any multiport POVM of amplitudes β_{jn}, by defining $\boldsymbol{B}_{s,\epsilon}$ to be the square matrix of these amplitudes $(\boldsymbol{B}_{s,\epsilon})_{jn} = \beta_{jn}$, the operator kets $|\Pi_j\rangle$ and the Fock kets $|n\rangle$ are then related by the simple linear system

$$\boldsymbol{V} = \boldsymbol{B}_{s,\epsilon}\boldsymbol{v}\,, \quad \boldsymbol{v} = \begin{pmatrix} \langle 0| \\ \vdots \\ \langle d-1| \end{pmatrix}, \tag{29}$$

where the column $\boldsymbol{V}$ of operator bras is as defined in Equation (8). The Fock bras can then be expressed in terms of the operator bras $|\Pi_j\rangle$ as $\boldsymbol{v} = \boldsymbol{B}_{s,\epsilon}^{-1}\boldsymbol{V}$. This compact form proves useful when evaluating the operator trace of the inverted frame operator $\mathcal{F}^{-1}$:

$$\mathrm{tr}\{\mathcal{F}^{-1}\} = \mathrm{tr}\{\mathcal{F}^{-1}\boldsymbol{v}^{\dagger}\boldsymbol{v}\} = \mathrm{tr}\{\mathcal{F}^{-1}\boldsymbol{V}^{\dagger}\boldsymbol{B}_{s,\epsilon}^{\mathrm{T}}{}^{-1}\boldsymbol{B}_{s,\epsilon}^{-1}\boldsymbol{V}\}\,. \tag{30}$$

At this stage, we emphasize the distinction between the operators (such as $\mathcal{F}$ and $|\Pi_j\rangle\langle\Pi_j|$) and the columns of (operator) kets (such as $\boldsymbol{v}$ and $\boldsymbol{V}$) to avoid confusion regarding the role of the operator trace $\mathrm{tr}\{\,\cdot\,\}$. With that, using the basic fact

$$\langle\Pi_j|\,\mathcal{F}^{-1}\,|\Pi_k\rangle = \mathrm{tr}\{\Pi_j\}\delta_{j,k} \tag{31}$$

for any minimal POVM, the answer

$$\mathrm{tr}\{\mathcal{F}^{-1}\} = \sum_{j=0}^{d-1}\mathrm{tr}\{\Pi_j\}\left(\boldsymbol{B}_{s,\epsilon}\boldsymbol{B}_{s,\epsilon}^{\mathrm{T}}\right)^{-1}_{jj} \tag{32}$$

is immediate and

$$\mathrm{TTF}_{\mathrm{multiport}}(s,\{\eta_j\}) = \frac{1}{d}\left[\sum_{j=0}^{d-1}\mathrm{tr}\{\Pi_j\}\left(\boldsymbol{B}_{s,\epsilon}\boldsymbol{B}_{s,\epsilon}^{\mathrm{T}}\right)^{-1}_{jj} - \frac{1}{\mathrm{tr}\{\Pi_{d-1}\}} - \frac{2(d-1)}{d+1}\right]. \tag{33}$$

5. Multiport Device of Equal Port Efficiencies and $s \to \infty$ Output Ports

To gain some physical insights from the structure of multiport devices, we begin with a systematic study of the special case where $\eta_j = \eta$. With this, Equation (4) immediately applies. Upon an introduction of the simple relation

$$[1-\eta(s-k)]^n = \left(\frac{\partial}{\partial t}\right)^n \mathrm{e}^{t\,[1-\eta(s-k)]}\bigg|_{t=0}, \tag{34}$$

subsequent analysis may be facilitated after rewriting the POVM amplitudes as

$$\beta_{jn} = \binom{s}{j}\left(\frac{\partial}{\partial t}\right)^n\left[\mathrm{e}^{t(1-\eta s)}\left(\mathrm{e}^{t\eta}-1\right)^j\right]\bigg|_{t=0}. \tag{35}$$

This formula, which is valid for any s and η, shall serve as a good starting point for deriving our main results.

5.1. Perfect Multiport Devices Without Losses

If the loss probability is zero ($\epsilon = 0$), the port efficiencies are then all equal to $\eta = 1/s$. It follows from Equation (35), that

$$\beta_{jn} = \binom{s}{j}\left(\frac{\partial}{\partial t}\right)^n \left(e^{\frac{t}{s}} - 1\right)^j\Big|_{t=0} = \frac{j!}{s^n}\binom{s}{j}\left\{ {n \atop j} \right\}, \tag{36}$$

after an invocation of the moment-generating formula

$$\frac{1}{a^n j!}\left(\frac{\partial}{\partial t}\right)^n \left(e^{at} - 1\right)^j\Big|_{t=0} = \left\{ {n \atop j} \right\} \tag{37}$$

for the Stirling number of the second kind $\left\{ {n \atop j} \right\}$. The combinatorial sum rule

$$\sum_{j=0}^{n} \frac{s!}{(s-j)!}\left\{ {n \atop j} \right\} = s^n, \tag{38}$$

guarantees the proper normalization of β_{jn} as it should.

For infinitely many ports ($s \to \infty$), the ratio $s!/(s-j)! \to s^j$ and the amplitudes

$$\beta_{jn} \to \frac{1}{s^{n-j}}\left\{ {n \atop j} \right\}\Big|_{s\to\infty} = \delta_{j,n} \tag{39}$$

become those of the Fock states. Put differently, as the multiport device grows in size, its functionality approaches that of the pure Fock-state measurement—indirect photon counting approaches direct photon counting in the large-s limit. As the matrix $\boldsymbol{B} \equiv \boldsymbol{B}_{s\to\infty,\epsilon=0}$ is simply the $d \times d$ identity matrix, the TTF takes the value

$$\mathrm{TTF}_{\mathrm{multiport}}\left(s \to \infty, \left\{\eta_j = \frac{1}{s}\right\}\right) = \frac{(d-1)^2}{d(d+1)}. \tag{40}$$

As d increases, the TTF approaches unity. It can be shown that the performance $\mathrm{TTF}_{\mathrm{multiport}}\left(s, \{\eta_j\}\right)$ of any arbitrary lossless multiport device is bounded from below by this Fock-state limit (see Appendix A).

5.2. Imperfect Multiport Devices with Losses

When photon losses are present [$\epsilon > 0, \eta = (1-\epsilon)/s$], Equation (35) gives

$$\beta_{jn} = \binom{s}{j}\left(\frac{\partial}{\partial t}\right)^n \left\{ e^{t\epsilon}\left[e^{\frac{t}{s}(1-\epsilon)} - 1\right]^j \right\}\Big|_{t=0}. \tag{41}$$

In the limit $s \to \infty$, the approximation $e^y \approx 1 + y$ for small y and $s!/(s-j)! \to s^j$ render

$$\begin{aligned} \beta_{jn} \;\to\; & \frac{(1-\epsilon)^j}{j!}\left(\frac{\partial}{\partial t}\right)^n \left(t^j e^{t\epsilon}\right)\Big|_{t=0} = \frac{(1-\epsilon)^j}{j!}\sum_{l=0}^{\infty}\frac{\epsilon^l}{l!}\underbrace{\left(\frac{\partial}{\partial t}\right)^n t^{j+l}\Big|_{t=0}}_{=\,n!\,\delta_{l,n-j}} \\ = \;& \binom{n}{j}(1-\epsilon)^j \epsilon^{n-j} = (\boldsymbol{B}_\epsilon)_{jn}. \end{aligned} \tag{42}$$

The expression for $\boldsymbol{B}_\epsilon$ was defined earlier in [26] as a separate consequence of multiport photon losses, the argument of which is independent of taking the limit $s \to \infty$. These amplitudes correspond to those of a POVM comprising binomial mixtures of Fock-state outcomes

$$\Pi_j = \sum_{m=j}^{d-1} |m\rangle \binom{m}{j} (1-\epsilon)^j \epsilon^{m-j} \langle m| \,, \tag{43}$$

which tends to the set of Fock states in the limits $\epsilon \to 0$ and $d \to \infty$. Thus for large multiport devices, the probabilities are primarily influenced by the number of detection and absorption events.

To calculate the TTF, we need the inverse of $\boldsymbol{B}_\epsilon \equiv \boldsymbol{B}_{s\to\infty,\epsilon}$, which can be deduced to be

$$\boldsymbol{B}_\epsilon^{-1} = \sum_{j=0}^{d-1}\sum_{n=0}^{d-1} e_j e_n \binom{n}{j} (1-\epsilon)^{-n}(-\epsilon)^{n-j} \tag{44}$$

by a reverse engineering of the binomial theorem. One can effortlessly verify the following obvious necessary property $\boldsymbol{B}_\epsilon^{-1}\boldsymbol{B}_\epsilon = \boldsymbol{1}$. The remaining task is to simply calculate the matrix elements of $(\boldsymbol{B}_\epsilon \boldsymbol{B}_\epsilon^{\mathrm{T}})^{-1}$:

$$\begin{aligned}(\boldsymbol{B}_\epsilon \boldsymbol{B}_\epsilon^{\mathrm{T}})^{-1}_{jj'} &= \sum_{n=0}^{d-1} \binom{j}{n}(1-\epsilon)^{-j}(-\epsilon)^{j-n}\binom{j'}{n}(1-\epsilon)^{-j'}(-\epsilon)^{j'-n} \\ &= \left(-\frac{\epsilon}{1-\epsilon}\right)^{j'+j} \sum_{n=0}^{d-1}\binom{j}{n}\binom{j'}{n}\frac{1}{\epsilon^{2n}} \\ &= \left(-\frac{\epsilon}{1-\epsilon}\right)^{j'+j} {}_2F_1\left(\begin{matrix}-j & -j' \\ & 1\end{matrix}; \frac{1}{\epsilon^2}\right),\end{aligned} \tag{45}$$

where we have considered a definition

$$\sum_{n=0}^{j_<} \binom{j}{n}\binom{j'}{n} y^n = {}_2F_1\left(\begin{matrix}-j & -j' \\ & 1\end{matrix}; y\right), \quad j_< = \min\{j, j'\}, \tag{46}$$

for the special case of the Gaussian hypergeometric function ${}_2F_1\left(\begin{matrix}a_1 & a_2 \\ & b_1\end{matrix}; y\right)$.

Furthermore, in terms of the regularized incomplete beta function $\mathrm{I}_z(a,b)$, the operator traces for this multiport POVM

$$\mathrm{tr}\{\Pi_j\} = \frac{1}{1-\epsilon}\left[1 - \mathrm{I}_\epsilon(d-j, j+1)\right]. \tag{47}$$

Notably, we have $\mathrm{tr}\{\Pi_{d-1}\} = (1-\epsilon)^{d-1}$, which can be obtained either by

$${}_2F_1\left(\begin{matrix}1 & d+1 \\ & 2\end{matrix}; y\right) = \frac{1}{dy}\left[\frac{1}{(1-y)^d} - 1\right] \tag{48}$$

or the simple physical reasoning that the registration of j clicks must at least originate from the presence of j photons ($j \leq n$). A substitution of this final piece of information as well as Equation (45) into Equation (33) leads to

$$\begin{aligned}&\mathrm{TTF}_{\text{multiport}}\left(s\to\infty, \left\{\eta_j = \frac{1-\epsilon}{s}\right\}\right) \\ =\;& \frac{1}{d}\left[\sum_{j=0}^{d-1}\mathrm{tr}\{\Pi_j\}\left(\frac{\epsilon}{1-\epsilon}\right)^{2j} {}_2F_1\left(\begin{matrix}-j & -j \\ & 1\end{matrix}; \frac{1}{\epsilon^2}\right) - \frac{1}{(1-\epsilon)^{d-1}} - \frac{2(d-1)}{d+1}\right].\end{aligned} \tag{49}$$

It is clear that $\epsilon = 0$ brings us back to the optimal result stated in Equation (40) by noting that

$$\epsilon^{2j}\,{}_2F_1\left(\begin{matrix}-j & -j \\ & 1\end{matrix};\frac{1}{\epsilon^2}\right)\Bigg|_{\epsilon=0} = 1 \tag{50}$$

that arises from the definition in Equation (46).

Figure 1 demonstrates the fit between the theoretically predicted TTF values with Equation (49) and the numerically calculated ones after performing Monte–Carlo averaging of the inverse of the Fisher operator $F(\rho)^{-1}$ (see Equation (19)) over the Haar measure of pure states. To this average numerically, it is sufficient to generate a sufficiently large number of random pure states $\{\rho_j = \mathcal{A}_j^\dagger\mathcal{A}_j/\mathrm{tr}\{\mathcal{A}_j^\dagger\mathcal{A}_j\}\}$ parametrized by the random complex auxiliary rank-one operators $\mathcal{A}_j$ that follow the standard Gaussian distribution and use them to compute the average of $F(\rho)^{-1}$.

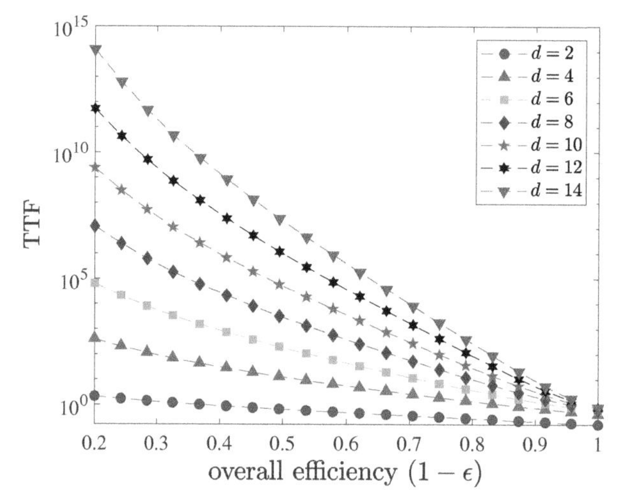

Figure 1. Numerical (colored markers) and theoretical (colored dashed curves) values of TTF (logarithmically scaled) for infinitely large multiport devices of various ϵ and Hilbert-space dimensions d. A total of 1000 random pure states were used to evaluate each numerical plot point. The convergence to the optimal TTF in Equation (40) at $\epsilon = 1$ is as expected. It therefore comes as no surprise that losses monotonically lowers reconstruction accuracy.

5.3. *Noisy Photon-Number Resolution of Multiport Devices with $s \to \infty$ and $\epsilon > 0$*

In the hypothetical situation where the photon-loss rate $\epsilon = 0$, the multiport device is capable of resolving photon numbers in an optical signal described by ρ of any arbitrary dimension d (recall that $d-1$ is then the maximum number of photons in the signal. We say that the d-dimensional Hilbert space is resolvable. Therefore any subspace of dimension $d_{\mathrm{res}} \leq d$ is by definition also resolvable. In real experiments however, a nonzero photon-loss rate directly limits the number of photons resolvable. The key relation that governs this restriction for $s \to \infty$ is Equation (47). For a fixed d, $\mathrm{tr}\{\Pi_j\}$ (or Π_j) becomes essentially zero above certain threshold $j = j_{\mathrm{thres}}$. This threshold value defines the dimension of the maximally resolvable subspace—$d_{\mathrm{res}} \leq j_{\mathrm{thres}+1}$.

More specifically, we may define j_{thres} as the largest integer for which

$$1 - \mathrm{I}_\epsilon(d - j_{\mathrm{thres}}, j_{\mathrm{thres}} + 1) > \mu_{\mathrm{thres}} \approx 0\,, \tag{51}$$

where μ_{thres} is a very small positive number close to zero. While this equation has no general analytical solution for finite d, we note that for large d, $\mathrm{I}_\epsilon(d-j, j+1) \approx \frac{1}{2} + \frac{1}{2}\tanh(j - d(1-\epsilon))$ is a remarkably good approximation. We may then use this to derive the simplified and approximate photon-number resolvability restriction

$$d_{\rm res} \leq d(1-\epsilon) + \tanh^{-1}(1-2\mu_{\rm thres})\,. \tag{52}$$

This observation impacts how we should perform asymptotic TTF analyses for multiport devices in the large d-limit. Unlike the ideal case where one simply takes $d \to \infty$ with Equation (40) to arrive at the finite value 1, this naive limit results in the divergence of $\mathrm{TTF}_{\rm multiport}$ for any finite ϵ. A careful thought reveals that indeed, for the tomography of photon-number distributions for dimension d to be IC, we require the necessary condition that ϵ be no greater than some critical value beyond which the inequality $1 - \mathrm{I}_\epsilon(1,d) > \mu_{\rm thres}$ becomes in valid. This condition may be approximately written as

$$\epsilon \leq \tanh^{-1}(1-2\mu_{\rm thres})/d \tag{53}$$

for sufficiently large d following Equation (52). Figures 2 and 3 show the important plots that characterize the informational completeness of any given (infinitely large) multiport device of nonzero photon-loss rate ϵ.

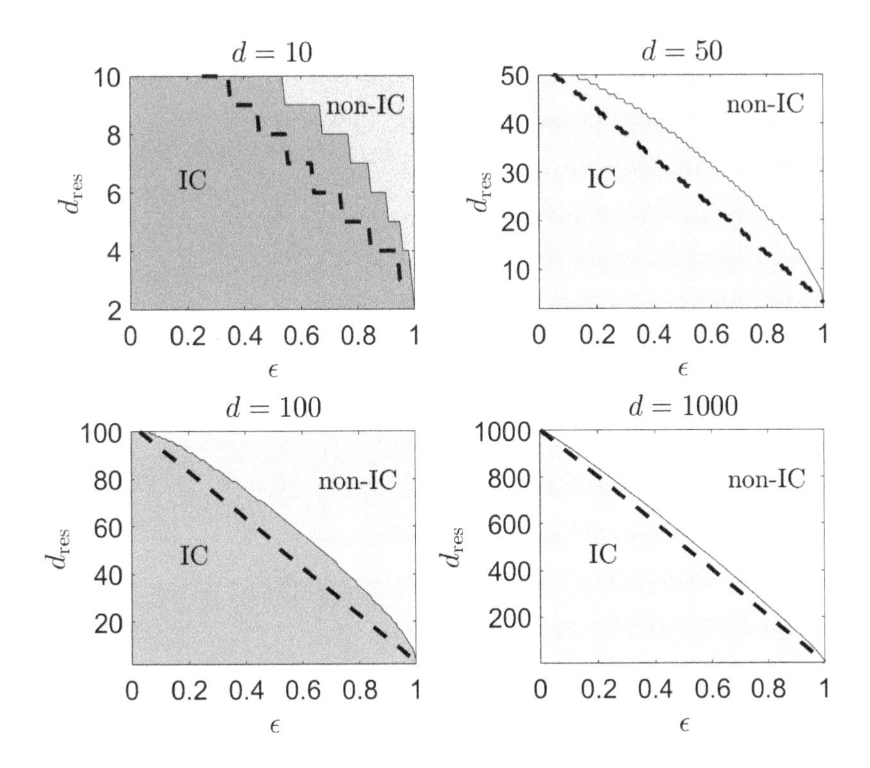

Figure 2. Informational completeness phase diagrams for various d in the $d_{\rm res}$-ϵ plane with $\mu_{\rm thres} = 10^{-3}$. Subspaces of dimensions below the boundary are resolvable, and hence render the multiport device of $s \to \infty$ and $\eta_j = (1-\epsilon)/s$ IC. Those of dimensions above the boundary are unresolvable with such a multiport device. The thick dashed curves represent the analytically calculated boundaries using the approximation in (52), which provide conservative underestimates for the maximum $d_{\rm res}$ compared to the numerically computed boundaries. Clearly, the range of ϵ for which the entire d-dimensional Hilbert space is completely resolvable reduces as d increases.

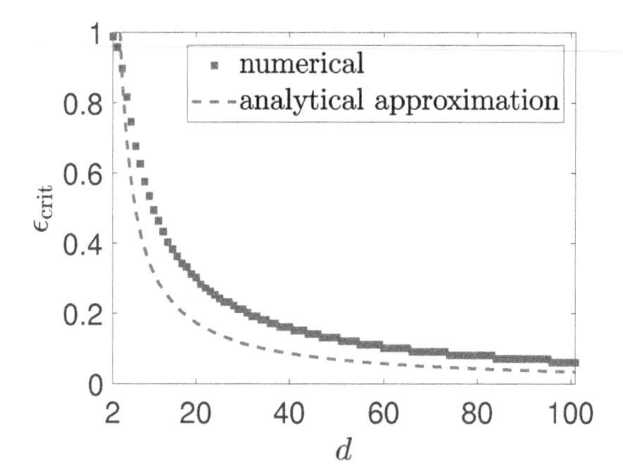

Figure 3. A plot of the critical ϵ value (ϵ_{crit}) against d that shows the maximum amount of photon losses a multiport device can tolerate before losing its informational completeness property. Here, $\mu_{\text{thres}} = 10^{-3}$. The simple $\epsilon_{\text{crit}} \sim 1/d$ behavior serves as a back-of-the-envelope solution for designing such devices.

6. Multiport Device of Equal Port Efficiencies and s Output Ports

6.1. *Perfect Multiport Devices without Losses*

The consideration of a finite-size multiport device with s output ports more closely resembles the real physical situation in the laboratory in which the resources that go into its implementation are limited. Even in this case, one can still easily compute the TTF for the $\epsilon = 0$ case where losses are absent in the device. Starting with the POVM amplitudes β_{jn} in Equation (36), we find that the inverse of their corresponding $\boldsymbol{B}_s \equiv \boldsymbol{B}_{s,\epsilon=0}$ amplitude matrix is simply given by

$$\boldsymbol{B}_s^{-1} = \sum_{j=0}^{d-1}\sum_{n=0}^{d-1} \boldsymbol{e}_j\boldsymbol{e}_n \frac{(s-n)!}{s!} s^j (-1)^{n-j} \begin{bmatrix} n \\ j \end{bmatrix} \tag{54}$$

and we owe this simple inversion formula to the existence of the (unsigned) Stirling number of the first kind $\begin{bmatrix} n \\ j \end{bmatrix}$ that is orthogonal to the Stirling number of the second kind $\begin{Bmatrix} n \\ j \end{Bmatrix}$ in the sense that

$$\sum_{n=j}^{k}(-1)^{n-k}\begin{Bmatrix} n \\ j \end{Bmatrix}\begin{bmatrix} k \\ n \end{bmatrix} = \sum_{n=j}^{k}(-1)^{n-j}\begin{Bmatrix} k \\ n \end{Bmatrix}\begin{bmatrix} n \\ j \end{bmatrix} = \delta_{j,k}\,. \tag{55}$$

This means that

$$\begin{aligned}\left(\boldsymbol{B}_s\boldsymbol{B}_s^{\mathrm{T}}\right)^{-1}_{jj'} &= (-1)^{j+j'}\frac{(s-j)!\,(s-j')!}{s!^2}\sum_{n=0}^{d-1} s^{2n}\begin{bmatrix} j \\ n \end{bmatrix}\begin{bmatrix} j' \\ n \end{bmatrix} \\ &= (-1)^{j+j'}\frac{(s-j)!\,(s-j')!}{s!^2}\,{}_2F_1^{(\mathrm{S1})}\left(\begin{matrix} -j & -j' \\ & 1 \end{matrix};s^2\right),\end{aligned} \tag{56}$$

where we have defined the *Stirling–Gaussian hypergeometric function of the first kind*

$$\sum_{n=0}^{j_<}\begin{bmatrix} j \\ n \end{bmatrix}\begin{bmatrix} j' \\ n \end{bmatrix} y^n = {}_2F_1^{(\mathrm{S1})}\left(\begin{matrix} -j & -j' \\ & 1 \end{matrix};y\right) \tag{57}$$

that is of analogous form to the usual Gaussian hypergeometric function in Equation (46). Accordingly, the Stirling–Gaussian hypergeometric function of the second kind ${}_2F_1^{(\mathrm{S2})}\left(\begin{matrix}-j & -j'\\ & 1\end{matrix};y\right)$ would then simply involve the Stirling numbers of the second kind.

The resulting performance certifier

$$\begin{aligned}&\mathrm{TTF}_{\mathrm{multiport}}\left(s,\left\{\eta_j=\frac{1}{s}\right\}\right)\\&=\frac{1}{d}\left\{\sum_{j=0}^{d-1}\frac{(s-j)!}{s!}\,{}_2F_1^{(\mathrm{S1})}\left(\begin{matrix}-j & -j\\ & 1\end{matrix};s^2\right)\sum_{n'=j}^{d-1}\frac{1}{s^{n'}}\left\{\begin{matrix}n'\\ j\end{matrix}\right\}-\frac{s^{d-1}(s-d+1)!}{s!}-\frac{2(d-1)}{d+1}\right\}\end{aligned}\tag{58}$$

allows us to evaluate the reconstruction accuracy for a finite-size multiport device of equal port efficiencies. As a verification of the validity of Equation (58), we compare it with numerically computed TTF for a sufficiently large set of random pure states distributed to the Haar measure (see Figure 4). Specifically, we note that for $d=2$, the TTF is a constant value of $1/6$, which tells us that for effective single-photon sources a two-port device functions exactly like a Fock-state measurement. This can be easily understood in hindsight by realizing that the only POVM outcomes that matter in this subspace are the vacuum and $n=1$ Fock states in the absence of losses. All other $s-1$ outcomes are not measured.

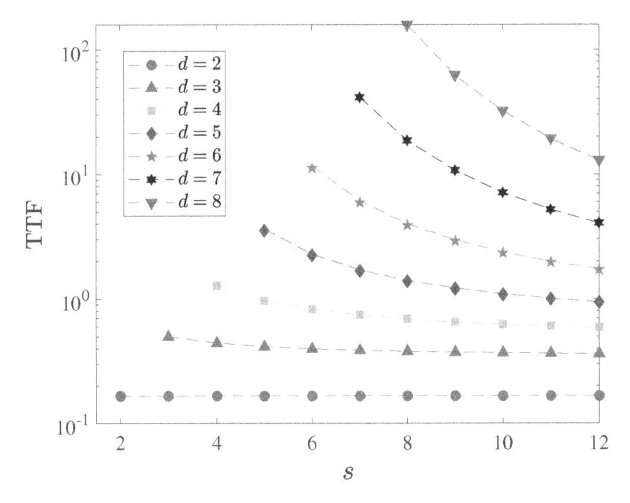

Figure 4. Numerical (colored markers) and theoretical (colored dashed curves) values of the TTF (logarithmically scaled) for finite-size multiport devices of various s values and Hilbert-space dimensions d. A total of 2000 random pure states were used to evaluate each numerical plot point. The $s \geq d$ regime illustrates the TTF for IC multiport POVMs only, which is the regime an observer would be interested in for the purpose of photon-number-distribution tomography.

6.2. Imperfect Multiport Devices with Losses

The rather specialized physical and mathematical structures of multiport devices permit us to obtain an analytical expression for the TTF even in the most general case where $s<\infty$ and $\epsilon>0$. The corresponding $\boldsymbol{B}_{s,\epsilon}$ for such multiport POVMs can again be inverted by the observation that $\boldsymbol{B}_{s,\epsilon}=\boldsymbol{B}_s\boldsymbol{B}_\epsilon$. In other words, a finite-size photoabsorptive multiport device is a device convolution of a perfect finite-size multiport device and photoabsorption losses. This is because

$$
\begin{aligned}
(\boldsymbol{B}_s\boldsymbol{B}_\epsilon)_{jn} &= \epsilon^n \binom{s}{j} \sum_{n'=0}^{n} \binom{n}{n'} \left(\frac{1}{\epsilon}\frac{\partial}{\partial t}\right)^{n'} \left[\mathrm{e}^{\frac{t(1-\epsilon)}{s}} - 1\right]^j \Bigg|_{t=0} \\
&= (-1)^j \binom{s}{j} \sum_{k=0}^{j} \binom{j}{k} (-1)^k \left(\epsilon + \frac{1-\epsilon}{s}k\right)^n \\
&= (-1)^j \binom{s}{j} \sum_{k=0}^{j} \binom{j}{k} (-1)^k \left[1 - \eta(s-k)\right]^n = (\boldsymbol{B}_{s,\epsilon})_{jn} \,. \qquad (59)
\end{aligned}
$$

This decomposition implies that $\boldsymbol{B}_{s,\epsilon}^{-1} = \boldsymbol{B}_\epsilon^{-1}\boldsymbol{B}_s^{-1}$, so that utilizing the results from Equations (44) and (54) for the two respective components, we can summarize the expressions for the performance measure:

$$
\begin{aligned}
\mathrm{TTF}_{\mathrm{multiport}}\left(s, \left\{\eta_j = \frac{1-\epsilon}{s}\right\}\right) &= \frac{1}{d}\left[\sum_{j=0}^{d-1} \mathrm{tr}\{\Pi_j\} \left(\boldsymbol{W}_{s,\epsilon}^{\mathrm{T}}\boldsymbol{W}_{s,\epsilon}\right)_{jj} - \frac{1}{\mathrm{tr}\{\Pi_{d-1}\}} - \frac{2(d-1)}{d+1}\right], \\
\boldsymbol{W}_{s,\epsilon\;jn} &= (-1)^{n-j}\epsilon^{-j}\frac{(s-n)!}{s!}\sum_{l=j}^{n}\left(\frac{\epsilon s}{1-\epsilon}\right)^l \binom{l}{j} \begin{bmatrix} n \\ l \end{bmatrix} . \qquad (60)
\end{aligned}
$$

Once more, we notice the constant TTF for $d = 2$ with a value of $(1+2\epsilon)/(6-6\epsilon)$ due to the s-independent multiport POVM consisting of the outcomes

$$
\begin{aligned}
\Pi_0 &= |0\rangle\langle 0| + |1\rangle\,\epsilon\,\langle 1| \,, \\
\Pi_1 &= |1\rangle\,(1-\epsilon)\,\langle 1| \,. \qquad (61)
\end{aligned}
$$

Figure 5 gives the comparison between theory and numerical computations for a sample ϵ.

6.3. Noisy Photon-Number Resolution of Multiport Devices with $s < \infty$ and $\epsilon > 0$

As with the case of $s \to \infty$ in Section 5.3, a nonzero photon-loss rate ϵ for a finite-size multiport device also reduces the number of photons that can be resolved. Therefore, ϵ should again be smaller than some critical value in order for the multiport device to characterize the complete d-dimensional photon-number distribution. This critical value may be computed, for every given value of s, according to the constraint $\sum_{n=0}^{d-1}(\boldsymbol{B}_{s,\epsilon})_{d-1,n} > \mu_{\mathrm{thres}} \approx 0$ that is to be satisfied by the largest value of ϵ.

In general, the critical value of ϵ has no easy analytical form. It is however numerically efficient to plot graphs of the critical values with respect to d for any physically reasonable s. Figure 6 shows some sample plots.

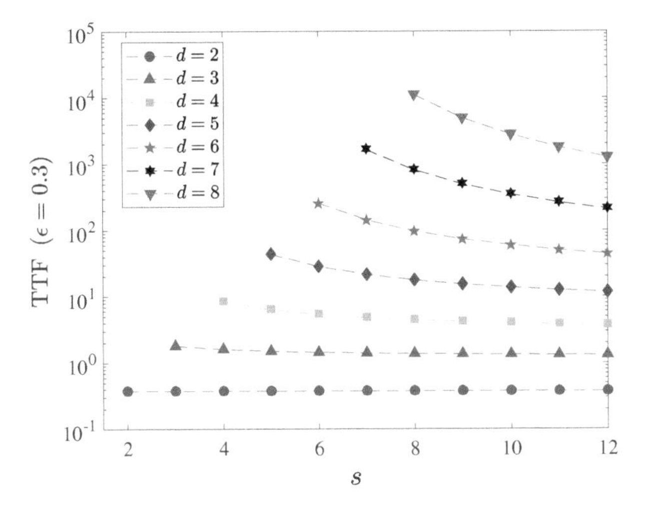

Figure 5. Numerical (colored markers) and theoretical (colored dashed curves) values of the TTF (logarithmically scaled) for finite-size multiport devices of various s values, a fixed $\epsilon = 0.3$, and Hilbert-space dimensions d. A total of 2000 random pure states were used to evaluate each numerical plot point. As in the case of $\epsilon = 0$, the TTF for $d = 2$ takes a constant value of 0.3809 for this particular ϵ value. The worsening of the tomographic performance with a finite loss probability is clearly manifested as an overall increase in the TTF values.

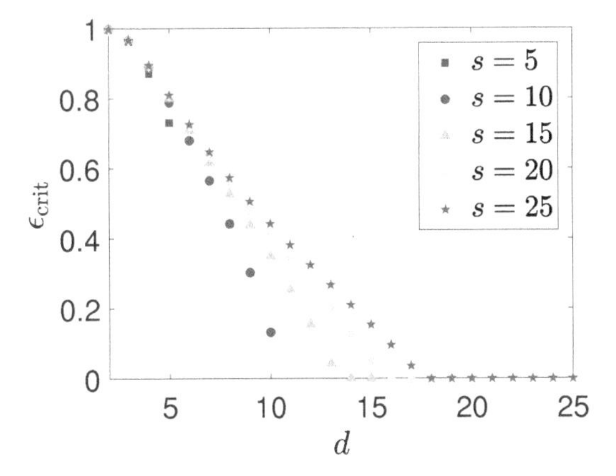

Figure 6. A plot of the critical ϵ value (ϵ_{crit}) against d for various number of outputs s of the multiport device. The threshold $\mu_{\mathrm{thres}} = 10^{-3}$ is chosen. For a given d-dimensional Hilbert subspace, increasing s also raises ϵ_{crit}, although for reasonable values of s such an increase is not dramatic even when $d \ll s$.

7. Discussion

We present a short series of studies related to the performance of multiport devices on photon-number distribution tomography. The central measure of performance is the quantum tomographic transfer function—the uniform average of the inverse Fisher information over all photon-number distributions in the probability simplex.

The mathematical framework for calculating the transfer function introduced in this article allows us to conclude that sufficiently-large multiport devices of equal transmissivity for each output port function either like a Fock-state measurement or binomial mixtures of Fock-state measurements respectively in the absence and presence of photon losses. These are followed by analytical treatments for finite-size multiport devices. In the presence of photon losses, we have studied and mapped out conditions concerning the photon-number resolving power of noisy multiport devices of both infinite and finite sizes. We show that devices of high photon losses possess weak photon-number resolving power and increasing the number of output ports may help only to a certain limited extent. The optimization of photodetectors and other optical components, especially for the purpose of curbing photon losses, is therefore crucial for building realistic multiport devices for indirect photon counting.

Author Contributions: Conceptualization, Y.S.T. and J.Ř.; methodology, Y.S.T.; software, Y.S.T.; validation, Y.S.T. and J.Ř.; formal analysis, Y.S.T.; investigation, Y.S.T.; resources, Y.S.T.; data curation, Y.S.T.; writing—original draft preparation, Y.S.T.; writing—review and editing, Y.S.T., H.J., J.Ř., Z.H., L.L.S.-S., and C.S.; visualization, Y.S.T.; supervision, Y.S.T., H.J., L.L.S.-S., and C.S.; project administration, Y.S.T., H.J., and L.L.S.-S.; funding acquisition, Y.S.T., H.J., J.Ř., Z.H., and L.L.S.-S.

Funding: We acknowledge financial support from the BK21 Plus Program (Grant No. 21A20131111123) funded by the Ministry of Education (MOE, Korea) and National Research Foundation of Korea (NRF), the NRF grant funded by the Korea government (MSIP) (Grant No. NRF-2019R1H1A3079890), the Spanish MINECO (Grant No. FIS2015-67963-P and PGC2018-099183-B-I00), the Grant Agency of the Czech Republic (Grant No. 18-04291S), and the IGA Project of the Palacký University (Grant No. IGA PrF 2019-007).

Conflicts of Interest: The authors declare no conflict of interest. The funders had no role in the design of the study; in the collection, analyses, or interpretation of data; in the writing of the manuscript, or in the decision to publish the results.

Appendix A. Optimality of the Fock-State Measurement for Noiseless Multiport Devices

We expect the commuting Fock-state measurement to be the optimal noiseless measurement for photon-number-distribution reconstruction. This expectation can be confirmed by showing that the value of the TTF in Equation (40) is indeed the optimal limit for all multiport devices. To this end, we exploit the inequalities

$$\begin{aligned} \mathrm{tr}\{AB\} &\leq \mathrm{tr}\{A\}\mathrm{tr}\{B\} \quad \text{for } A \geq 0 \text{ and } B \geq 0\,, && \text{(A1)} \\ \mathrm{tr}\{A\}\mathrm{tr}\{A^{-1}\} &\geq \dim\{A\}^2 \quad \text{for any invertible } A\,, && \text{(A2)} \end{aligned}$$

and remind ourselves that the TTF expression in Equation (33) holds for any s, $\{\eta_j\}$ and ϵ. As the operator trace

$$\mathrm{tr}\{\mathcal{F}\} = \sum_{j=0}^{d-1} \frac{\mathrm{tr}\{\Pi_j^2\}}{\mathrm{tr}\{\Pi_j\}} \leq \sum_{j=0}^{d-1} \mathrm{tr}\{\Pi_j\} = \mathrm{tr}\{1\} = d \tag{A3}$$

of the general frame operator is bounded from above according to (A1), the inequality in (A2) implies that

$$\mathrm{tr}\{\mathcal{F}^{-1}\} \geq d\,. \tag{A4}$$

Together with the obvious fact that the probability of detecting all available photons from the input signal never exceeds one ($\mathrm{tr}\{\Pi_{d-1}\} \leq 1$), we have the general inequality

$$\mathrm{TTF}_{\mathrm{multiport}}(s, \{\eta_j\}) \geq \mathrm{TTF}_{\mathrm{multiport}}\left(s \to \infty, \left\{\eta_j = \frac{1}{s}\right\}\right) \tag{A5}$$

to confirm that the Fock-state measurement condition $\{s \to \infty, \{\eta_j = 1/s\}\}$ is indeed optimal.

Appendix B. Averages over the Probability Simplex

We shall give simple derivations of the identities in (24). The general m-moment integral of interest in our context takes the form

$$I_m = \int_0^1 \mathrm{d}p_0 \cdots \int_0^1 \mathrm{d}p_{d-1}\, \delta\left(1 - \sum_{l=0}^{d-1} p_l\right) p_j^m, \tag{A6}$$

in which the simplex constraint $\sum_{l=0}^{d-1} p_l = 1$ is obeyed. Using the integral representation

$$\delta(x) = \int \frac{\mathrm{d}k}{2\pi}\, \mathrm{e}^{\mathrm{i}kx} \tag{A7}$$

for the delta function,

$$\begin{aligned} I_m &= \int \frac{\mathrm{d}k}{2\pi}\, \mathrm{e}^{\mathrm{i}k} \int_0^1 \mathrm{d}p_0 \cdots \int_0^1 \mathrm{d}p_{d-1}\, \mathrm{e}^{-\mathrm{i}k(p_0 + \cdots + p_{d-1})}\, p_j^m \\ &= \int \frac{\mathrm{d}k}{2\pi}\, \mathrm{e}^{\mathrm{i}k} \frac{\left(1 - \mathrm{e}^{-\mathrm{i}k}\right)^{d-1}}{(\mathrm{i}k)^{d-1}} \int_0^1 \mathrm{d}p_j\, \mathrm{e}^{-\mathrm{i}kp_j}\, p_j^m \\ &= \int \frac{\mathrm{d}k}{2\pi}\, \mathrm{e}^{\mathrm{i}k} \frac{\left(1 - \mathrm{e}^{-\mathrm{i}k}\right)^{d-1}}{(\mathrm{i}k)^{d-1}} \left(\mathrm{i}\frac{\mathrm{d}}{\mathrm{d}k}\right)^m \left[\frac{1}{\mathrm{i}k}\left(1 - \mathrm{e}^{-\mathrm{i}k}\right)\right], \end{aligned} \tag{A8}$$

where if $y \equiv \mathrm{i}k$,

$$\begin{aligned} \left(\mathrm{i}\frac{\mathrm{d}}{\mathrm{d}k}\right)^m \left[\frac{1}{\mathrm{i}k}\left(1 - \mathrm{e}^{-\mathrm{i}k}\right)\right] &= (-1)^m \left(\frac{\mathrm{d}}{\mathrm{d}y}\right)^m \left[y^{-1}(1 - \mathrm{e}^{-y})\right] \\ &= m!\, y^{-m-1}(1 - \mathrm{e}^{-y}) - \sum_{n=1}^{m} \frac{m!}{n!}\, y^{-m+n-1}\, \mathrm{e}^{-y}, \end{aligned} \tag{A9}$$

so that

$$I_m = m! \int \frac{\mathrm{d}k}{2\pi}\, \mathrm{e}^{\mathrm{i}k} \frac{(1 - \mathrm{e}^{-\mathrm{i}k})^d}{(\mathrm{i}k)^{m+d}} - \sum_{n=1}^{m} \frac{m!}{n!} \int \frac{\mathrm{d}k}{2\pi} \frac{\left(1 - \mathrm{e}^{-\mathrm{i}k}\right)^{d-1}}{(\mathrm{i}k)^{m-n+d}}. \tag{A10}$$

The first term can be evaluated using the identity

$$\frac{1}{y^{m+1}} = \frac{1}{m!} \int_0^\infty \mathrm{d}t\, t^m\, \mathrm{e}^{-yt}. \tag{A11}$$

This gives

$$\begin{aligned} m! \int \frac{\mathrm{d}k}{2\pi}\, \mathrm{e}^{\mathrm{i}k} \frac{(1 - \mathrm{e}^{-\mathrm{i}k})^d}{(\mathrm{i}k)^{m+d}} &= \frac{m!}{(m+d-1)!} \sum_{n=0}^{d} \binom{d}{n} (-1)^n \int_0^\infty \mathrm{d}t\, t^{m+d-1} \underbrace{\int \frac{\mathrm{d}k}{2\pi}\, \mathrm{e}^{\mathrm{i}k(1-n-t)}}_{=\delta(1-n-t)} \\ &= \frac{m!}{(m+d-1)!} \sum_{n=0}^{d} \binom{d}{n} (-1)^n\, \delta_{n,0} \\ &= \frac{m!}{(m+d-1)!}. \end{aligned} \tag{A12}$$

Next, it is possible to argue that the second term of (A10) is zero, since upon repeating the same exercise, we arrive at $\delta(n+t)$ instead of $\delta(1-n-t)$ as in the first line of (A12).

Therefore, the mth moment of p_j over the probability simplex is

$$\mathbb{E}_{|\rho\rangle}[p_j^m] = \frac{I_m}{I_0} = \binom{m+d-1}{m}^{-1}. \tag{A13}$$

In the regime of $d \gg m$, we then have $\mathbb{E}_{|\rho\rangle}[p_j^m] \approx \frac{\sqrt{2\pi m}}{(ed/m)^m} = O\left(\frac{1}{d^m}\right)$.

References

1. Wildfeuer, C.F.; Pearlman, A.J.; Chen, J.; Fan, J.; Migdall, A.; Dowling, J.P. Resolution and sensitivity of a Fabry-Perot interferometer with a photon-number-resolving detector. Quantum Limits in Optical Interferometry. *Phys. Rev. A* **2009**, *80*, 043822. [CrossRef]
2. Demkowicz-Dobrzański, R.; Jarzyna, M.; Kołodyński, J. Resolution and sensitivity of a Fabry-Perot interferometer with a photon-number-resolving detector. *Prog. Opt.* **2015**, *60*, 345.
3. Von Helversen, M.; Böhm, J.; Schmidt, M.; Gschrey, M.; Schulze, J.-H.; Strittmatter, A.; Rodt, S.; Beyer, J.; Heindel, T.; Reitzenstein, S. Quantum metrology of solid-state single-photon sources using photon-number-resolving detectors. *New J. Phys.* **2019**, *21*, 035007. [CrossRef]
4. Wu, J-.Y.; Toda, N.; Hofmann, H.F. Observation of squeezed states with strong photon-number oscillations. *Phys. Rev. A* **2009**, *100*, 013814.
5. Cattaneo, M.; Paris, M.G.A.; Olivares, S. Hybrid quantum key distribution using coherent states and photon-number-resolving detectors. *Phys. Rev. A* **2018**, *98*, 012333. [CrossRef]
6. Stucki, D.; Ribordy, G.; Stefanov, A.; Zbinden, H.; Rarity, J.G.; Wall, T. Photon counting for quantum key distribution with peltier cooled ingaas/inp apds. *J. Mod. Opt.* **2001**, *48*, 1967. [CrossRef]
7. Kilmer, T.; Guha, S. Boosting linear-optical Bell measurement success probability with predetection squeezing and imperfect photon-number-resolving detectors. *Phys. Rev. A* **2019**, *99*, 032302. [CrossRef]
8. Ren, M.; Wu, E.; Liang, Y.; Jian, Y.; Wu, G.; Zeng, H. Quantum random-number generator based on a photon-number-resolving detector. *Phys. Rev. A* **2011**, *83*, 023820. [CrossRef]
9. Applegate, M.J. Efficient and robust quantum random number generation by photon number detection. *Appl. Phys. Lett.* **2015**, *107*, 071106. [CrossRef]
10. Eisaman, M.D.; Fan, J.; Migdall, A.; Polyakov, S.V. Single-photon sources and detectors. *Rev. Sci. Instrum.* **2011**, *82*, 071101. [CrossRef]
11. Jönsson, M.; Björk, G. Evaluating the performance of photon-number-resolving detectors. *Phys. Rev. A* **2019**, *99*, 043822. [CrossRef]
12. Mirin, R.P.; Nam, S.W.; Itzler, M.A. Single-Photon and Photon-Number-Resolving Detectors. *IEEE Photonics J.* **2011**, *4*, 629. [CrossRef]
13. Marsili, F.; Bitauld, D.; Gaggero, A.; Jahanmirinejad, S.; Leoni, R.; Mattioli, F.; Fiore, A. Physics and application of photon number resolving detectors based on superconducting parallel nanowires. *New J. Phys.* **2009**, *11*, 045022. [CrossRef]
14. Marsili, F.; Najafi, F.; Dauler, E.; Bellei, F.; Hu, X.; Csete, M.; Molnar, R.J.; Berggren, K.K. Single-Photon Detectors Based on Ultranarrow Superconducting Nanowires. *Nano Lett.* **2011**, *11*, 2048. [CrossRef] [PubMed]
15. Jahanmirinejad, S.; Frucci, G.; Mattioli, F.; Sahin, D.; Gaggero, A.; Leoni, R.; Fiore, A. Photon-number resolving detector based on a series array of superconducting nanowires. *Appl. Phys. Lett.* **2012**, *101*, 072602. [CrossRef]
16. Matekole, E.S.; Vaidyanathan, D.; Arai, K.W.; Glasser, R.T.; Lee, H.; Dowling, J.P. Room-temperature photon-number-resolved detection using a two-mode squeezer. *Phys. Rev. A* **2017**, *96*, 053815. [CrossRef]
17. Ma, J.; Masoodian, S.; Starkey, D.A.; Fossum, E.R. Photon-number-resolving megapixel image sensor at room temperature without avalanche gain. *Optica* **2017**, *4*, 1474. [CrossRef]
18. Zolotov, P.; Divochiy, A.; Vakhtomin, Y.; Moshkova, M.; Morozov, P.; Seleznev, V.; Smirnov, K. Photon-number-resolving SSPDs with system detection efficiency over 50 at telecom range. *AIP Conf. Proc.* **2018**, *1936*, 020019.
19. Cai, Y.; Chen, Y.; Chen, X.; Ma, J.; Xu, G.; Wu, Y.; Xu, A.; Wu, E. Metamodelling for Design of Mechatronic and Cyber-Physical Systems. *Appl. Sci.* **2019**, *9*, 2638. [CrossRef]

20. Paul, H.; Törma, P.; Kiss, T.; Jex, I. Photon Chopping: New Way to Measure the Quantum State of Light. *Phys. Rev. Lett.* **1996**, *76*, 2464. [CrossRef]
21. Kok, P.; Braunstein, S.L. Detection devices in entanglement-based optical state preparation. *Phys. Rev. A* **2001**, *63*, 033812. [CrossRef]
22. Rohde, P.P.; Webb, J.G.; Huntington, E.H.; Ralph, T.C. Photon number projection using non-number-resolving detectors. *New J. Phys.* **2007**, *9*, 233. [CrossRef]
23. Řeháček, J.; Hradil, Z.; Haderka, O.; Peřina J., Jr.; Hamar, M. Multiple-photon resolving fiber-loop detector. *Phys. Rev. A* **2003**, *67*, 061801(R). [CrossRef]
24. Banaszek, K.; Walmsley, I.A. Fiber-assisted detection with photon number resolution. *Opt. Lett.* **2003**, *28*, 52. [CrossRef] [PubMed]
25. Fitch, M.J.; Jacobs, B.C.; Pittman, T.B.; Franson, J.D. Photon-number resolution using time-multiplexed single-photon detectors. *Phys. Rev. A* **2003**, *68*, 043814. [CrossRef]
26. Achilles, D.; Silberhorn, C.; Sliwa, C.; Banaszek, K.; Walmsley, I.A.; Fitch, M.J.; Jacobs, B.C.; Pittman, T.B.; Franson, J.D. Photon-number-resolving detection using time-multiplexing. *J. Mod. Opt.* **2004**, *51*, 1499. [CrossRef]
27. Avenhaus, M.; Laiho, K.; Chekhova, M.V.; Silberhorn, C. Accessing Higher Order Correlations in Quantum Optical States by Time Multiplexing. *Phys. Rev. Lett.* **2010** *104*, 063602. [CrossRef]
28. Kruse, R.; Tiedau, J.; Bartley, T.J.; Barkhofen, S.; Silberhorn, C. Limits of the time-multiplexed photon-counting method. *Phys. Rev. A* **2017**, *95*, 023815. [CrossRef]
29. Teo, Y.S. *Introduction to Quantum-State Estimation*; World Scientific Publishing Co.: Singapore, 2015.
30. Řeháček J.; Teo, Y.S.; Hradil, Z. Determining which quantum measurement performs better for state estimation. *Phys. Rev. A* **2015**, *92*, 012108. [CrossRef]

quantum reports

Article

Superposition Principle and Born's Rule in the Probability Representation of Quantum States †

Igor Ya. Doskoch and Margarita A. Man'ko *

Lebedev Physical Institute, Leninskii Prospect 53, Moscow 119991, Russia; iggor@lebedev.ru
* Correspondence: mankoma@lebedev.ru

† Based on the talk presented by Margarita A. Man'ko at the 16th International Conference on Squeezed States and Uncertainty Relations {ICSSUR} (Universidad Complutense de Madrid, Spain, 17–21 June 2019).

Received: 5 September 2019; Accepted: 20 September 2019; Published: 26 September 2019

Abstract: The basic notion of physical system states is different in classical statistical mechanics and in quantum mechanics. In classical mechanics, the particle system state is determined by its position and momentum; in the case of fluctuations, due to the motion in environment, it is determined by the probability density in the particle phase space. In quantum mechanics, the particle state is determined either by the wave function (state vector in the Hilbert space) or by the density operator. Recently, the tomographic-probability representation of quantum states was proposed, where the quantum system states were identified with fair probability distributions (tomograms). In view of the probability-distribution formalism of quantum mechanics, we formulate the superposition principle of wave functions as interference of qubit states expressed in terms of the nonlinear addition rule for the probabilities identified with the states. Additionally, we formulate the probability given by Born's rule in terms of symplectic tomographic probability distribution determining the photon states.

Keywords: entanglement; interference phenomenon; superposition of quantum states; quantum tomograms

---

1. Introduction

The superposition principle of quantum states, identified with the wave functions introduced by Schrödinger [1], is the fundamental property of quantum systems. Its formulation provides the possibility to associate with two wave functions corresponding to the two pure states of the system, the third function which is an arbitrary linear combination of the two wave functions. A property of the quantum world is that this combination also corresponds to a physical state of the system. An analogous formulation takes place for two state vectors in a Hilbert space and the linear combination of these vectors [2]. The foundation of quantum mechanics and the role of superposition principle were also discussed for quantum states [3,4], which are associated with the density matrices of density operators acting in the Hilbert space, introduced in [5,6]. The foundations of quantum optics were developed in connection with coherence properties of the electromagnetic-field states in [7,8] as well as in [9].

The aim of this paper is to demonstrate how such a phenomenon as interference of quantum states, described by the superposition principle of complex wave functions, is considered in the probability representation of quantum mechanics [10–21]. In the probability picture, the interference is described by the nonlinear addition rule of the probabilities describing the superposed states; the rule provides the probabilities describing the superposition state, and this corresponds to the nonlinear addition rule of the projectors. We demonstrate this rule on an example of the superposition of qubit states.

The superposition principle describes the interference phenomenon in quantum mechanics. Since the density operators $\hat{\rho}_{\psi_1} = \mid \psi_1\rangle\langle\psi_1 \mid$ and $\hat{\rho}_{\psi_2} = \mid \psi_2\rangle\langle\psi_2 \mid$ provide the expression for the density operator of superposition state $\mid \psi\rangle\langle\psi \mid$, where vector $\mid \psi\rangle$ is a linear combination of vectors $\mid \psi_1\rangle$ and $\mid \psi_2\rangle$, the superposition principle can be formulated as an addition rule of the probabilities (quantum-state tomograms). We provide explicitly an example of such addition rule for a particular system, namely, for the spin-1/2 particle or the two-level atom.

The other goal of this paper is to formulate Born's rule for probabilities given by an overlap of two state wave functions in the form of expression providing the probability $P_\psi(\varphi) = |\langle\psi \mid \varphi\rangle|^2$ as a function of probabilities determining the states given by the density operators $\hat{\rho}_\psi$ and $\hat{\rho}_\varphi$. The result obtained shows that such a complete quantum phenomenon as the interference of wave functions can be formulated by applying only classical tools like probabilities used in classical statistics.

This paper is organized as follows:

Section 2 is ad memoriam of Roy Glauber and George Sudarshan, where we give a short review of their fundamental results in quantum physics. In Section 3, we provide a brief review of the scientific results of our colleague Viktor Dodonov, in connection with his 70th anniversary. In Section 4, we discuss the qubit-state density matrix in the probability representation using classical-like probability distributions of dichotomic variables [22]. Section 5 is devoted to the superposition principle in the probability representation, while the quantum suprematism picture of qubit states is presented in Section 6 and Born's rule is formulated as a nonlinear addition rule of probabilities in Section 7. We discuss quantum tomography of continuous variables and quantum tomography of coherent states in Sections 8 and 9, respectively. Born's rule for oscillator states are considered in Section 10. Finally, our results and prospectives are given in Section 11.

2. Ad Memoriam of Roy Glauber and George Sudarshan

Last year we suffered great losses. Professor Roy J. Glauber and Professor E. C. George Sudarshan, founders of quantum optics, passed away in 2018. We are extremely sad about this. (Ad Memoriam of Roy Glauber and George Sudarshan was also presented in the talks of Margarita A. Man'ko at the 26th Central European Workshop on Quantum Optics {CEWQO} (Paderborn University, Germany, 3–7 June 2019) [23] and the 18th International Symposium "Symmetries in Sciences" (Gasthof Hotel Lamm, Bregenz, Vorarlberg, Austria, 4–9 August 2019) [24] and will be published in the *Journal of Russian Laser Research* (2019) and the *Journal of Physics: Conference Series* (2020), respectively.)

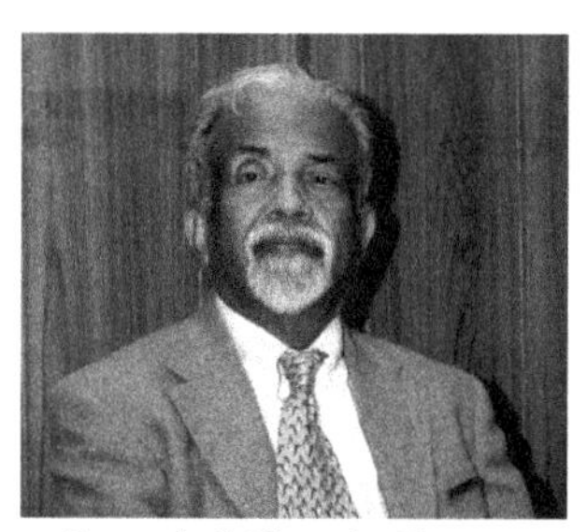

Ennackal Chandy George Sudarshan (Texas University) 1931–2018

These distinguished scientists made famous discoveries in the foundations of quantum physics that provide the possibility today to raise quantum optics and quantum mechanics to a level of understanding such that quantum technologies can now be developed. The operation of such devices as lasers is based on understanding the coherence properties of radiation and realizing how to achieve the conditions for obtaining such properties.

Roy Jay Glauber
(Harvard University)
1925–2018

In 1963, the notion of the coherent state of electromagnetic-field oscillations as well as the terminology "coherent state" of an arbitrary oscillator were introduced. Roy Glauber and George Sudarshan simultaneously published the papers [7,8] where the properties of coherent states were discussed. These publications are cited in the majority of papers where quantum optics and quantum information technologies are discussed.

The general linear positive map of the density matrix to the other density matrices for finite-dimensional systems was presented in the form [25], which later on was generalized in [26] for arbitrary systems. In addition, we should point out that the new evolution equation for open quantum systems, which generalizes the Schrödinger equation for the wave function and the von Neumann equation for the density matrix considered in the case of unitary evolution to the case of nonunitary evolution, was obtained in [27,28] and called the GKSL equation.

The pioneer scientific results obtained by these distinguished researchers play an important role in developing quantum optics. Studies of the properties of quantum states, especially the coherence and squeezing phenomena, correlations, unitary evolution, and the evolution of the electromagnetic fields in the presence of dissipation are based on the original results of Glauber and Sudarshan. All applications of quantum optics discussed in connection with the development of quantum technologies, motivated by the attempts to construct quantum computers and quantum information devices, are based on the theoretical notion and foundations of quantum mechanics associated with the results obtained by Glauber and Sudarshan.

A specific phase-space quasidistribution representation of quantum states, which employs the basis of coherent states, was introduced independently by Glauber and Sudarshan—it is called the Glauber–Sudarshan P-representation. This representation plays an important role in discussing the properties of quantum systems analogously to the Wigner quasidistribution function [29] and the Husimi–Kano quasidistribution [30,31]. These famous contributions play an important role in the foundations of quantum optics and quantum information, and in developing future quantum technologies.

Substantial developments in laser physics were made at the Lebedev Physical Institute in Moscow by Nikolay G. Basov, who won the Nobel Prize in 1964 together with Aleksandr M. Prokhorov and Charles H. Townes due to their revolutionary work in the invention of masers and lasers (as well as laser physics based on quantum mechanics and quantum optics). The official formulation reads that the Nobel Prize was given "for fundamental work in the field of quantum electronics, which has led to the construction of oscillators and amplifiers based on the maser–laser principle".

Roy J. Glauber received the Nobel Prize in 2005 "for his contribution to the quantum theory of optical coherence", along with John L. Hall and Theodor W. Hansch who received the Nobel Prize "for their contributions to the development of laser-based precision spectroscopy, including the optical frequency comb technique". Roy Glauber made an important contribution to the nuclear physics providing a rigorous analysis of the scattering theory. Additionally, the important results were obtained by Glauber in the theory of correlation functions in quantum optics. The correlation functions play a substantial role on the analysis of classical and quantum phenomena in radiation fields.

Professor Glauber exercised substantial influence on the development of quantum optics in European Scientific Centers—he would invite young researchers to Harvard University for collaborations, among which we can list Fritz Haake from Germany, Paolo Tombesi from Italy, Vladimir Man'ko from the Soviet Union, and Stig Stenholm from Finland.

Since the entire scientific life of M. A. Man'ko was connected with the Lebedev Physical Institute, she was a witness and participant in the collaboration of Roy Glauber with Lebedev scientists. The results of these collaborations were published in such series as the Proceedings of the Lebedev Institute [32] and Journal of Experimental and Theoretical Physics [33]. Among the four Nobel-prize papers, there was a paper by G. Schrade, V. I. Man'ko, W. P. Schleich, and R. Glauber [34], where the collaboration of Harvard University, the Lebedev Institute (Moscow), and University of Ulm was mentioned.

As for George Sudarshan, the great scientist, we are happy to recall that he was in Moscow many times due to the International Workshops on Squeezed States, Group Theoretical Colloquium, and the International Conference on Squeezed States and Uncertainty Relations. Particular mention should be made of the University Federico II of Naples and especially Professor Giuseppe Marmo who enabled many European scientists to collaborate with Professor Sudarshan in Italy.

There exist other representations, like the probability representation of quantum states [35–39], where states are identified with fair probability distributions. The state superposition is expressed in this representation using the addition rule for the probabilities. We are happy to let the readers of the Journal of Russian Laser Research know that famous results concerning the study of entanglement phenomenon, quantum tomography, and superposition principle in terms of density operators were published by Sudarshan with coauthors in [3,4,40,41].

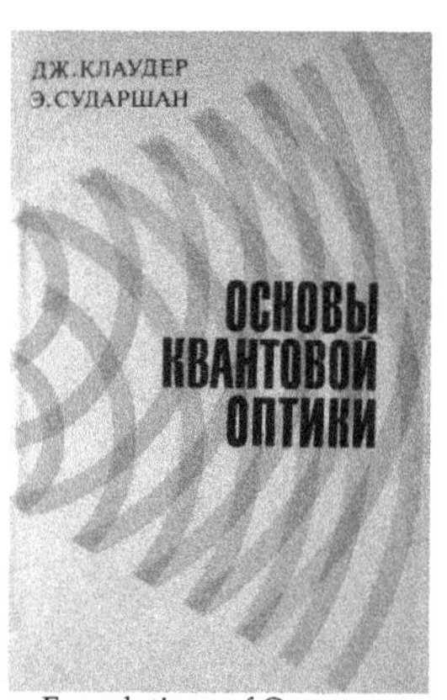

Foundations of Quantum Optics – Russian edition.

The famous book—*Foundations of Quantum Optics* [9], written by John Klauder and George Sudarshan and translated into Russian—is an excellent textbook on the foundation of quantum theory; it is used in all universities of the world, including universities in Russia, by students and professors to obtain basic knowledge in developing new quantum technologies in the future.

The results of Glauber and Sudarshan mentioned above provided the theoretical basis in studies of the evolution of open quantum systems, theory of quantum channels, and applications of these approaches in future quantum technologies. Therefore, we are extremely pleased to be able to witness these deep international connections today, and we show below a kaleidoscope of pictures taken recently and a long time ago connected with participants of the ICSSUR series (with Professor Young Suh Kim, the Founder) and the CEWQO series (with Professor Jozsef Janszky, the Founder), as well as some other meetings.

3. Professor Viktor V. Dodonov: On the Occasion of His 70th Birthday

On 26 November 2018, Viktor V. Dodonov, a recognized Russian physicist and current professor at the National University of Brazil (Brazilia), turned 70. In addition to his University academic activity, Professor Dodonov is an Editorial board member of our Journal of Russian Laser Research published by Springer in New York, and we celebrate this date with all our friends and colleagues, see also [42].

Professor Viktor V. Dodonov—in addition to his scientific activity and obtaining many interesting results in foundations and applications of quantum theory—participated together with us during all his scientific life in organizing conferences, workshops, and publications of the Proceedings of the conferences, and is the Editorial Board member of Journal of Russian Laser Research. Together with Vladimir Man'ko, he has issued the book *Invariants and the Evolution of Nonstationary Quantum Systems*, published by Nauka in Moscow as volume No. 183 of the *Lebedev Institute Proceedings* and translated into English by Nova Science Publishers, Commack, New York in 1983 [43], where many of his results on time-dependent integrals of motion, parametric oscillators, even and odd coherent states, and general theory of uncertainty relations were presented.

Professor Dodonov is the coauthor of the paper [44] where developing the approach of Glauber and Sudarshan for describing the superposition of coherent states modeling Schrödinger cat states, and where the even and odd coherent states were introduced. He is also the coauthor of papers [45,46], where the idea of experimental checking of the *nonstationary Casimir effect* (later on called in the literature the *dynamical Casimir effect*) using the devices based on Josephson junctions was suggested. The photons created due to the dynamical Casimir effect were predicted to be in squeezed states. Experimental results demonstrating the existence of this effect were obtained in [47].

A substantial contribution of foundations of quantum mechanics was done by Professor Dodonov, who extended the approach for analyzing the uncertainty relations [48,49]. The new kinds of uncertainty relations for different physical observables including the position and momentum, spin variables, as well as entropic inequalities play an important role in discussing quantum-information devices and quantum-channel technique.

In times of the Soviet Union, Viktor Dodonov together with the Lebedev people organized several international meetings in Moscow, Moscow Region (Zvenigorod), Urmala (Latvia), Baky (Azerbaijan), Tambov (his native place), and so on. Scientific Programs of the conferences and publication of the Proceedings were always a prerogative of Dodonov—he did this job perfectly.

We are absolutely happy to have Professor Dodonov as the Editorial Board Member of Journal of Russian Laser Research, as permanent referee of Physica Scripta, and hope that he will help us to produce referee's estimation of contributions to Quantum Reports, where we announced to publish the Proceedings of the 16th ICSSUR.

We wish Viktor V. Dodonov long scientific activity and continuation of his work with us!

G. Sudarshan (1), Y. S. Kim (2), F. Haake (3), and M. Nieto (4) among the participants of ICSSUR 1992 in Moscow.

P. Tombesi, M. A. Man'ko, R. J. Glauber, Y. S. Kim, and D. Han at ICSSUR 1992 in Moscow.

M. A. Man'ko and V. I. Man'ko visited E. Wigner in his house in Princeton, New Jersey in 1990.

Y. S. Kim, M. A. Man'ko, Mrs. Kim, and V. I. Man'ko at ICSSUR 1991 in College Park, Maryland.

E.S. Fradkin, V.I. Man'ko, F. Gürsey, M.A. Markov, F. Iachello, and A. Bohm at Group Theory Colloquium 1990 in Moscow.

V. I. Man'ko, M. A. Man'ko, and Y. S. Kim at ICSSUR 1992 in Moscow.

V. A. Isakov, M. A. Man'ko, O. V. Man'ko, and A. S. Chirkin at ICSSUR 2001 in Boston, Massachusetts.

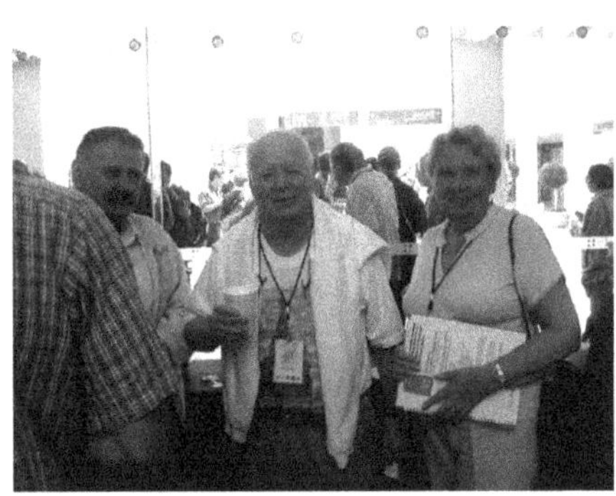

J. Janszky, J. Klauder, and M. A. Man'ko at ICSSUR 2003 in Puebla, Mexico.

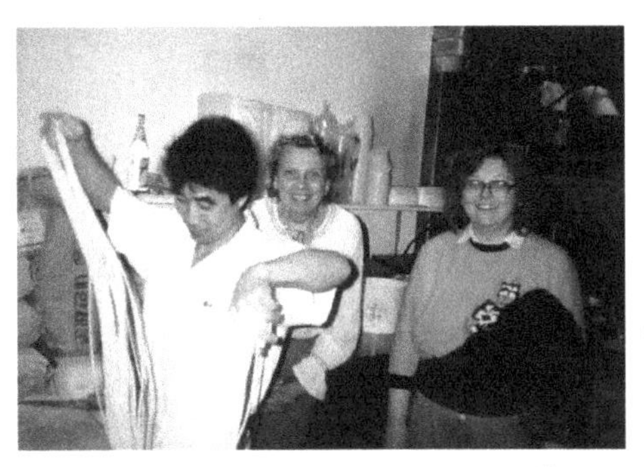

M. A. Man'ko and O. V. Man'ko at Harmonic Oscillators Workshop 1992 in College Park, Maryland.

V. I. Man'ko (1), M. A. Man'ko (2), R. Kerner (3), and A. Solomon (4) at Quantum Theory and Symmetries 2011 in Prague.

O.V. Man'ko (1), S. Biedenharn (3), Mrs. Kim (3), M.A. Man'ko (4), and E. Moshinsky (5) at ICSSUR 1992 in Moscow.

Y. S. Kim, V. I. Man'ko, and M. A. Man'ko at Group Theory Colloquium 1992 in Salamanca, Spain.

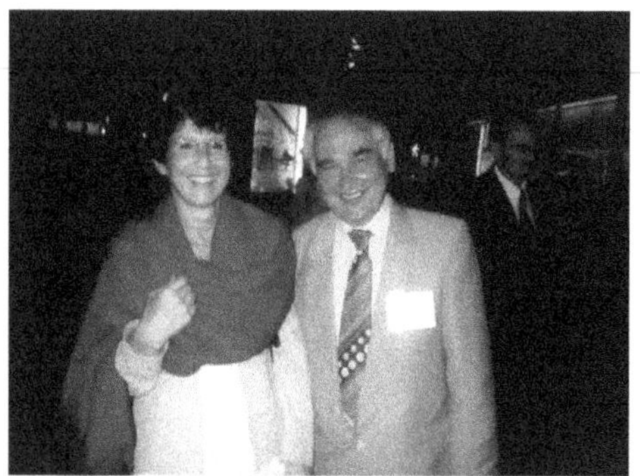

E. Giacobino and V. I. Man'ko
at Group Theory Colloquium 2002 in Paris.

Y. S. Kim (1), J. Janszky (2), and V. I. Man'ko (3)
at ICSSUR 2003 in Puebla, Mexico.

ICSSUR 2003 in Puebla, Mexico.

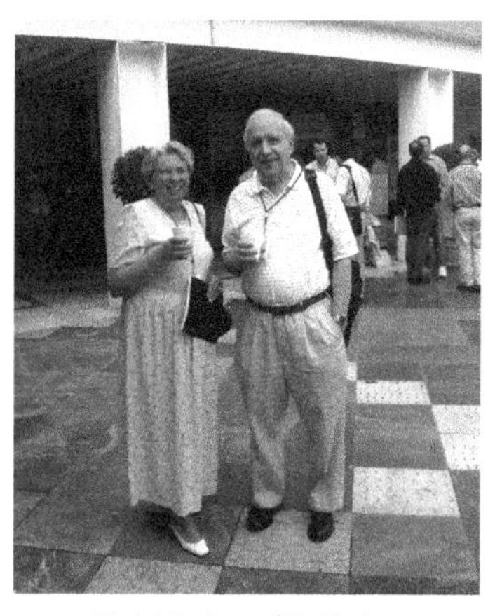

M. A. Man'ko and R. Glauber
at ICSSUR 2003 in Puebla, Mexico.

S. Mizrahi, L. Dodonova, V. V. Dodonov, and M. A. Man'ko
at ICSSUR 2005 in Besancon, France.

ICSSUR 2005 in Besancon, France.

W. Schleich (1), T. Kramer (2), M. Kleber (3), P. Kramer (4), M.A. Man'ko (5), V.I. Man'ko (6), V.V. Dodonov (7), and M. Scully (8) among the participants of Quantum Nonstationary Systems 2007 in Blaubeuren, Germany.

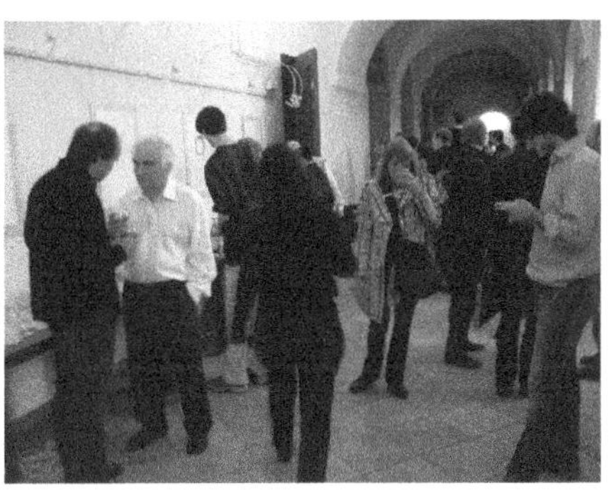

W. Vogel and V. I. Man'ko among the participants of ICSSUR 2009 in Olomouc, Czech Republic.

A. Messina, M. A. Man'ko, V. V. Dodonov, V. I. Man'ko, Y. S. Kim, D. M. Gitman, and B. Militello at ICSSUR 2011 in Foz do Iguacu, Brazil.

V. I. Man'ko and V. V. Dodonov at ICSSUR 2011 in Foz do Iguacu, Brazil.

ICSSUR 2011 in Foz do Iguacu, Brazil.

V. I. Man'ko, M. A. Man'ko, and S. Mizrahi at ICSSUR 2011 in Foz do Iguacu, Brazil.

A. Solomon, V. V. Dodonov, Mrs. Kim,
and Y. S. Kim at ICSSUR 2011 in Foz do Iguacu, Brazil.

A. Vourdas and A. Solomon
at ICSSUR 2011 in Foz do Iguacu, Brazil.

A. Messina, L. Dodonova, V. I. Man'ko, and
B. Militello at ICSSUR 2011 in Foz do Iguacu, Brazil.

O. V. Man'ko (1), Y. S. Kim (2), and S. Mizrahi (3)
with the participants of ICSSUR 2011 in Foz do Iguacu, Brazil.

R. J. Glauber and M. A. Man'ko
at ICSSUR 1992 in Moscow.

O. V. Man'ko and V. I. Man'ko
at ICSSUR 2003 in Puebla, Mexico.

A. Zeilinger, H. Rauch, and S. Stenholm
at CEWQO 2009 in Turku, Finland.

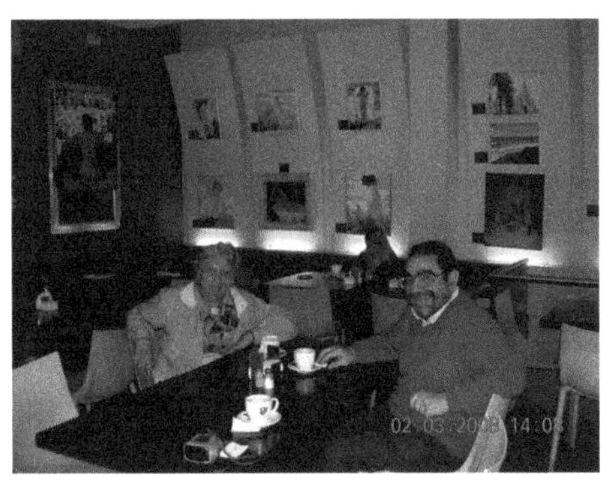

M. A. Man'ko and G. Marmo
at Universidad Carlos III de Madrid in 2008.

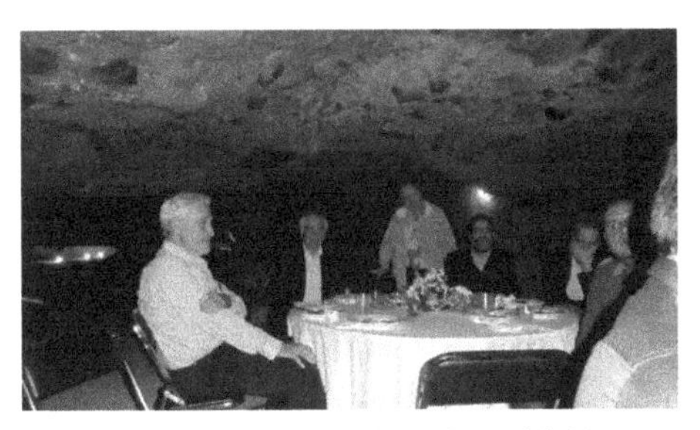

T. Seligman, V. I. Man'ko, M. A. Man'ko, and G. Marmo
at *Problems of Mathematical and Quantum Physics*
(75 + 75 years of Margarita and Vladimir Man'ko),
Cuernavaca, Morelos, Mexico, 2015.

V. V. Dodonov, L. Dodonova, V. I. Man'ko, M. A. Man'ko,
M. K. Atakishiyeva, and N. M. Atakishiyev
at *Problems of Mathematical and Quantum Physics*
(75 + 75 years of Margarita and Vladimir Man'ko),
Cuernavaca, Morelos, Mexico, 2015.

4. Qubit States in the Probability Representation

The density matrix of the spin-1/2 state is 2×2 matrix $\rho_{mm'}$; for spin projections on the z axis $m, m' = \pm 1/2$, it reads

$$\rho = \begin{pmatrix} \rho_{1/2,1/2} & \rho_{1/2,-1/2} \\ \rho_{-1/2,1/2} & \rho_{-1/2,-1/2} \end{pmatrix}, \tag{1}$$

where $1 \geq \rho_{1/2,1/2}, \rho_{-1/2,-1/2} \geq 0$ are probabilities to have spin projections $+1/2$ and $-1/2$ on the z axis. Off-diagonal matrix elements $\rho_{1/2,-1/2} = \rho^*_{-1/2,1/2}$ provide the condition of hermiticity of the density matrix $\rho^\dagger = \rho$. The non-negativity of the eigenvalues of this matrix gives $\det \rho \geq 0$, i.e., $\rho_{1/2,1/2} \cdot \rho_{-1/2,-1/2} - |\rho_{1/2,-1/2}|^2 \geq 0$.

Recently [50,51], the probability representation of the density matrix was found to be

$$\rho = \begin{pmatrix} p_3 & p_1 - (1/2) - i(p_2 - 1/2) \\ p_1 - (1/2) + i(p_2 - 1/2) & 1 - p_3 \end{pmatrix}. \tag{2}$$

One can check that the number $p_3 = \rho_{11}$, $1 \geq p_3 \geq 0$, is the probability to have the spin projection on the z axis equal to $+1/2$ in the state with the density matrix ρ. This means that, in view of Born's rule, the probability p_3 reads

$$p_3 = \mathrm{Tr}\left[\begin{pmatrix} \rho_{1/2,1/2} & \rho_{1/2,-1/2} \\ \rho_{-1/2,1/2} & \rho_{-1/2,-1/2} \end{pmatrix} \rho_3\right]. \tag{3}$$

The state with the density matrix $\rho_3 = \begin{pmatrix} 1 & 0 \\ 0 & 0 \end{pmatrix}$ is the pure state $\begin{pmatrix} 1 \\ 0 \end{pmatrix} (1\ 0) = \rho_3$ with the state vector $\mid +1/2\rangle_z = \begin{pmatrix} 1 \\ 0 \end{pmatrix}$, which is the eigenstate of the Pauli matrix $\sigma_z = \begin{pmatrix} 1 & 0 \\ 0 & -1 \end{pmatrix}$ determining the spin-projection operator $s_z = \sigma_z/2$ on the z-axis.

The number p_1 is given by an analogous relation

$$p_1 = \mathrm{Tr}\left[\begin{pmatrix} \rho_{1/2,1/2} & \rho_{1/2,-1/2} \\ \rho_{-1/2,1/2} & \rho_{-1/2,-1/2} \end{pmatrix} \rho_1\right], \tag{4}$$

where the density matrix $\rho_1 = \begin{pmatrix} 1/2 & 1/2 \\ 1/2 & 1/2 \end{pmatrix} = \begin{pmatrix} 1\sqrt{2} \\ 1/\sqrt{2} \end{pmatrix} (1/\sqrt{2}\, 1/\sqrt{2})$, with the state vector $\mid +1/2\rangle_x = \begin{pmatrix} 1/\sqrt{2} \\ 1/\sqrt{2} \end{pmatrix}$. It is the eigenvector of the Pauli matrix $\sigma_x = \begin{pmatrix} 0 & 1 \\ 1 & 0 \end{pmatrix}$ determining the spin-projection operator $s_x = \sigma_x/2$ on the x-axis.

One can easily check that, due to Born's rule, the number

$$p_2 = \mathrm{Tr}\left[\begin{pmatrix} \rho_{1/2,1/2} & \rho_{1/2,-1/2} \\ \rho_{-1/2,1/2} & \rho_{-1/2,-1/2} \end{pmatrix} \rho_2\right] \tag{5}$$

is the probability to have the spin projection on the y-axis equal to $+1/2$ in the state with the density matrix $\rho_2 = \begin{pmatrix} 1/2 & -i/2 \\ i/2 & 1/2 \end{pmatrix}$, since $\begin{pmatrix} 1/2 & -i/2 \\ i/2 & 1/2 \end{pmatrix} = \begin{pmatrix} 1/\sqrt{2} \\ i/\sqrt{2} \end{pmatrix} (1/\sqrt{2}\ -i/\sqrt{2})$, and the state vector $\mid +1/2\rangle_y = \begin{pmatrix} 1/\sqrt{2} \\ i/\sqrt{2} \end{pmatrix}$ is the eigenvector of the Pauli matrix $\sigma_y = \begin{pmatrix} 0 & -i \\ i & 0 \end{pmatrix}$ determining the spin-projection operator $s_y = \sigma_y/2$ on the y-axis.

The state of spin equal to $1/2$ can be determined by three probability distributions $(p_1,\ 1-p_1)$, $(p_2,\ 1-p_2)$, and $(p_3,\ 1-p_3)$, which can be considered as probabilities for three different nonideal classical coins in such a game as coin flipping, coin tossing, or heads or tails, which is the practice of throwing a coin in the air and checking which side is showing when it lands, in order to choose between two alternatives p_k or $(1-p_k)$; $k = 1,2,3$.

For quantum states, there exist quantum correlations given by the condition $\det\rho \geq 0$ or

$$(p_1 - 1/2)^2 + (p_2 - 1/2)^2 + (p_3 - 1/2)^2 \leq 1/4.$$

This condition does not take place for classical coins.

5. Pure States of Qubits and Their Superposition

The classical condition for these probabilities, e.g., if $p_1 = p_2 = p_3 = 1$, reads

$$(p_1 - 1/2)^2 + (p_2 - 1/2)^2 + (p_3 - 1/2)^2 \leq 3/4. \tag{6}$$

One can check that the density matrix of the pure state $\mid \psi\rangle$, providing the density matrix

$$\rho_\psi = \mid \psi\rangle\langle\psi \mid \tag{7}$$

of the form (2), where

$$| \psi\rangle = \begin{pmatrix} \sqrt{p_3} \\ \dfrac{p_1 - 1/2}{\sqrt{p_3}} + i\,\dfrac{p_2 - 1/2}{\sqrt{p_3}} \end{pmatrix} \tag{8}$$

represents a pure state if and only if

$$(p_1 - 1/2)^2 + (p_2 - 1/2)^2 + (p_3 - 1/2)^2 = 1/4. \tag{9}$$

The inequalities

$$1/4 < (p_1 - 1/2)^2 + (p_2 - 1/2)^2 + (p_3 - 1/2)^2 \leq 3/4 \tag{10}$$

are forbidden for "quantum" coins or for spin-1/2 projection probabilities in the state with any density matrix ρ.

Thus, we show that the pure state of spin equal to 1/2 is determined by three probabilities $0 \leq p_1, p_2, p_3 \leq 1$ to have spin projection $+1/2$ on the three directions x, y, and z, respectively. This means that all properties of spin-1/2 states, including the superposition principle, can be formulated in terms of these probabilities.

In view of gauge invariance of wave functions, we assume the first component of the Pauli spinor $| \psi\rangle$ to be a real and non-negative number; the second component can be given as a complex number $\sqrt{1 - p_3}\, e^{i\varphi}$, where the phase φ is the function of the probabilities, i.e,

$$\cos\varphi = \frac{p_1 - 1/2}{\sqrt{p_3(1 - p_3)}}, \qquad \sin\varphi = \frac{p_2 - 1/2}{\sqrt{p_3(1 - p_3)}}. \tag{11}$$

We point out that new trigonometric formulas (11) are expressed in terms of probabilities [50] describing dichotomic random variables.

Thus, for two arbitrary pure spin states $| \psi_1\rangle$ and $| \psi_2\rangle$, the superposition principle means the following:

The Pauli spinors $| \psi\rangle = C_1 \mid \psi_1\rangle + C_2 \mid \psi_2\rangle$, such that $\langle\psi \mid \psi\rangle = 1$ and $\langle\psi_1 \mid \psi_2\rangle = 0$, the numbers $C_1 = \sqrt{\Pi_3}$ and $C_2 = \dfrac{\Pi_1 - 1/2}{\sqrt{\Pi_3}} + \dfrac{i(\Pi_2 - 1/2)}{\sqrt{\Pi_3}}$ are expressed in terms of probabilities $0 \leq \Pi_1, \Pi_2, \Pi_3 \leq 1$ satisfying the equality $(\Pi_1 - 1/2)^2 + (\Pi_2 - 1/2)^2 = \Pi_3(1 - \Pi_3)$; they also have the form

$$| \psi\rangle = \begin{pmatrix} \sqrt{P_3} \\ \dfrac{P_1 - 1/2}{\sqrt{P_3}} + \dfrac{i(P_2 - 1/2)}{\sqrt{P_3}} \end{pmatrix} e^{i\varphi}. \tag{12}$$

Due to gauge invariance of quantum-state vectors, these spinors are determined up to the phase factors by the three probabilities P_1, P_2, and P_3. The density matrix $\rho_\psi = \dfrac{| \psi\rangle\langle\psi |}{\langle\psi \mid \psi\rangle}$ expressed in terms of the numbers P_1, P_2, and P_3, provides the addition rule of the probabilities p_1, p_2, p_3 and $\mathcal{P}_1, \mathcal{P}_2, \mathcal{P}_3$, determining the nonorthogonal vectors $| \psi_1\rangle$ and $| \psi_2\rangle$ yielding the explicit relations

$$P_3 = (1/\mathcal{T}) \left\{\Pi_3 p_3 + (1 - \Pi_3)\mathcal{P}_3 + 2\sqrt{p_3\mathcal{P}_3}\,(\Pi_1 - 1/2)\right\}, \tag{13}$$

$$\begin{aligned} P_1 - 1/2 = (1/\mathcal{T})\Big\{&\Pi_3(p_1 - 1/2) + (\mathcal{P}_1 - 1/2)(1 - \Pi_3) \\ &+ [(\Pi_1 - 1/2)(p_1 - 1/2) + (\Pi_2 - 1/2)(p_2 - 1/2)]\,\sqrt{\mathcal{P}_3/p_3} \\ &+ [(\Pi_1 - 1/2)(\mathcal{P}_1 - 1/2) - (\Pi_2 - 1/2)(\mathcal{P}_2 - 1/2)]\,\sqrt{p_3/\mathcal{P}_3}\Big\}, \end{aligned} \tag{14}$$

where

$$\mathcal{T}=1+\frac{2}{\sqrt{p_3\mathcal{P}_3}}\{(\Pi_1-1/2)\left[(p_1-1/2)(\mathcal{P}_1-1/2)+(\mathcal{P}_2-1/2)(p_2-1/2)+p_3\mathcal{P}_3\right] \\ +(\Pi_2-1/2)\left[(p_2-1/2)(\mathcal{P}_1-1/2)-(p_1-1/2)(\mathcal{P}_2-1/2)\right]\}. \quad (15)$$

Equations (13)–(15) provide the addition rule for probabilities determining the pure states $\mid\psi_1\rangle$ and $\mid\psi_2\rangle$ in the general case where these states are not orthogonal. Here, we use the same parameters Π_1, Π_2, and Π_3 defining the complex numbers C_1 and C_2.

At $\langle\psi_1\mid\psi_2\rangle=0$, the above number $\mathcal{T}=1$. The relative phase χ in the superposition of states $\mid\psi_1\rangle$ and $\mid\psi_2\rangle$ is determined by the phase of the complex number C_2, $\cos\chi=\dfrac{\Pi_1-1/2}{\sqrt{\Pi_3(1-\Pi_3)}}$. The obtained rule of probability addition corresponds to the formula for superposition principle obtained in [3]—for $\hat{\rho}_{\psi_1}\hat{\rho}_{\psi_2}=0$, it reads

$$\hat{\rho}_\psi=\lambda_1\hat{\rho}_{\psi_1}+\lambda_2\hat{\rho}_{\psi_2}+\sqrt{\lambda_1\lambda_2}\,\frac{\hat{\rho}_{\psi_1}\hat{\rho}_0\hat{\rho}_{\psi_2}+\hat{\rho}_{\psi_2}\hat{\rho}_0\hat{\rho}_{\psi_1}}{\sqrt{\operatorname{Tr}\rho_{\psi_1}\rho_0\rho_{\psi_2}\rho_0}}, \quad (16)$$

where $\hat{\rho}_0$ is an arbitrary projector, and $0\geq\lambda_1,\,\lambda_2\geq1$ are the probabilities such that $\lambda_1+\lambda_2=1$.

Taking the trace of Equation (16) yields

$$\operatorname{Tr}\left\{\hat{\rho}_{\psi_1}\hat{\rho}_0\hat{\rho}_{\psi_2}+\hat{\rho}_{\psi_2}\hat{\rho}_0\hat{\rho}_{\psi_1}\right\}=0, \quad (17)$$

which does not hold in general, but holds if $\langle\psi_1\mid\psi_2\rangle=0$.

Thus, we see that pure and mixed states of the spin-1/2 system can be identified with three probability distributions $(p_1,1-p_1)$, $(p_2,1-p_2)$, and $(p_3,1-p_3)$.

Assume that we have three nonideal classical coins in such a game as coin flipping, coin tossing, or heads ("UP") or tails ("DOWN"), which is the practice of throwing a coin in the air and checking which side is showing when it lands, in order to choose between three alternatives p_k or $(1-p_k)$; $k=1,2,3$. We interpret the numbers p_1, p_2, and p_3 as "coin" probabilities to be in the position "UP" for each coin, and $(1-p_1)$, $(1-p_2)$, and $(1-p_3)$, to be in the position "DOWN".

For pure states, there is the rule of nonlinear addition of the distributions that is equivalent to the superposition of Pauli spinors. This rule can be illustrated by addition of two Triadas of Malevich's squares determined by "coin" probabilities p_1, p_2, p_3, $\mathcal{P}_1$, $\mathcal{P}_2$, $\mathcal{P}_3$, and Π_1, Π_2, Π_3, and providing the probabilities P_1, P_2, P_3 as a result. This rule can be illustrated by the formula for adding the probabilities

$$\begin{pmatrix}p_1\\p_2\\p_3\end{pmatrix}\oplus_{\vec{\Pi}}\begin{pmatrix}\mathcal{P}_1\\\mathcal{P}_2\\\mathcal{P}_3\end{pmatrix}=\begin{pmatrix}P_1\\P_2\\P_3\end{pmatrix}, \quad (18)$$

where symbol $\oplus_{\vec{\Pi}}$ of the addition means that we use the probabilities Π_1, Π_2, and Π_3 associated with complex numbers C_1 and C_2 determining the superposition $\mid\psi\rangle=C_1\mid\psi_1\rangle+C_2\mid\psi_2\rangle$ of states $\mid\psi_1\rangle$ and $\mid\psi_2\rangle$.

6. Illustration of the Qubit State by Triada of Malevich's Squares

Now, we explain how to illustrate qubit states by Triada of Malevich's squares [50,52–56].

Given the triangle $A_1A_2A_3$ with sides y_k; $k=1,2,3$,

$$y_k=\left(2+2p_k^2-6p_k+2p_{k+1}^2+2p_kp_{k+1}\right)^{1/2}; \quad (19)$$

in the above equation, the $k+1$ addition is modulo 3. We construct three squares with sides y_k associated with triangle $A_1A_2A_3$ as shown in Figure 1 and called *Triada of Malevich's squares*. The sum of the areas of these three squares is expressed in terms of the three probabilities p_k as

$$S = y_1^2 + y_2^2 + y_3^2 = 2\left[3\,(1 - p_1 - p_2 - p_3) + 2p_1^2 + 2p_2^2 + 2p_3^2 + p_1p_2 + p_2p_3 + p_3p_1\right]. \tag{20}$$

The three squares constructed, using the sides of the triangle, are analogs of the *Triada of Malevich's squares* in art. The properties of area S associated with the triada, given by Equation (20), are different for the classical system states and for the quantum system states, namely, for three classical coins and for qubit states.

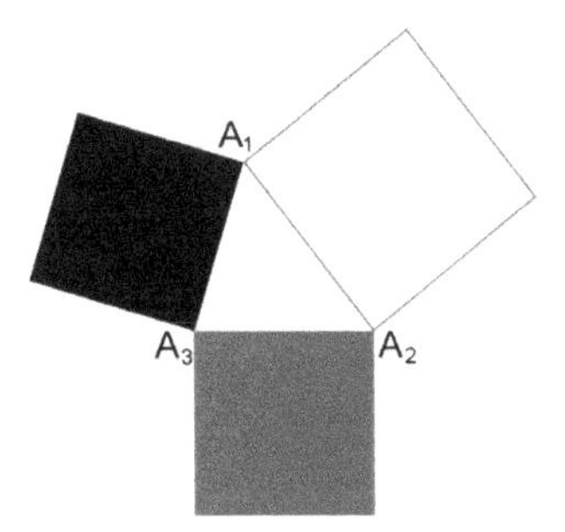

Figure 1. *Triada of Malevich's squares* corresponding to the spin-1/2 state and determined by the triangle $A_1A_2A_3$.

For classical coins, the numbers p_1, p_2, and p_3 take any values in the domains $0 \leq p_k \leq 1$—this means that for statistics of classical coins the area of the *Triada of Malevich's squares* satisfies the inequality $3/2 \leq S \leq 6$.

For qubit states, the probabilities $0 \leq p_k \leq 1$ to have spin projections $m = +1/2$ along three orthogonal directions satisfy the non-negativity condition of the density matrix—this provides the maximum value of the sum of areas of three Malevich's squares $S = 3$.

The formulas obtained correspond to the addition rule for two *Triadas of Malevich's squares*, illustrated by Figure 2.

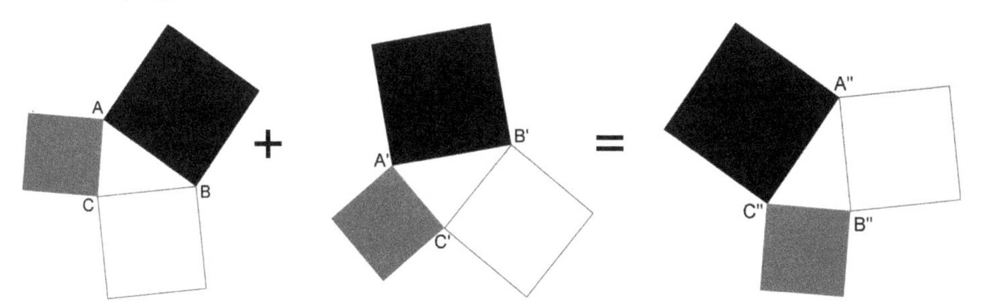

Figure 2. The superposition principle for two pure spin-1/2 states as a result of addition of two *Triadas of Malevich's squares*, which yields the *Triada of Malevich's squares* associated with the addition rule.

7. Born's Rule for Qubits as a Quadratic Form of Probabilities

Born's rule provides the probability w_{12} to obtain the properties of the state with the density matrix ρ_2

$$\rho_2 = \begin{pmatrix} p_3^{(2)} & p_1^{(2)} - (1/2) - i(p_2^{(2)} - 1/2) \\ p_1^{(2)} - (1/2) + i(p_2^{(2)} - 1/2) & 1 - p_3^{(2)} \end{pmatrix}, \tag{21}$$

if one measures these properties in the state with the density matrix ρ_1

$$\rho_1 = \begin{pmatrix} p_3^{(1)} & p_1^{(1)} - (1/2) - i(p_2^{(1)} - 1/2) \\ p_1^{(1)} - (1/2) + i(p_2^{(1)} - 1/2) & 1 - p_3^{(1)} \end{pmatrix}. \tag{22}$$

According to Born's rule $w_{12} = \text{Tr}\,(\rho_2\rho_1)$, we have the equality $w_{12} = w_{21}$.

Thus, the probabilities $p_1^{(k)}, p_2^{(k)}, p_3^{(k)}; k = 1,2,3$, due to Born's rule, provide the probability w_{12} which, as the function of probabilities $p_1^{(1)}, p_2^{(1)}, p_3^{(1)}, p_1^{(2)}, p_2^{(2)}$, and $p_3^{(2)}$, reads

$$w_{12} = p_3^{(1)} p_3^{(2)} + \left(1 - p_3^{(1)}\right)\left(1 - p_3^{(2)}\right) + 2\left[\left(p_1^{(1)} - 1/2\right)\left(p_1^{(2)} - 1/2\right) + \left(p_2^{(1)} - 1/2\right)\left(p_2^{(2)} - 1/2\right)\right],$$

i.e., we expressed the probability w_{12} given by Born's rule in terms of probabilities determining two states of qubit. It is worth noting that the above formula provides the probability given by a quadratic form of other probabilities; such quadratic forms were not considered in the literature related to probability theory. Such expression of Born's probability w_{12} is a generic property of qubit and qudit states, which can be given by the quadratic form, we will prove this in a future publication.

8. Probability in Quantum Mechanics and Quantum Optics for States with Continuous Variables

Now, we formulate, for continuous variables, the superposition principle of quantum-state wave functions in the probability representation of quantum mechanics, where the states are identified with probability distributions.

In standard formulation of quantum mechanics, the pure states are associated with complex wave functions $\psi(x) = |\psi(x)| \exp[i\varphi(x)]$. The mixed states are associated with density matrices $\rho(x,x')$, which are Hermitian matrices $\rho = \rho^\dagger$ with unit trace $\text{Tr}\,\rho = 1$ and non-negative eigenvalues.

Recently [35,37,38,57], the probability representation of quantum states for both continuous variables like quantum-oscillator states and for discrete variables, like spin states or N-level atom states, was introduced. The review and development of the tomographic-probability distribution of quantum states are given in [39,51,58,59]. On the other hand, the classical-particle states are determined either by their position and momentum (if there is no fluctuations) or by the probability-distribution functions $f(q,p) \geq 0$, such that $\int f(p,q)\,dp\,dq = 1$ in the case of presence of fluctuations like, e.g., thermal fluctuations.

In quantum mechanics, there exist different quasiprobability distribution representations of quantum states like the Wigner function $W(q,p)$ [29], which is similar to the classical probability distribution $f(q,p)$ in the phase space, but it can take negative values and, due to this fact, it is not the probability distribution.

Nevertheless, there exists the bijective map of the density matrices (density operators $\hat{\rho}$) onto fair probability distributions both for continuous variables [37] given by the Radon transform of the Wigner function and determined by the trace of the product of operator $\hat{\rho}$ and Dirac delta-function of a specific operator, it is

$$w(X,\mu,\nu) = \text{Tr}\left[\hat{\rho}\,\delta\left(X\hat{\mathbf{1}} - \mu\hat{q} - \nu\hat{p}\right)\right]. \tag{23}$$

Operators $\hat{q}$ and $\hat{p}$ are the position and momentum operators; $\hat{\mathbf{1}}$ is the identity operator. Here, the Dirac delta-function of operator $(X\hat{\mathbf{1}} - \mu\hat{q} - \nu\hat{p})$ is given by $\delta\left(X\hat{\mathbf{1}} - \mu\hat{q} - \nu\hat{p}\right) = \frac{1}{2\pi}\int e^{ik(X\hat{\mathbf{1}} - \mu\hat{q} - \nu\hat{p})}dk$, the Fourier transform with $-\infty \leq X \leq \infty$ being photon quadrature (oscillator's position) and μ and ν, real parameters determining the reference frame in the phase space where the oscillator's position X is considered.

For $\mu = \cos\theta$ and $\nu = \sin\theta$, the probability distribution of variable X is called the optical tomogram of quantum radiation state. The inverse transform reads

$$\hat{\rho} = \frac{1}{2\pi}\int w(X,\mu,\nu)\exp\left[i\left(X\hat{\mathbf{1}} - \mu\hat{q} - \nu\hat{p}\right)\right] dX\,d\mu\,d\nu. \tag{24}$$

Thus, the state (photon state, oscillator's state) can be described either by the density operator $\hat{\rho}$ or by the tomographic-probability distribution $w(X,\mu,\nu)$, since they contain the same information on the state.

The superposition principle of the wave functions means that, for two given wave functions $\psi_1(x)$ and $\psi_2(x)$ describing two different pure states $\mid\psi_1\rangle$ and $\mid\psi_2\rangle$, the wave function reads $\psi(x) = C_1\psi_1(x) + C_2\psi_2(x)$ and the state vector is $\mid\psi\rangle = C_1 \mid\psi_1\rangle + C_2 \mid\psi_2\rangle$, where C_1 and C_2 are complex coefficients, also describes the physical system state.

9. Coherent State Superposition

We consider nonlinear addition of tomographic-probability distributions corresponding to the superposition of coherent states $\mid\alpha\rangle$, i.e., normalized eigenstates of the oscillator annihilation operator $\hat{a}$, $(\hat{a}\mid\alpha\rangle = \alpha\mid\alpha\rangle)$ introduced and studied in quantum optics by Roy Glauber and George Sudarshan [7,8]. The wave function of the coherent state in the position representation reads

$$\psi_\alpha(x) = \pi^{-1/4}\exp\left(-\frac{x^2}{2} - \frac{|\alpha|^2}{2} - \frac{\alpha^2}{2} + \sqrt{2}\alpha x\right); \qquad \text{we assume} \qquad \hbar = m = \omega = 1. \tag{25}$$

The tomogram of the state with the wave function $\psi(y)$ is given by its fractional Fourier transform [60]

$$w(X,\mu,\nu) = \frac{1}{2\pi|\nu|}\left|\int \psi(y)\exp\left[\frac{i\mu}{2\nu}y^2 - \frac{iXy}{\nu}\right]dy\right|^2. \tag{26}$$

This formula provides tomogram of the coherent state in the form of Gaussian distribution

$$w_\alpha(X,\mu,\nu) = \frac{1}{\sqrt{2\pi\sigma^2}}\exp\left(-\frac{(X-\bar{X})^2}{2\sigma^2}\right), \tag{27}$$

where $\sigma^2 = (\mu^2+\nu^2)/2$ and $\bar{X} = \mu\sqrt{2}\,\mathrm{Re}\,\alpha + \nu\sqrt{2}\,\mathrm{Im}\,\alpha$.

The superposition of states $\mid\alpha\rangle_\pm = N_\pm\,(\mid\alpha\rangle \pm \mid -\alpha\rangle)$ introduced and called even and odd coherent states in [44] provides the addition rule for tomographic-probability distributions $w_\alpha(X,\mu,\nu)$ and $w_{-\alpha}(X,\mu,\nu)$, it is given by the formula

$$w_{\alpha_\pm}(X,\mu,\nu) = \frac{N_\pm^2}{2\pi|\nu|}\left|\int\left[\psi_\alpha(y) \pm \psi_{-\alpha}(y)\right]\exp\left[\frac{i\mu}{2\nu} - \frac{iXy}{\nu}\right]dy\right|^2.$$

Thus, the first two terms of the tomographic-probability distributions are contributions of states $\mid\alpha\rangle$ and $\mid -\alpha\rangle$, while the other two terms describe the contribution of the interference term into the tomographic-probability distribution.

In terms of state vectors $\mid\alpha\rangle$ and $\mid -\alpha\rangle$, we have the addition rule of the form

$$w_{\alpha\pm}(X,\mu,\nu) = N_\pm^2\ \mathrm{Tr}\left[\delta\left(X - \mu\hat{q} - \nu\hat{p}\right)\left(\mid\alpha\rangle\langle\alpha\mid + \mid -\alpha\rangle\langle -\alpha\mid \pm \mid\alpha\rangle\langle -\alpha\mid \pm \mid -\alpha\rangle\langle\alpha\mid\right)\right]. \tag{28}$$

Thus, we can formulate the addition rule of the tomographic-probability distributions as follows:

$$w_{\alpha\pm}(X,\mu,\nu) = N_\pm^2\left[w_\alpha(X,\mu,\nu) + w_{-\alpha}(X,\mu,\nu) + F(X,\mu,\nu)\right], \tag{29}$$

where

$$F(X,\mu,\nu) = \mathrm{Tr}\left[\delta\left(X - \mu\hat{q} - \nu\hat{p}\right)\left(\pm\mid\alpha\rangle\langle -\alpha\mid \pm \mid -\alpha\rangle\langle\alpha\mid\right)\right]. \tag{30}$$

10. Born's Rule for Continuous Variables

In the general case, Born's rule provides the probability W_{12} to have the properties of a system state with density operator $\hat{\rho}_1$ if these properties are measured in the system with the density operator $\hat{\rho}_2$, given as $W_{12} = \mathrm{Tr}\,(\hat{\rho}_1\hat{\rho}_2)$. If one has the operator $\hat{U}(x)$ and the operator $\hat{D}(x)$, such that one maps an operator $\hat{A}$ onto the function $f_A(x) = \mathrm{Tr}\,\hat{A}\hat{U}(x)$, and the function $f_A(x)$ is mapped onto the operator $\hat{A} = \int f_A(x)\hat{D}_A(x)\,dx$, the generic formula for Born's rule reads

$$W_{12} = \int K_B(x_1x_2)f_{A_1}(x_1)f_{A_2}(x_2)\,dx_1\,dx_2, \tag{31}$$

where $K_B(x_1x_2) = \mathrm{Tr}\left(\hat{D}(x_1)\hat{D}(x_2)\right)$ is the kernel providing the probability W_{12} in the formalism of the used operators $\hat{U}(x)$ and $\hat{D}(x)$.

As we discussed for symplectic-tomography scheme, $x = (X, \mu, \nu)$, we have that

$$\hat{U}(X,\mu,\nu) = \delta(X\hat{\mathbf{1}} - \mu\hat{q} - \nu\hat{p}), \qquad \hat{D}(X,\mu,\nu) = \frac{1}{2\pi}\exp i(X\hat{\mathbf{1}} - \mu\hat{q} - \nu\hat{p}).$$

Thus, the kernel $K_B(x_1, x_2) = K_B(X_1, \mu_1, \nu_1, X_2, \mu_2, \nu_2)$ is given by the formula

$$K_B(X_1,\mu_1,\nu_1,X_2,\mu_2,\nu_2) = \frac{e^{i(X_1+X_2)}}{4\pi^2}\,\mathrm{Tr}\left(e^{i(\mu_1\hat{q}+\nu_1\hat{p})}\cdot e^{i(\mu_2\hat{q}+\nu_2\hat{p})}\right). \tag{32}$$

As operator $\hat{A}$, we used the density operators $\hat{\rho}_1$ and $\hat{\rho}_2$. Applying this kernel, we obtain Born's rule for symplectic tomograms.

In fact, one has the relation for probabilities $|\langle\psi_1 \mid \psi_2\rangle|^2 = W_{12}$ (Born's rule) in terms of tomograms $w_1(X, \mu\nu)$ and $w_2(X, \mu\nu)$, where

$$w_k(X,\mu,\nu) = \mathrm{Tr}\,\hat{\rho}_k\delta\left(X\mathbf{1} - \mu\hat{q} - \nu\hat{p}\right), \quad k = 1,2, \tag{33}$$

i.e.,

$$W_{12} = \frac{1}{2\pi}\int w_1(X_1,\mu,\nu)w_2(X_2,-\mu,-\nu)e^{i(X_1+X_2)}\,dX_1\,dX_2\,d\mu\,d\nu. \tag{34}$$

Thus, the tomographic-probability distributions determining the oscillator states $\mid\psi_1\rangle$ and $\mid\psi_2\rangle$ provide the probability W_{12} to have the properties of the pure state $\mid\psi_2\rangle$ if they are measured in the pure state $\mid\psi_1\rangle$.

We expressed the probability given by Born's rule in terms of tomographic-probability distributions determining the photon states. It is worth noting that Born's rule (34) is given by functional quadratic form of tomographic probabilities determined by the parameters μ and ν, characterizing the reference frames analogous to the qubit case.

11. Conclusions

On an example of spin - 1/2, we showed that for quantum systems considered in the probability representation of quantum mechanics, the superposition principle can be expressed as a specific nonlinear addition rule of the probabilities determining pure quantum states. From the viewpoint of simplex theory, the new nonlinear expressions obtained in the form of polynomials of several variables (probabilities), which again yield the probabilities, can be obtained using the discussed qubit-system properties.

We found an explicit expression for the probability determined by Born's rule in terms of the probabilities determining the quantum states.

Additionally, we considered the superposition principle of quantum system states with continuous variables, like oscillator's ones, and discussed the addition rule of the tomographic-probability distributions on the example of even and odd coherent states, which demonstrate that generic properties of addition of probabilities for arbitrary state vectors for both discrete (spin) variables and continuous variables (oscillator) can be formulated as the nonlinear addition rule of probability distributions describing the states.

Author Contributions: Conceptualization, M.A.M.; writing–original draft preparation, I.Y.D. and M.A.M.; writing–review and editing, I.Y.D. and M.A.M.; supervision, M.A.M.

Funding: This research received no external funding.

Acknowledgments: M.A.M. thanks the Organizers of the 16th International Conference on Squeezed States and Uncertainty Relations (Madrid, 17–21 June 2019) and especially Luis Sanchez-Soto and Alberto Ibort for their invitation and kind hospitality. This paper is the talk of M.A.M. at the 16th International Conference on Squeezed States and Uncertainty Relations (Universidad Complutense de Madrid, Spain, 17–21 June 2019). Section *Ad Memoriam of Roy Glauber and George Sudarshan* was also presented in the talks of M.A.M. at the 26th Central European Workshop on Quantum Optics (Paderborn University, Germany, 3–7 June 2019) and the 18th International Symposium "Symmetries in Sciences" (Gasthof Hotel Lamm, Bregenz, Voralberg, Austria, 4–9 August 2019) and will be published in the *Journal of Russian Laser Research* (2019) and the *Journal of Physics: Conference Series* (2020), respectively.

Conflicts of Interest: The authors declare no conflict of interest.

References

1. Schrödinger, E. Quantisierung als Eigenwertproblem (Zweite Mitteilung). *Ann. Phys.* **1926**, *384*, 361–376; 489–527. [CrossRef]
2. Dirac, P.A.M. *The Principles of Quantum Mechanics*; Clarendon Press: Oxford, UK, 1981; ISBN 9780198520115.
3. Man'ko, V.I.; Marmo, G.; Sudarshan, E.C.G.; Zaccaria, F. On the Relation between Schrödinger and von Neumann Equations. *J. Russ. Laser Res.* **1999**, *20*, 421–437. [CrossRef]
4. Sudarshan, E.C.G. Search for Purity and Entanglement. *J. Russ. Laser Res.* **2003**, *24*, 195–203. [CrossRef]
5. Landau, L. Das Dämpfungsproblem in der Wellenmechanik. *Z. Phys.* **1927**, *45*, 430–441. [CrossRef]
6. Von Neumann, J. Wahrscheinlichkeitstheoretischer Aufbau der Quantenmechanik. *Gött. Nach.* **1927**, *1927*, 245–272.
7. Glauber, R.J. Photon Correlations. *Phys. Rev. Lett.* **1963**, *10*, 84–86. [CrossRef]
8. Sudarshan, E.C.G. Equivalence of Semiclassical and Quantum Mechanical Descriptions of Statistical Light Beams. *Phys. Rev. Lett.* **1963**, *10*, 277–279. [CrossRef]
9. Klauder, J.R.; Sudarshan, E.C.G. *Fundamentals of Quantum Optics*; Benjamin: New York, NY, USA, 1968; ISBN 978-0486450087; Russian translation, Mir: Moscow, USSR, 1970.
10. Man'ko, M.A. Probability Representation of Spin States and Inequalities for Unitary Matrices. *Theor. Math. Phys.* **2011**, *168*, 985–993. [CrossRef]
11. Man'ko, M.A.; Man'ko, V.I. Probability Description and Entropy of Classical and Quantum Systems. *Found. Phys.* **2011**, *41*, 330–344. [CrossRef]
12. Man'ko, M.A.; Man'ko, V.I. Dynamic Symmetries and Entropic Inequalities in the Probability Representation of Quantum Mechanics. *AIP Conf. Proc.* **2011**, *1334*, 217–248. [CrossRef]
13. Man'ko, M.A.; Man'ko, V.I. The Probability Representation as a New Formulation of Quantum Mechanics. *J. Phys. Conf. Ser.* **2012**, *380*, 012005:1–012005:14. [CrossRef]
14. Man'ko, M.A.; Man'ko, V.I. Tomographic Entropic Inequalities in the Probability Representation of Quantum Mechanics. *AIP Conf. Proc.* **2012**, *1488*, 110–121. [Beauty in Physics: Theory and Experiment: In Honor of Francesco Iachello on the Occasion of His 70th Birthday (Bijker, R., Ed.; Hacienda Cocoyoc, Mexico, 14–18 May 2012)]. [CrossRef]
15. Man'ko, M.A.; Man'ko, V.I. Statistics of Observables in the Probability Representation of Quantum and Classical System States. *AIP Conf. Proc.* **2012**, *1424*, 234–245. [Quantum Theory: Reconsideration of Foundations-6 (Adenier, G., Khrennikov, A.Yu., Eds.)]. [CrossRef]
16. Man'ko, M.A.; Man'ko, V.I.; Marmo, G.; Simoni, A.; Ventriglia, F. Introduction to Tomography, Classical and Quantum. *Nuovo Cimento C* **2013**, *36*, 163–182. [CrossRef]
17. Man'ko, M.A. Tomographic Rényi Entropy of Multimode Gaussian States. *Phys. Scr.* **2013**, *87*, 038113:1–038113:3. [CrossRef]
18. Man'ko, M.A. Joint Probability Distributions and Conditional Probabilities in the Tomographic Representation of Quantum States. *Phys. Scr.* **2013**, *T153*. [CrossRef]
19. Man'ko, M.A.; Man'ko, V.I. Entanglement and Other Quantum Correlations of a Single Qudit State. *Int. J. Quantum Inf.* **2014**, *12*, 156006. [CrossRef]
20. Man'ko, M.A.; Man'ko, V.I. Hidden Correlations in Indivisible Qudits as a Resource for Quantum Technologies on Examples of Superconducting Circuits. *J. Phys. Conf. Ser.* **2016**, *698*. [CrossRef]
21. Man'ko, M.A. Conditional Information and Hidden Correlations in Single-Qudit States. *J. Russ. Laser Res.* **2017**, *38*, 211–222. [CrossRef]
22. Man'ko, M.A.; Man'ko, V.I.; Marmo, G.; Ventriglia, F.; Vitale, P. Dichotomic Probability Representation of Quantum States. *arXiv* **2019**, arXiv:1905.10561.

23. 26th Central European Workshop on Quantum Optics. Available online: https://cewqo2019.uni-paderborn.de/ (accessed on 24 September 2019).
24. The Legendary Symmetries in Science Symposia. Available online: https://itp.uni-frankfurt.de/symmetries-in-science/ (accessed on 24 September 2019).
25. Sudarshan, E.C.G.; Mathews, P.M.; Rau, J. Stochastic Dynamics of Quantum-Mechanical Systems. *Phys. Rev.* **1961**, *121*, 920–924. [CrossRef]
26. Kraus, K. *States, Effects and Operations. Fundamental Notions of Quantum Theory*; Böhm, A., Dollard, J.D., Wootters, W.H., Eds.; Springer: Berlin, Germany, 1983; ISBN 978-3-540-12732-1.
27. Gorini, V.; Kossakowski, A.; Sudarshan, E.C.G. Completely Positive Dynamical Semigroups of N-Level Systems. *J. Math. Phys.* **1976**, *17*, 821–825. [CrossRef]
28. Lindblad, G. On the Generators of Quantum Dynamical Semigroups. *Commun. Math. Phys.* **1976**, *48*, 119–130. [CrossRef]
29. Wigner, E. On the Quantum Correction For Thermodynamic Equilibrium. *Phys. Rev.* **1932**, *40*, 749–759. [CrossRef]
30. Husimi, K. Some Formal Properties of the Density Matrix. *Proc. Phys. Math. Soc. Jpn.* **1940**, *22*, 264–314._264. [CrossRef]
31. Kano, Y. A New Phase-Space Distribution Function in the Statistical Theory of the Electromagnetic Field. *J. Math. Phys.* **1965**, *6*, 1913–1915. [CrossRef]
32. Glauber, R.; Man'ko, V.I. Damping and Fluctuations in the Systems of Two Entangled Quantum Oscillators. In *Group Theory, Gravitation, and Physics of Elementary Particles, Proceedings of the Lebedev Physical Institute* [in Russian]; Komar, A.A., Ed.; Nauka: Moscow, Russia, 1986; Volume 167. [English translation by Nova Science: Commack, New York, 1987; Volume 167, ISBN 9780941743020].
33. Glauber, R.; Man'ko, V.I. Damping and Fluctuations in the Systems of Two Entangled Quantum Oscillators. *Zh. Eksp. Teor. Fiz.* **1984**, *87*, 790–804. [*Sov. Phys. JETP* **1984**, *60*, 450–457].
34. Schrade, V. I.; Man'ko, V.I.; Schleich, W.P.; Glauber, R. Wigner Functions in the Paul Trap. *Quantum Semiclass. Opt.* **1995**, *7*, 307–325. [CrossRef]
35. Man'ko, V.I.; Man'ko, O.V. Spin State Tomography. *J. Exp. Theor. Phys.* **1997**, *85*, 430–434. [CrossRef]
36. Chernega, V.N.; Man'ko, O.V.; Man'ko, V.I. Subadditivity Condition for Spin Tomograms and Density Matrices of Arbitrary Composite and Noncomposite Qudit Systems. *J. Russ. Laser Res.* **2014**, *35*, 278–290. [CrossRef]
37. Mancini, S.; Man'ko, V.I.; Tombesi, P. Symplectic Tomography as Classical Approach to Quantum Systems. *Phys. Lett. A* **1996**, *213*, 1–6. [CrossRef]
38. Dodonov, V.V.; Man'ko, V.I. Positive Distribution Description for Spin States. *Phys. Lett. A* **1997**, *229*, 335–339. [CrossRef]
39. Asorey, M.; Ibort, A.; Marmo, G.; Ventriglia, F. Quantum Tomography Twenty Years Later. *Phys. Scr.* **2015**, *90*, 074031:1–074031:17. [CrossRef]
40. Man'ko, V.I.; Marmo, G.; Sudarshan, E.C.G.; Zaccaria, F. Photon Distribution in Nonlinear Coherent States. *J. Russ. Laser Res.* **2000**, *21*, 305–316. [CrossRef]
41. Man'ko, V.I.; Marmo, G.; Sudarshan, E.C.G.; Zaccaria, F. Entanglement Structure of the Adjoint Representation of the Unitary Group and Tomography of Quantum States. *J. Russ. Laser Res.* **2003**, *24*, 507–543.:JORR.0000004166.55179.aa. [CrossRef]
42. Klimov, A.B.; Man'ko, O.V.; Man'ko, V. I. Professor Viktor V. Dodonov: On the Occasion of His 70th Birthday. *J. Russ. Laser Res.* **2019**, *40*, 105–106. [CrossRef]
43. Dodonov, V.V.; Man'ko, V.I. Invariants and the Evolution of Nonstationary Quantum Systems. In *Proceedings of the Lebedev Physical Institute*; Markov, M.A., Ed.; Nauka: Moscow, Russia, 1987; Volume 183; ISSN 0203-5820. [English translation by Nova Science: Commack, New York, USA, 1989, Volume 183].
44. Dodonov, V.V.; Malkin, I.A.; Man'ko, V.I. Even and Odd Coherent States and Excitations of a Singular Oscillator. *Physica* **1974**, *72*, 597–615. [CrossRef]
45. Dodonov, V.V.; Man'ko, O.V.; Man'ko, V.I. Correlated States in Quantum Electronics (Resonant Circuit). *J. Sov. Laser Res.* **1989**, *10*, 413–420. [CrossRef]
46. Dodonov, V.V.; Klimov, A.B.; Man'ko, V.I. Nonstationary Casimir Effect and Oscillator Energy Level Shift. *Phys. Lett. A* **1989**, *142*, 511–513. [CrossRef]

47. Wilson, C.M.; Johansson, G.; Pourkabirian, A.; Simoen, M.; Johansson, J.R.; Duty, T.; Nori F.; Delsing, P. Observation of the Dynamical Casimir Effect in a Superconducting Circuit. *Nature* **2011**, *479*, 376–379. [CrossRef]
48. Dodonov, V.V. 'Nonclassical' States in Quantum Optics: A 'Squeezed' Review of the First 75 Years. *J. Opt. B Quantum Semiclass.* **2002**, *4*. R1–R33. [CrossRef]
49. Dessano, H.; Dodonov, A.V. One- and Three-Photon Dynamical Casimir Effects Using a Nonstationary Cyclic Qutrit. *Phys. Rev. A* **2018**, *98*, 022520:1–022520:7. [CrossRef]
50. Chernega, V.N.; Man'ko, O.V.; Man'ko, V.I. Triangle Geometry of the Qubit State in the Probability Representation Expressed in Terms of the Triada of Malevich's Squares. *J. Russ. Laser Res.* **2017**, *38*, 141–149. [CrossRef]
51. Man'ko, V.I.; Marmo, G.; Ventriglia, F.; Vitale, P. Metric on the space of quantum states from relative entropy. Tomographic reconstruction. *J. Phys. A Math. Theor.* **2017**, *50*, 335302:1–335302:29. [CrossRef]
52. Chernega, V.N.; Man'ko, O.V.; Man'ko, V.I. Quantum Suprematism Picture of Triada of Malevich's Squares for Spin States and the Parametric Oscillator Evolution in the Probability Representation of Quantum Mechanics. *J. Phys. Conf. Ser.* **2018**, *1071*, 012008:1–012008:13. [CrossRef]
53. Man'ko, M.A. Hidden Correlations in Quantum Optics and Quantum Information. *J. Phys. Conf. Ser.* **2018**, *1071*, 012015:1–012015:15. [CrossRef]
54. Man'ko, M.A.; Man'ko, V.I. New Entropic Inequalities and Hidden Correlations in Quantum Suprematism Picture of Qudit States. *Entropy* **2018**, *20*, 692. [CrossRef]
55. Lopez-Saldivar, J.A.; Castaños, O.; Nahmad-Achar, E.; López-Peña, R.; Man'ko, M.A.; Man'ko, V.I. Geometry and Entanglement of Two-Qubit States in the Quantum Probabilistic Representation. *Entropy* **2018**, *20*, 630. [CrossRef]
56. Lopez-Saldivar, J.A.; Castaños, O.; Man'ko, M.A.; Man'ko, V.I. A New Mechanism of Open System Evolution and Its Entropy Using Unitary Transformations in Noncomposite Qudit Systems. *Entropy* **2019**, *21*, 736. [CrossRef]
57. Man'ko, M.A.; Man'ko, V.I. Properties of Nonnegative Hermitian Matrices and New Entropic Inequalities for Noncomposite Quantum Systems. *Entropy* **2015**, *17*, 2876–2894. [CrossRef]
58. Asorey, M.; Facchi, P.; Man'ko, V.I.; Marmo, G.; Pascazio, S.; Sudarshan, E.G.C. Radon Transform on the Cylinder and Tomography of a Particle on the Circle. *Phys. Rev. A* **2007**, *76*. [CrossRef]
59. Ibort, A.; Man'ko, V.I.; Marmo, G.; Simoni, A.; Ventriglia, F. An Introduction to the Tomographic Picture of Quantum Mechanics. *Phys. Scr.* **2009**, *79*. [CrossRef]
60. Man'ko, V.I.; Mendes, R.V. Non-Commutative Time-Frequency Tomography. *Phys. Lett.* **1999**, *A263*, 53–61. [CrossRef]

Article

Selective Engineering for Preparing Entangled Steady States in Cavity QED Setup

Emilio H. S. Sousa and J. A. Roversi *

Instituto de Fisica "Gleb Wataghin", Universidade Estadual de Campinas, 13083970 Campinas, SP, Brazil
* Correspondence: roversi@ifi.unicamp.br

Received: 31 May 2019; Accepted: 3 July 2019; Published: 8 July 2019

Abstract: We propose a dissipative scheme to prepare maximally entangled steady states in cavity QED setup, consisting of two two-level atoms interacting with the two counter-propagating whispering-gallery modes (WGMs) of a microtoroidal resonator. Using spontaneous emission and cavity decay as the dissipative quantum dynamical source, we show that the steady state of this system can be steered into a two-atom single state as well as into a two-mode single state. We probed the compound system with weak field coupled to the system via a tapered fiber waveguide, finding it is possible to determine whether the two atoms or two modes are driven to a maximally entangled state. Through the transmission and reflection measurements, without disturbing the atomic state, when the cavity modes are being driven, or without disturbing the cavity field state, when a single atom being driven, one can get the information about the maximal entanglement. We also investigated for both subsystem, two-atom and two-mode states, the entanglement generation and under what conditions one can transfer entanglement from one subsystem to the other. Our scheme can be selectively used to prepare both maximally entangled atomic state as well as maximally entangled cavity-modes state, providing an efficient method for quantum information processing.

Keywords: entangled states; two atoms; two-modes; cavity QED setup

---

1. Introduction

The preparation and control of entangled states via dissipative engineering have attracted great interest in the last several years [1,2]. Different from the unitary evolution based schemes, these schemes use decoherence as a powerful resource in the state preparation process without destroying the quantum entanglement. In [3], the authors showed that the cavity decay plays an integral part in preparing a maximally entangled state of two Λ atoms trapped in an optical cavity. Stannigel et al. showed that the driven-dissipative preparation of entangled states can be obtained in cascaded quantum-optical networks between individual nodes [4]. In [5], the authors used the energy relaxation of the single superconducting qubit coupled to two spatially separated transmission line resonators for generating a two-mode entangled state. In addition, the generation of a two-mode entangled states has been investigated by quantum reservoir engineering [6]. Other schemes based on the waveguide QED configuration are presented in [7–12].

On the other hand, the identification of a quantum state is usually achieved by quantum state tomography [13]. However, this method directly performs a series of projective measurements on many identical copies of the quantum state, inevitably disturbing the state of the system. To circumvent this problem, quantum nondemolition measurements [14] are projected to prevent the back action of the measurement on the detected observable. Recently, proposals to realize a quantum nondemolition measurements of a superconducting flux qubit [15], pair of atomic samples [16] and nonclassical state of a massive object [17] have been demonstrated. However, an interesting question is whether we can use an experimentally feasible method to detect an entangled state of two atoms or two modes, selectively, in a single quantum system via quantum nondemolition measurements.

Here, we report a dissipative scheme to prepare maximally entangled steady states in cavity QED setup, which consists of two two-level atoms interacting with the two counter-propagating whispering-gallery modes (WGMs) of a microtoroidal resonator. The steady state of this system can be steered into an entangled atomic state or into an entangled field state by the dissipative quantum dynamical process. In this scheme, both atomic spontaneous emission and cavity decay are utilized as a resource to engineer the targeted state. In addition, the dissipative steady-state production requires neither precise time control nor initial state preparation. To probe these entangled states without disturbing them, we performed transmission and reflection measurements through the incident weak field of two ways: (i) driving the cavity mode; and (ii) driving a single atom. For this propose, both the microtoroidal cavity and the atom were coupled to a tapered fiber waveguide and detectors. By injecting and controlling a weak field into a tapered fiber, we could determine whether the two atoms (driving the cavity mode) or two modes (driving a single atom) are in a maximally entangled state by a single click on the detector, without disturbing the atomic state or cavity field state. Thus, one of the detectors acted as a witness of the preparation of the entangled atomic state and the other of the entangled field state. Note that our goal was not to find a way to simultaneously prepare two entangled states, but to explore the possibility to get a maximally entangled state between two atoms or two modes, under certain conditions. Compared with previous proposals, the present scheme indicates a possibility of preparing an entangled atomic state as well as an entangled field state [18] as well as using the dissipation as a powerful resource to engineer in those states [19]. We also investigated the time evolution of the entanglement for both subsystem, i.e. two atoms and two modes, and under what conditions one can transfer an entangled state of two qubits from one subsystem to the other.

2. Model

Our system consists of a pair of identical two-level atoms that interact with the evanescent fields of a microtoroidal cavity, as shown in Figure 1. We denote the atoms by label $i = 1, 2$ with frequency ω_{eg} and the atomic ground and excited states by $|g\rangle_i$ and $|e\rangle_i$, respectively. The cavity supports two WGMs at frequency ω_c and with annihilation (creation) operators $\hat{a}$ $(\hat{a}^\dagger)$ and $\hat{b}$ $(\hat{b}^\dagger)$. These two modes have an intrinsic loss rate κ_{in} and are coupled to each other with coupling strength J. Each atom is coupled simultaneously with the two WGMs via evanescent field with a coherent coupling strength described by g_i. dipole–dipole interaction of strength Ω is also included. The Hamiltonian of the whole system can be written in the form (in unit $\hbar$) [20–22]:

$$\begin{aligned} H &= \omega_{eg}(\hat{\sigma}_1^+\hat{\sigma}_1^- + \hat{\sigma}_2^+\hat{\sigma}_2^-) + \omega_c(\hat{a}^\dagger\hat{a} + \hat{b}^\dagger\hat{b}) + J(\hat{a}^\dagger\hat{b} + \hat{a}\hat{b}^\dagger) + \Omega(\hat{\sigma}_1^+\hat{\sigma}_2^- + \hat{\sigma}_2^+\hat{\sigma}_1^-) \\ & (g_1^*\hat{a}^\dagger\hat{\sigma}_1^- + g_1\hat{a}\hat{\sigma}_1^\dagger) + (g_1\hat{b}^\dagger\hat{\sigma}_1^- + g_1^*\hat{b}\hat{\sigma}_1^\dagger) + (g_2^*\hat{a}^\dagger\hat{\sigma}_2^- + g_2\hat{a}\hat{\sigma}_2^\dagger) + (g_2\hat{b}^\dagger\hat{\sigma}_2^- + g_2^*\hat{b}\hat{\sigma}_2^\dagger) \end{aligned} \quad (1)$$

where $\hat{\sigma}_i^+ = |e\rangle_i\langle g|$ and $\hat{\sigma}_i^- = |g\rangle_i\langle e|$ are the raising and lowering operators of the atom i.

Introducing dissipation, the dynamics of the system is governed by the master equation [21,22]:

$$\dot{\rho}(t) = -i[H, \rho(t)] + (\gamma/2)\sum_{i=1}^{2}\mathcal{D}[\hat{\sigma}_-^i]\rho(t) + \kappa\mathcal{D}[\hat{a}]\rho(t) + \kappa\mathcal{D}[\hat{b}]\rho(t), \quad (2)$$

where γ is the spontaneous emission rate of the atoms, κ is the cavity decay rate and $\mathcal{D}[\mathcal{O}]\rho(t) \equiv 2\mathcal{O}\rho(t)\mathcal{O}^\dagger - \mathcal{O}^\dagger\mathcal{O}\rho(t) - \rho(t)\mathcal{O}^\dagger\mathcal{O}$. The spectrum of the system, i.e., its allowed states, are represented by the eigenvalues and eigenvectors of H, in the ideal case (note that, including system dissipation, the eigenvectors with zero eigenvalues are still the same stationary entangled states present in the ideal case, whereas the others two decay in time), are given by

$$E_0^0 = 0; \qquad |E_0^0\rangle = |gg00\rangle$$

$$E_1^1 = 0; \qquad |E_1^1\rangle = \frac{1}{\sqrt{g_1^2 + g_2^2}}(g_1|eg00\rangle - g_2|ge00\rangle)$$

$$E_1^2 = 0; \qquad |E_1^2\rangle = \frac{1}{\sqrt{g_1^2 + g_2^2}}(g_1|gg10\rangle - g_2|gg01\rangle)$$

$$E_1^{\pm} = \pm i\sqrt{2}\sqrt{g_1^2 + g_2^2}; \qquad |E_1^{\pm}\rangle = \frac{1}{\sqrt{2}}\Big\{\frac{1}{\sqrt{2}}(|gg10\rangle + |gg01\rangle) \pm \frac{1}{\sqrt{g_1^2 + g_2^2}}(g_1|eg00\rangle + g_2|ge00\rangle)\Big\}$$

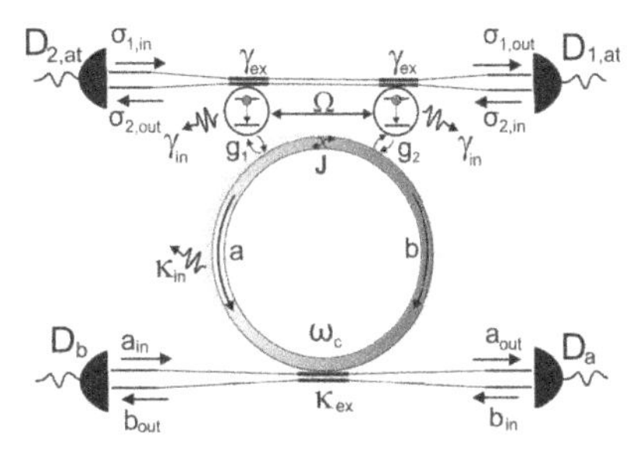

Figure 1. Experimental scheme of the atom–cavity system. The cavity consists of two WGMs coupled simultaneously to a pair of two-level atoms interacting via dipole–dipole interaction Ω. In this case, both the cavity and the atom are coupled to a tapered optical fiber in an overcoupled regime [21]. The modes of fiber are described by $\{\hat{a}_{in}, \hat{a}_{out}, \hat{b}_{in}, \hat{b}_{out}\}$ coupled to the cavity and $\{\hat{\sigma}_{1,in}, \hat{\sigma}_{1,out}, \hat{\sigma}_{2,in}, \hat{\sigma}_{2,out}\}$ coupled to the atom, in terms of the input–output fields to a detectors.

Note that, for $g_1 = g_2$, the eigenvector correspondent to the eigenvalue $E_1^1 = 0$ is a tensorial product between the maximally entangled states of the atoms and the cavity in a vacuum state and for $E_1^2 = 0$ is a tensorial product between the atoms in ground state and a maximally entangled states between the field modes,

$$|E_1^1\rangle_{g_1=g_2} = \frac{1}{\sqrt{2}}(|eg\rangle - |ge\rangle) \otimes |00\rangle = |\psi^-, V\rangle$$
$$|E_1^2\rangle_{g_1=g_2} = |gg\rangle \otimes \frac{1}{\sqrt{2}}(|10\rangle - |01\rangle) = |G, \phi^-\rangle$$

where $|G\rangle = |gg\rangle$, $|V\rangle = |00\rangle$, $|\psi^-\rangle = \frac{1}{\sqrt{2}}(|eg\rangle - |ge\rangle)$ and $|\phi^-\rangle = \frac{1}{\sqrt{2}}(|10\rangle - |01\rangle)$. The energy-level diagram of the atom–toroid system with the transitions and decay rates, via Fermi's golden rule (without spontaneous emission ($\gamma = 0$), collective decay rates of the modes are given by $\Gamma_c = \kappa/2$ e $\Gamma_c^{\pm} = \kappa$, while, without cavity loss ($\kappa = 0$), the collective decay rates of the atoms are given by $\Gamma_a = \gamma/4$ e $\Gamma_a^{\pm} = \gamma/2$) [23], and the probe fields (which are discussed further below) are shown in Figure 2. In this study, we only considered the lowest energy of the system, since we were interested in its steady state, which is a mixture of these states for any initial state of the system. Thus, the steady state of the system is a mixture of the lower energy states of each subspace:

$$\rho_{ss} = (1 - P_{spont} - P_{cav})|G,00\rangle\langle G,00| + P_{spont}|\phi^-,00\rangle\langle\phi^-,00| + P_{cav}|\psi^-,00\rangle\langle\psi^-,00| \tag{3}$$

where P_{spont} is the projection of the initial state of the atom in the state $|\psi^-\rangle$ and P_{cav} is the projection of the initial state of the fields in the state $|\phi^-\rangle$. Equation (3) can be obtained directly from Equation (2) to $t \to 0$.

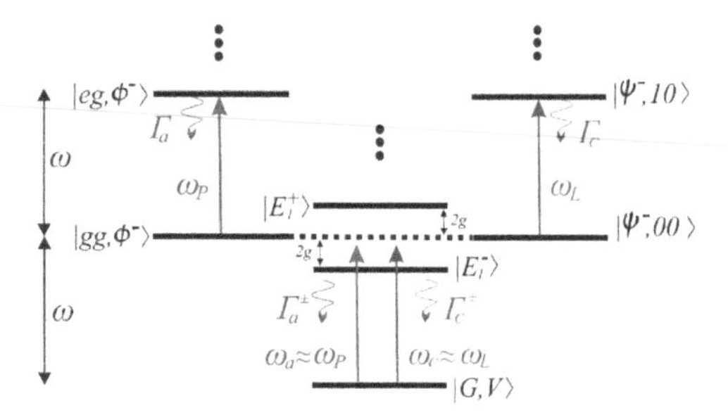

Figure 2. Energy levels diagram of the atom-cavity system with collective decay rates and probe fields where (blue) driving the cavity mode with strength ϵ and frequency ω_L and (red) driving the single atom with strength η and frequency ω_P.

Following the method described in [18], to discriminate these states without disturbing them, we must monitor the system using a weak probe field, keeping the system still with a single excitation. In our case, we used two distinct procedures to monitor the atom–cavity system via probe field: (i) drive the cavity mode to distinguish between the atomic states $|G\rangle$ and $|\psi^-\rangle$, restricted to the time interval $\kappa/g^2 \ll t < 1/\gamma$ [24]; and (ii) drive a single atom to distinguish between the states of the fields $|V\rangle$ and $|\phi-\rangle$ in the time interval $\kappa t \ll 1$ and $g^2 t/\gamma \gg 1$. In both cases, such discrimination of states consisted in measuring the coefficients of transmission and reflection of the incident field, without disturbing the atomic system (Case *i*) or the cavity modes (Case *ii*). In addition, using the formalism of input–output theory [25], the output fields are given by

$$a_{out}(t) = a_{in}(t) + \sqrt{2\kappa_{ex}} a(t) \tag{4}$$

$$b_{out}(t) = b_{in}(t) + \sqrt{2\kappa_{ex}} b(t) \tag{5}$$

where the coherent amplitudes of the input fields are given by $\langle a_{in}\rangle = \frac{i\epsilon}{\sqrt{2\kappa_{ex}}}$ and $\langle b_{in}\rangle = 0$ [22]. In this way, the transmission and reflection coefficients of the modes are defined as

$$T_c = \frac{\langle a^\dagger_{out} a_{out}\rangle_{ss}}{|\epsilon|^2/(2\kappa_{ex})} \tag{6}$$

$$R_c = \frac{\langle b^\dagger_{out} b_{out}\rangle_{ss}}{|\epsilon|^2/(2\kappa_{ex})}. \tag{7}$$

Similarly, the transmission and reflection coefficients from the atoms are given by

$$T_a = \frac{\langle \sigma^+_{1,out}\sigma^-_{1,out}\rangle_{ss}}{|\eta|^2/(2\gamma_{ex})} \tag{8}$$

$$R_a = \frac{\langle \sigma^+_{2,out}\sigma^-_{2,out}\rangle_{ss}}{|\eta|^2/(2\gamma_{ex})}. \tag{9}$$

3. Monitoring the Atom–Cavity System by Driven the Cavity Mode

Our purpose in this section is to probe stationary atomic states, that is, to distinguish an uncorrelated atomic state ($|G\rangle$) from a maximally entangled state ($|\psi^-\rangle$) without disturbing them, e.g., by performing a quantum nondemolition measurement. In this case, the mode a of the cavity is driven by a probe field given by $H_L = \epsilon(\hat{a} e^{i\omega_L t} + \hat{a}^\dagger e^{-i\omega_L t})$, where ϵ and ω_L are the strength and the frequency of the probe field, respectively. In Figure 2, we can observe that, if the system is in the state $|\psi-, 00\rangle$, the external field is capable of promoting the transition $|\psi^-, 00\rangle \to |\psi^-, 10\rangle$, resonantly. Note that $|\psi^-\rangle$ is a dark state and, in this case, the atoms do not "see" the cavity modes and this would be the same as if imposing $g \simeq 0$, thus, obtaining $T_c = 1$ and $R_c = 0$. On the other hand, if the system is in the state $|G,V\rangle$, the external field leads to the transition $|G,V\rangle \rightarrow |E_1^{\pm}\rangle$ having detuning (Δ_L) between the frequencies of the probe and atom–field system given by $\pm 2g$, resulting in $T_c = 0$ and $R_c = 1$, as explained below. This occurs because the two atoms are strongly coupled to the cavity, i.e., $|g| \gg \{\kappa_{in}, J, \Omega, |\Delta_L|, \Gamma_c\}$, resulting in an intracavity field redistribution between counter-propagating modes a and b [21]. Thus, the radiated field is dynamically controlled by atomic polarization which produces a destructive interference in the output field a_{out} between the components a_{in} and $\sqrt{2}\kappa_{ex}$ resulting in a transmission $T_c \rightarrow 0$, and, consequently, the incident field is fully reflected in the fiber via the output field b_{out}, resulting in $R_c \rightarrow 1$ (note that $\mathcal{T}_c + \mathcal{R}_c < 1$ due to the losses in the system) [21,26]. Therefore, the transmission and reflection of the probe field applied to the cavity mode can be used as witness of the atomic states, without disturbing them so that, $T_c = 1$ and $R_c = 0$ implies that $\rho_{ss}^{at} \rightarrow |\psi^-\rangle\langle\psi^-|$, and $T_c = 0$ and $R_c = 1$ implies that $\rho_{ss}^{at} \rightarrow |G\rangle\langle G|$. In a compact way, the steady state of the atomic system, tracing over the modes of the cavity, is given by:

$$Tr_c[\rho_{ss}] \rightarrow \rho_{ss}^{at} = (1 - P_{spont})|G\rangle\langle G| + P_{spont}|\psi^-\rangle\langle\psi^-|. \tag{10}$$

Figure 3 shows the transmission and reflection of the cavity modes as a function of the detuning ($\Delta_L = \omega_c - \omega_L$) between the frequencies of the probe field and the atom–toroid system in overcoupled regime, for $\{\epsilon, \kappa_{ex}, \kappa_{in}, \Omega, J, \Gamma_c, g\}/2\pi = \{10, 20, 0.2, 0, 0, 5.2, 45\}$MHz. In Figure 3, the dashed line correspond to the case where $\rho_{ss}^{at} \rightarrow |\psi^-\rangle\langle\psi^-|$, so that the incident field is fully transmitted ($T_c = 1$ and $R_c = 0$), equivalent to empty cavity type behavior ($g = 0$). On the other hand, when $\rho_{ss}^{at} \rightarrow |G\rangle\langle G|$ (solid line), it was observed that, near the resonance ($\Delta_L = 0$), the strong coupling between the atoms and the cavity modes, that is, $g > \kappa_{ex}$, causes the incident field to be fully reflected ($T_c = 0$ and $R_c = 1$) and, in this case, the atoms act as a "mirror" for the incident field.

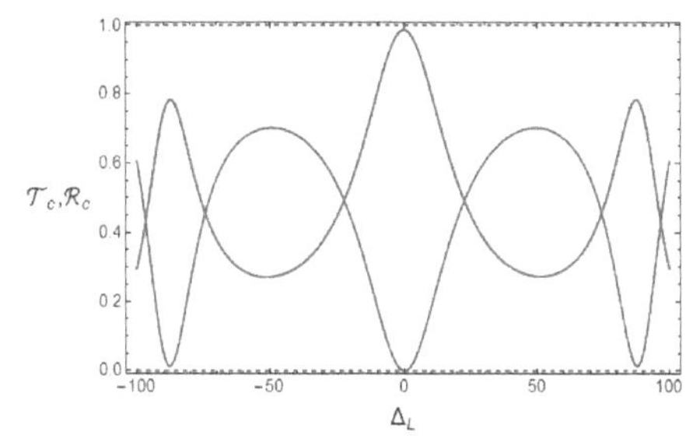

Figure 3. Transmission (blue) and reflection (red) of the cavity modes as a function of the detuning between the frequencies of the probe field and the atom–cavity system in overcoupled regime $\kappa_{ex} \gg \{\kappa_{in}, J\}$ and $\omega_c = \omega_a$. The chosen parameters were $\{\epsilon, \kappa_{ex}, \kappa_{in}, \Omega, J, \Gamma_c, g\}/2\pi = \{10, 20, 0.2, 0, 0, 5.2, 45\}$ MHz. The dashed line corresponds to $\rho_{ss}^{at} \rightarrow |\psi^-\rangle\langle\psi^-|$ and the solid line for $\rho_{ss}^{at} \rightarrow |G\rangle\langle G|$.

4. Monitoring the Atom–Cavity System by Driven the Single Atom

In this section, we probe the steady states of cavity modes, that is, differentiate an uncorrelated state ($|V\rangle$) from an entangled state between two modes ($|\phi\rangle$) by driving the single atom, without disturbing the states of the cavity modes. In this setup, the incident field is given by $H_P = \eta(\sigma_1^- e^{i\omega_P t} + \sigma_1^+ e^{-i\omega_P t})$, where η and ω_P are the strength and the frequency of the probe field, respectively. Note that the external field was applied to atom 1, as shown in Figure 2. It was also observed that the probe field is capable of promoting, resonantly, the transition $|gg,\phi^-\rangle \rightarrow |eg,\phi^-\rangle$. Similar to the previous discussion, $|\phi^-\rangle$ is a dark state and, in this case, the cavity modes do not "see" atoms and this would be the same as if imposing $g \simeq 0$, resulting in $T_a = 1$ and $R_a = 0$. However, for $|gg,V\rangle$, the incident field is capable of promoting the transition, of resonance, $|gg,V\rangle \rightarrow |E_{\pm}\rangle$ with detuning $\pm 2g$, resulting in $T_a = 0$ and $R_a = 1$. In this case, the incident field is fully reflected due to destructive interference between the components $\sigma_{1,in}^-$ and $\sqrt{2\gamma_{ex}}\sigma_1^-$ resulting in $T_a \rightarrow 0$.

Therefore, the transmission and reflection of the probe field applied on atom 1 can be used as witness of the state of the cavity modes without disturbing them, so that, $T_a = 1$ and $R_a = 0$ implies that $\rho_{ss}^c \rightarrow |\phi^-\rangle\langle\phi^-|$ and $T_a = 0$ and $R_a = 1$ implies that $\rho_{ss}^c \rightarrow |V\rangle\langle V|$. The steady state of the cavity modes, tracing over the atomic states, is given by:

$$Tr_{at}[\rho_{ss}] \rightarrow \rho_{ss}^c = (1 - P_{cav})|V\rangle\langle V| + P_{cav}|\phi^-\rangle\langle\phi^-|. \tag{11}$$

Figure 4 shows the transmission and reflection of the atoms as a function of the detuning between the frequencies of the probe field and the atom–toroid system $\Delta_P = \omega_a - \omega_P$ in an overcoupled regime, for $\{\eta, \gamma_{ex}, \gamma_{in}, \Omega, J, \Gamma_a, g\}/2\pi = \{10, 40, 0.2, 0, 0, 5.2, 45\}$ MHz.

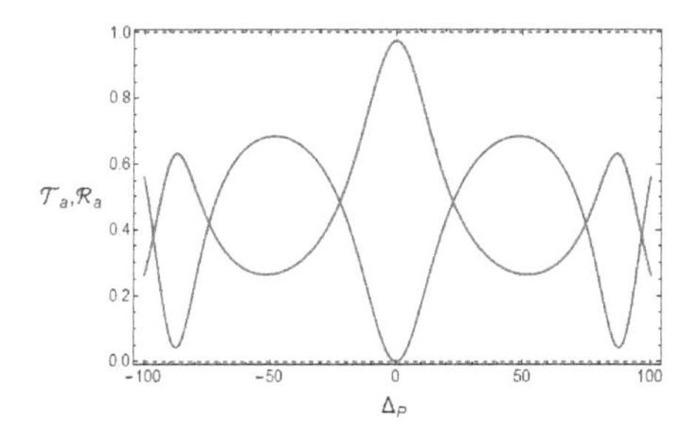

Figure 4. Transmission (blue) and reflection (red) of the atoms as a function of the detuning between the frequencies of the probe field and the atom–cavity system in an overcoupled regime $\gamma_{ex} \gg \{\gamma_{in}, \Omega\}$ and $\omega_c = \omega_a$. The chosen parameters were $\{\gamma_{ex}, \gamma_{in}, \Omega, J, \Gamma_c\}/2\pi = \{40, 0.2, 0, 0, 5.2\}$ MHz. The dashed line corresponds to $\rho_{ss}^c \rightarrow |\phi^-\rangle\langle\phi^-|$ and the solid line for $\rho_{ss}^c \rightarrow |V\rangle\langle V|$.

In Figure 4, the dashed lines correspond to the case that $\rho_{ss}^c \rightarrow |\phi-\rangle\langle\phi-|$, where the probe field is fully transmitted ($T_a = 1$ and $R_a = 0$), because, in this state, the cavity modes do not "see" atoms ($g = 0$). However, when $\rho_{ss}^c \rightarrow |V\rangle\langle V|$ (solid lines), it is noted that, around the resonance ($\Delta_P = 0$), the strong coupling between the modes and the atoms ($g > \gamma_{ex}$) induces a complete reflection of the incident field ($T_a = 0$ and $R_a = 1$).

5. Transfer of Entanglement between Two Atoms and Two Modes

In this section, we investigate the case when the system is initially prepared in an entangled state and under what conditions it is possible to obtain the transfer of entanglement between the two subsystems, e.g., from the two atoms to the two modes. In this case, we consider the following two initial states: $|\phi(0)\rangle_1 = \frac{1}{\sqrt{2}}(|eg\rangle + |ge\rangle) \otimes |00\rangle$ and $|\phi(0)\rangle_2 = \frac{1}{\sqrt{2}}(|10\rangle + |01\rangle) \otimes |gg\rangle$. For this purpose, we use negativity [27] as measure of the degree of entanglement between the two atoms ($\mathcal{N}_a$) and the two-modes ($\mathcal{N}_c$).

Figure 5 shows the time evolution of the negativity between the atoms (blue lines) and between the fields (red lines) as a function of the scaled time gt for the initial states $|\phi(0)\rangle_1$ (left) and $|\phi(0)\rangle_2$ (right) with γ fixed and different values of κ. It was observed that, in both initial states, it is possible to completely transfer the entanglement between the two subsystems when $\kappa = \gamma = 0.01g$ (next to the ideal case), as expected. For small values of gt and higher values of κ, the degradation of the entanglement between atoms is more robust when compared to the entanglement between the modes.

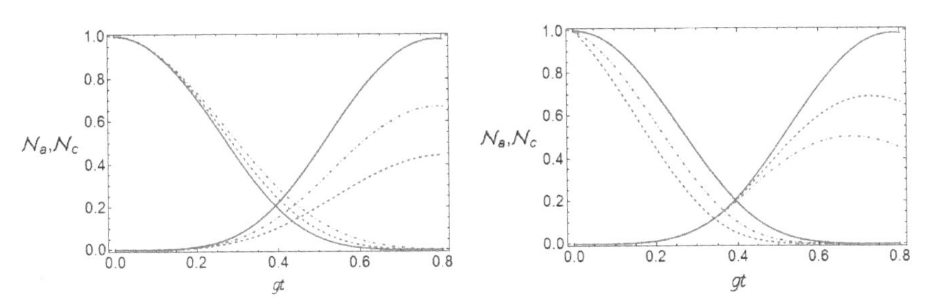

Figure 5. Time evolution of the negativity between the atoms (blue) and fields (red) as a function of time scaled gt for the initial states: (**left**) $|\phi(0)\rangle_1 = \frac{1}{\sqrt{2}}(|eg00\rangle + |ge00\rangle)$; and (**right**) $|\phi(0)\rangle_2 = \frac{1}{\sqrt{2}}(|gg10\rangle + |gg01\rangle)$. The chosen parameters were $\gamma = 0.01g$ and $\Omega = J = 0$ for $\kappa = 0.01g$ (solid line), $\kappa = 0.5g$ (dot-dashed line) and $\kappa = 1.0g$ (dashed line).

6. Conclusions

In conclusion, using a dissipative atom-microtorid system, we have shown that it is possible to prepare a selectively maximally entangled state between two atoms as well as a maximally entangled state between two modes. In this case, the results were obtained by measuring the transmission and reflection coefficients of a weak probe field applied to the atom–cavity system, so as not to disturb the atomic state (drive the cavity mode) or the state of the cavity field (drive the single atom) by performing a quantum nondemolition measurement. In addition, we have shown that it is possible to transfer an entangled state of two qubits from the two atoms to the two modes under certain conditions. Therefore, our results may contribute to a better understanding of the preparation and transfer of entangled states that are of great interest to the quantum information processing.

Author Contributions: All authors contributed substantially to the research.

Funding: This research was funded by Conselho Nacional de Desenvolvimento Científico e Tecnológico (CNPq) grant number 140039/2016-3 (http://www.carloschargas.cnpq.br), Coordenação de Aperfeiçoamento de Pessoal de Nível Superior (CAPES) grant number 1247693/2013 (https://www.capes.gov.br), Fundação de Apoio a Pesquisa do Estado de São Paulo (FAPESP) (http://www.fapesp.br) and Instituto Nacional de Ciência e Tecnologia (INCT) (http://www.inct-iq@if.ufrj.br).

Acknowledgments: The authors would like to thank Mohinder Paul Sharma for the valuable comments and careful English revision of the manuscript.

Conflicts of Interest: The authors declare no conflict of interest.

References

1. Horn, K.P.; Reiter, F.; Lin, Y.; Leibfried, D.; Koch, C.P. Quantum optimal control of the dissipative production of a maximally entangled state. *New J. Phys.* **2018**, *20*, 123010. [CrossRef]
2. Jin, Z.; Su, S.L.; Zhu, A.D.; Wang, H.F.; Zhang, S. Dissipative preparation of distributed steady entanglement: An approach of unilateral qubit driving. *Opt. Express* **2017**, *25*, 88–101. [CrossRef] [PubMed]
3. Kastoryano, M.J.; Reiter, F.; Sørensen, A.S. Dissipative Preparation of Entanglement in Optical Cavities. *Phys. Rev. Lett.* **2011**, *106*, 090502. [CrossRef] [PubMed]
4. Stannigel, K.; Rabl, P.; Zoller, P. Driven-dissipative preparation of entangled states in cascaded quantum-optical networks. *New J. Phys.* **2012**, *14*, 063014. [CrossRef]
5. Ma, S.L.; Li, Z.; Fang, A.P.; Li, P.B.; Gao, S.Y.; Li, F.L. Controllable generation of two-mode-entangled states in two-resonator circuit QED with a single gap-tunable superconducting qubit. *Phys. Rev. A* **2014**, *90*, 062342. [CrossRef]
6. Arenz, C.; Cormick, C.; Vitali, D.; Morigi, G. Generation of two-mode entangled states by quantum reservoir engineering. *J. Phys. B Atomic Mol. Opt. Phys.* **2013**, *46*, 224001. [CrossRef]
7. Petiziol, F.; Dive, B.; Carretta, S.; Mannella, R.; Mintert, F.; Wimberger, S. Accelerating adiabatic protocols for entangling two qubits in circuit QED. *Phys. Rev. A* **2019**, *99*, 042315. [CrossRef]

8. Mirza, I.M.; Schotland, J.C. Two-photon entanglement in multiqubit bidirectional-waveguide QED. *Phys. Rev. A* **2016**, *94*, 012309. [CrossRef]
9. Macrì, V.; Nori, F.; Kockum, A.F. Simple preparation of Bell and Greenberger-Horne-Zeilinger states using ultrastrong-coupling circuit QED. *Phys. Rev. A* **2018**, *98*, 062327. [CrossRef]
10. Mirza, I.M.; Schotland, J.C. Multiqubit entanglement in bidirectional-chiral-waveguide QED. *Phys. Rev. A* **2016**, *94*, 012302. [CrossRef]
11. Mirza, I.M. Controlling tripartite entanglement among optical cavities by reservoir engineering. *J. Mod. Opt.* **2015**, *62*, 1048–1060. [CrossRef]
12. Mirza, I.M. Bi- and uni-photon entanglement in two-way cascaded fiber-coupled atom–cavity systems. *Phys. Lett. A* **2015**, *379*, 1643–1648. [CrossRef]
13. Cramer, M.; Plenio, M.B.; Flammia, S.T.; Somma, R.; Gross, D.; Bartlett, S.D.; Landon-Cardinal, O.; Poulin, D.; Liu, Y.K. Efficient quantum state tomography. *Nat. Commun.* **2010**, *1*, 149. [CrossRef] [PubMed]
14. Grangier, P.; Levenson, J.A.; Poizat, J.P. Quantum non-demolition measurements in optics. *Nature* **1998**, *396*, 537. [CrossRef]
15. Takashima, K.; Nishida, M.; Matsuo, S.; Hatakenaka, N. Quantum Nondemolition Measurement of a Superconducting Flux Qubit. *AIP Conf. Proc.* **2006**, *850*, 945–946. [CrossRef]
16. Di Lisi, A.; Mølmer, K. Entanglement of two atomic samples by quantum-nondemolition measurements. *Phys. Rev. A* **2002**, *66*, 052303. [CrossRef]
17. Lecocq, F.; Clark, J.B.; Simmonds, R.W.; Aumentado, J.; Teufel, J.D. Quantum Nondemolition Measurement of a Nonclassical State of a Massive Object. *Phys. Rev. X* **2015**, *5*, 041037. [CrossRef] [PubMed]
18. Rossatto, D.Z.; Villas-Boas, C.J. Method for preparing two-atom entangled states in circuit QED and probing it via quantum nondemolition measurements. *Phys. Rev. A* **2013**, *88*, 042324. [CrossRef]
19. Li, P.B.; Gao, S.Y.; Li, F.L. Quantum-information transfer with nitrogen-vacancy centers coupled to a whispering-gallery microresonator. *Phys. Rev. A* **2011**, *83*, 054306. [CrossRef]
20. Jin, J.S.; Yu, C.S.; Pei, P.; Song, H.S. Positive effect of scattering strength of a microtoroidal cavity on atomic entanglement evolution. *Phys. Rev. A* **2010**, *81*, 042309. [CrossRef]
21. Dayan, B.; Parkins, A.; Aoki, T.; Ostby, E.; Vahala, K.; Kimble, H. A photon turnstile dynamically regulated by one atom. *Science* **2008**, *319*, 1062–1065. [CrossRef] [PubMed]
22. Aoki, T.; Parkins, A.S.; Alton, D.J.; Regal, C.A.; Dayan, B.; Ostby, E.; Vahala, K.J.; Kimble, H.J. Efficient Routing of Single Photons by One Atom and a Microtoroidal Cavity. *Phys. Rev. Lett.* **2009**, *102*, 083601. [CrossRef] [PubMed]
23. Cohen-Tannoudji, C.; Dupont-Roc, J.; Grynberg, G. *Atom-Photon Interactions: Basic Processes and Applications*; Wiley-VCH: Weinheim, Germany, 1998; p. 678, ISBN 0-471-29336-9.
24. Bullough, R.K. Photon, quantum and collective, effects from rydberg atoms in cavities. *Hyperfine Interact.* **1987**, *37*, 71–108. [CrossRef]
25. Gardiner, C.W.; Collett, M.J. Input and output in damped quantum systems: Quantum stochastic differential equations and the master equation. *Phys. Rev. A* **1985**, *31*, 3761–3774. [CrossRef]
26. Aghamalyan, D.; You, J.B.; Chu, H.S.; Png, C.E.; Krivitsky, L.; Kwek, L.C. Quantum transistor realized with a single Λ-level atom coupled to the microtoroidal cavity. *arXiv* **2019**, arXiv:1902.11052.
27. Vidal, G.; Werner, R.F. Computable measure of entanglement. *Phys. Rev. A* **2002**, *65*, 032314. [CrossRef]

 quantum reports

Review

Resource Theories of Nonclassical Light

Kok Chuan Tan * and Hyunseok Jeong *

Department of Physics and Astronomy, Seoul National University, Seoul 08826, Korea
* Correspondence: bbtankc@gmail.com (K.C.T.); h.jeong37@gmail.com (H.J.)

Received: 12 September 2019; Accepted: 24 September 2019; Published: 26 September 2019

Abstract: In this focused review we survey recent progress in the development of resource theories of nonclassical light. We introduce the resource theoretical approach, in particular how it pertains to bosonic/light fields, and discuss several different formulations of resource theories of nonclassical light.

Keywords: quantum optics; nonclassicality; quantum resource theories; non-Gaussianity

---

1. Introduction

In quantum optics, there is general agreement that the most classical states of light are the coherent states [1–4]. In classical physics, a system can be completely described by a single point in phase space. In quantum mechanics, this is not possible due to the uncertainty principle [5,6]. For this reason, the phase space description of quantum systems must necessarily be via some distribution, rather than a point, that strictly obeys the uncertainty principle. In this sense the coherent state, being a minimum uncertainty state with no preferred axial direction in phase space, is arguably the closest quantum mechanical equivalent to a classical single point description of a physical system. However, one has to bear in mind that coherent states still fundamentally represent a quantum mechanical system. For this reason, despite being the most classical among the quantum states, quantum mechanical properties may still be pried from it. This is demonstrated by their strict observance of the uncertainty principle, as well as by their continued usefulness in purely quantum protocols such as quantum key distribution [7].

Nonclassical states, which are quantum states whose properties cannot be purely interpreted as directly deriving from the set of coherent states, are of considerable theoretical and practical interest. Fundamentally, because the set of classical states occupies only a small corner of the full Hilbert space, we are motivated to fully explore the set of physically allowable quantum states and to study their properties. Practically, there are many applications where nonclassical states can demonstrate genuine superiority over classical ones. These include applications in quantum information [8], quantum metrology [4], biology [9] and imaging [10]. It is also frequently the case that strongly nonclassical states are more useful in many protocols than weakly nonclassical ones. We are therefore interested in formulating methods that will enable us to first of all identify correctly when a light field can be considered nonclassical, as well as quantitatively describe the extent of nonclassicality present in quantum systems.

A recent trend in quantum information circles is to study different notions of nonclassicality via the so-called resource theoretical framework [11]. Such frameworks remove some of the ambiguity and arbitrariness in earlier studies of nonclassicality, by interpreting nonclassicality as resources to overcome the limitations of some set of quantum operation. Any quantifier of nonclassicality within this framework therefore has to be consistent with this view. This article seeks to provide some background to this approach, and to summarize the recent developments in this arena.

2. Preliminaries

As mentioned, the most classical states of light are typically considered to be the coherent states. One way to define the coherent states is as the eigenvectors $|\alpha\rangle$ of the annihiliation operator a such that

$$a|\alpha\rangle = \alpha|\alpha\rangle,$$

where α is any complex number. If one accepts that among pure states $|\alpha\rangle$ is classical, then classical statistical mixtures of such states should also be classical. This suggests that if the density operator of a general mixed can be expressed in the form

$$\rho_{\text{cl}} = \int d^2\alpha P_{\text{cl}}(\alpha)\,|\alpha\rangle\langle\alpha|,$$

where $\int d^2\alpha P_{\text{cl}}(\alpha) = 1$ is some positive probability density function, then we can say that the quantum state ρ is classical. We see that for classical states, the quantum state is fully described by the distribution P_{cl}.

It turns out that for general mixed states, a similar representation of the state is also possible. Due to the seminal work of Glauber [2] and Sudarshan [3], one may show that every quantum state of light may be written in the form

$$\rho = \int d^2\alpha\, P(\alpha)\,|\alpha\rangle\langle\alpha|,$$

where $P(\alpha)$ is called the Glauber-Sudarshan P-function. Observe the close similarity to the definition of a classical state $\rho_{\text{cl}} = \int d^2\alpha P_{\text{cl}}(\alpha)\,|\alpha\rangle\langle\alpha|$. For general quantum states, however, the function $P(\alpha)$ is a quasiprobability distribution instead of a proper, positive probability density function. This means that $P(\alpha)$ is always properly normalized such that $\int d^2\alpha\, P(\alpha) = 1$, but may permit negative values. We see that when $P(\alpha)$ corresponds to a positive probability density function the state must be classical. Otherwise, we say that the state is nonclassical.

3. The Resource Theoretical Framework

A quantum resource theory [11] is a general framework for quantifying notions of nonclassicality. We stress that this approach is not limited to the study of nonclassicality in light fields. There are many different kinds of quantum resource theories that are currently being considered, such as the resource theories of entanglement [12] and coherence [13]. One benefit of studying different notions of nonclassicality under a unifying framework is that it also provides a method of considering the relationships between different quantum resources, rather than treating each different quantum resource as its own separate island.

While the underlying notion of nonclassicality may differ from resource theory to resource theory, the fundamental approach remains broadly the same. The idea is to cast different notions of quantumness as resources that are not freely available. If one has access of the full repertoire of quantum mechanical operations, such limitations are not possible. For this reason, a fundamental requirement of a resource theory is to first define a set of "free" quantum operations.

We now describe this in more precise terms. Let the set of classical states be $\mathcal{C}$, which is a strict subset of the Hilbert space. Any state that belongs to the complement of the set $\mathcal{C}$, i.e., does not belong to $\mathcal{C}$, is nonclassical by definition. Associated with the set of classical states $\mathcal{C}$, we will also define a set of operations $\mathcal{O}$. $\mathcal{O}$ must be a strict subset of the set of all possible quantum operations, with the only requirement being that if $\Phi \in \mathcal{O}$ and $\rho \in \mathcal{C}$, then $\Phi(\rho) \in \mathcal{C}$. In other words, $\mathcal{O}$ must be some collection of quantum operations that are unable to produce nonclassical states from classical ones.

Given the set of classical states $\mathcal{C}$ and the set of free operations $\mathcal{O}$, we are then in the position to formulate our resource theory. We shall say that a nonnegative measure $N(\rho)$ is a nonclassicality measure under a particular resource theory if it satisfies the following basic properties:

1. $N(\rho) = 0$ if $\rho \in \mathcal{C}$.
2. (Monotonicity) $N(\rho) \geq N(\Phi(\rho))$ if $\Phi \in \mathcal{O}$.
3. (Convexity), i.e., $N(\sum_i p_i \rho_i) \leq \sum_i p_i N(\rho_i)$.

Property 1 simply requires that the measure $N(\rho)$ returns positive values only when ρ is nonclassical. Property 3 imposes the condition that $N(\rho)$ must be a convex function of state. This is a natural requirement to ensure that you cannot increase nonclassicality by statistically mixing any two quantum states ρ and σ and considering $p\rho + (1-p)\sigma$. It is intuitively clear that such statistical mixing processes are inherently classical procedures, and so cannot be expected to produce additional nonclassicality in any reasonable measure N.

Property 2 is perhaps the primary property that defines the resource theoretical approach. It imposes the requirement that any nonclassicality measure $N(\rho)$ must always be a nonclassicality monotone with respect to the set of operations $\mathcal{O}$. This means that the nonclassicality measure $N(\rho)$ must always monotonically decrease under the quantum operation Φ, if $\Phi \in \mathcal{O}$. This property encapsulates the idea that one can neither freely produce nor increase nonclassicality if you are limited to performing only quantum operations that belong to the set $\mathcal{O}$. Suppose $N(\rho)$ is some physically relevant quantity that we are trying to maximize. There is clearly no way to do so by solely performing operations in $\mathcal{O}$ so the only way to increase $N(\rho)$ is by replacing the input state ρ with another nonclassical state σ such that $N(\sigma) > N(\rho)$. In this picture, sources of nonclassical quantum states may be interpreted as additional quantum resources that are required in order to overcome the limitations inherent to classical states $\mathcal{C}$ and operations $\mathcal{O}$.

We stress that the above is only a framework and does not constitute a physical result on its own. Physics comes into the picture by studying specific physical scenarios and then studying the properties of physically meaningful measures $N(\rho)$ under various conditions, so it is imperative that $N(\rho)$ has physically meaningful interpretations in the first place. What this framework does is only to imbue the quantity $N(\rho)$ with an additional resource interpretation, in the sense that we have described in the previous paragraph.

For the quantification of nonclassicality in light, the set of classical states is unambiguous: $\mathcal{C}$ must be the set of states with classical P-functions. The set of operations $\mathcal{O}$ therefore needs to be defined in order to formulate a resource theory of nonclassicality. The earliest known proposal to formulate a resource theory of nonclassicality for light is by Gehrke et al. [14,15]. In this resource theory, the set of free quantum operations is abstract. Subsequently, resource theories based on more physically relevant sets of free operations were also considered. We will discuss these various approaches in the subsequent sections.

4. Resource Theory Based on Abstract Free Operations

An early proposal to formulate a resource theory of nonclassicality for light is by Gehrke et al. [14,15]. In this proposal the set of free operations $\mathcal{O}$ is proposed to be the maximal set of quantum operations Φ that always maps a classical state to another classical state, i.e., $\Phi(\rho) \in \mathcal{C}$ if $\Phi \in \mathcal{O}$ for every $\rho \in \mathcal{C}$. This is the largest possible set of quantum operations that one can define under the resource theoretical approach.

One example of a nonclassicality measure under this resource theory is the nonclassicality degree. The nonclassicality degree is a discrete measure of nonclassicality, similar to the Schmidt number [16,17] in the resource theory of entanglement. The set of coherent states is well known to form an overcomplete basis. This means that any pure state can be written in terms of a superposition of coherent states. As a result, we can consider the minimum number of superpositions r such that

$$|\psi\rangle = \sum_{i=1}^{r} \lambda_i |\alpha_i\rangle ,$$

where $\{|\alpha_i\rangle\}$ is some set of coherent states. The nonclassicality degree [14] is then defined as

$$\kappa(|\psi\rangle) := r - 1.$$

Since the only classical pure states are the coherent states, for any nonclassical pure state we must have $\kappa(|\psi\rangle) \geq 0$.

For mixed states, we can extend the measure for pure states via a convex roof construction [18,19]. This means that we consider all possible pure state decompositions $\{p_i, |\psi_i\rangle\}$ such that $\rho = \sum_i p_i |\psi_i\rangle\langle\psi_i|$. The nonclassicality degree of some density operator ρ can then be defined as the minimax quantity

$$\kappa(\rho) := \min_{\{p_i, |\psi_i\rangle\}} \max_i \kappa(|\psi_i\rangle).$$

This is basically the largest nonclassicality degree $\kappa(|\psi_i\rangle)$, minimized over all possible pure state decompositions $\{p_i, |\psi_i\rangle\}$.

Another example of a nonclassicality measure under this resource theory is the nonclassicality distance. The nonclassicality distance is actually a family of geometric nonclassicality measures rather than a single measure. Suppose we have a distance measure $d(\rho_1, \rho_2)$ for any quantum states ρ_1 and ρ_2. We can define the nonclassicality distance as

$$\delta(\rho) := \inf_{\sigma_{\text{cl}}} d(\rho, \sigma_{\text{cl}}).$$

In particular, some distance measures are known to be contractive under arbitrary completely positive, trace preserving (CPTP) quantum maps Φ. This means that for every ρ_1, ρ_2 and CPTP map Φ, we have $d(\rho_1, \rho_2) \geq d(\Phi(\rho_1), \Phi(\rho_2))$. If the distance measure d has such a property then we see that if $\Phi \in \mathcal{O}$ is a free operation, then

$$\begin{aligned}
\delta(\rho) &:= \inf_{\sigma_{\text{cl}}} d(\rho, \sigma_{\text{cl}}) \\
&\geq \inf_{\sigma_{\text{cl}}} d(\Phi(\rho), \Phi(\sigma_{\text{cl}})) \\
&\geq \delta(\Phi(\rho)).
\end{aligned}$$

We are therefore guaranteed that $\delta(\rho)$ monotonically decreases under free operations. If the distance measure d is additionally convex with respect to its arguments, then Properties 1, 2, and 3 will always be satisfied. Examples of distance measures satisfying the required properties are the trace distance [20] and the Bures distance [21]. The trace distance is defined as

$$d_{\text{Tr}}(\rho_1, \rho_2) := \|\rho_1 - \rho_2\|_1,$$

where $\|A\|_1 := \text{Tr}\left(\sqrt{A^\dagger A}\right)$ is the trace norm. The Bures distance is defined as

$$d_{\text{BU}}(\rho, \sigma) := \sqrt{2 - 2F(\rho, \sigma)},$$

where $F(\rho, \sigma) := \text{Tr}\{\sqrt{\sqrt{\rho}\sigma\sqrt{\rho}}\}$ is the Bures fidelity. One notable example of a distance measure that does not possess the required properties is the Hilbert-Schmidt distance, which is known to be not contractive under general CPTP maps [22].

The primary issue with Gehrke et al.'s resource theory is how to interpret this abstract definition of the set of free operations $\mathcal{O}$. While simple to define, there is no known simple characterization of O and it is unclear what kind of operations they represent physically in the laboratory. For the resource theoretical approach, this may not be always desirable, since one of the primary motivations behind this approach is to consider nonclassicality in terms of resource states that overcomes the limitations of the free operations $\mathcal{O}$. From this perspective, there is no strong physical motivation in Gehrke et al.'s approach as to why one should be interested to utilize nonclassical resources to overcome such abstract definitions of $\mathcal{O}$, beyond the fact that it is technically allowable under the resource theoretical framework.

5. Resource Theory Based on Linear Optical Operations

Recently, Tan et al. [23], noting that nonlinear operations are required in order to produce nonclassical states, proposed a resource theory of nonclassicality based on the set of linear optical operations. This approach is valid since linear optical operations always map a classical state to another classical state. This resolves the main problem with Gehrke et al.'s resource theory by replacing an abstract definition of $\mathcal{O}$ with a physically relevant set of operations that can be performed in the laboratory, and which also permits a relatively simple characterization.

In simple terms, a linear optical operation is any quantum operation that is implementable using any combination of passive linear optical elements, displacement operations, plus interaction with classical ancillas. The most fundamental operation in a general linear optical operation is a unitary linear optical operation U_L representing any combination of beam splitters, mirrors, and phase shifters, supplemented by displacement operations. A general linear optical unitary operation U_L can be characterized rather simply. Let $a_k^\dagger$ be the creation operator of the kth mode, then U_L represents any transformation of the type

$$a_i^\dagger \to \sum_{k=1}^{K} \mu_k a_k^\dagger + \bigoplus_{k=1}^{K} \alpha_k I_k$$

where $\{\mu_k\}$ are any set of complex values satisfying $\sum_{k=1}^{K} |\mu_k|^2 = 1$, I_k are identity operators on the kth mode, and α_k are complex numbers. More generally, a linear optical map is defined to be any map Φ_L that can be represented by the expression

$$\Phi_L(\rho_A) = \mathrm{Tr}_E(U_L \rho_A \otimes \sigma_E U_L^\dagger), \tag{1}$$

where σ_E is some classical state. By defining $\mathcal{O}$ to be the set of linear optical maps, we can see that $\mathcal{O}$ is not only simple to define, it also has a clear physical interpretation.

One example of a measure under this resource theory is the amount of coherent superposition between the coherent states [23]. By decomposing a state $|\psi\rangle$ as a superposition of a carefully chosen set of coherent states $|\alpha_i\rangle$ such that $|\psi\rangle = \sum_i c_i |\alpha_i\rangle$, one can take a continuous coherence measure C from the resource theory of coherence and transform it to a nonclassicality measure $N_C(|\psi\rangle)$ which quantifies the amount of coherent superposition among the set of coherent states $|\alpha_i\rangle$ according to the coefficients c_i. In Ref. [23] $N_C(|\psi\rangle)$ is referred to as the α-coherence. One may show that the α-coherence satisfies the required Properties 1, 2 and 3 under the resource theory of Tan et al. so it is a valid nonclassicality quantifier. The measure N_C may be interpreted as a continuous extension of the nonclassical degree Ref. [14], which is a discrete measure that counts the minimum number of superpositions rather than the amount of superposition. The α-coherence therefore provides a neat interpretation of optical nonclassicality in terms of the resource theory of coherence [13]. In fact, it was also noted [23] that the nonclassicality of optical systems shares many points of similarities with quantum coherence. For instance, both nonclassicality [24–26] and coherence [27–30] may be converted into entanglement using only free operations. The relationship between entanglement, coherence and nonclassicality may be an interesting point of investigation going forward.

Another measure that belongs to the resource theory by Tan et al. is called the metrological power. It is based on the quantum Fisher information, which quantifies the usefulness of a quantum probe for parameter estimation tasks. This measure was considered independently in both Ref. [31] and Ref. [32] with the former focusing on pure states, and the latter with a greater focus on general mixed states.

We now introduce the quantum Fisher information and discuss its relationship to parameter estimation. Suppose we have some unitary $U_\theta := e^{-i\theta G}$, where θ is a real number representing some physical parameter, and G is some Hermitian operator called the generator. Under this unitary evolution, an initial state ρ will evolve according to $\rho \to \rho_\theta = U_\theta \rho U_\theta^\dagger$. The primary goal of parameter estimation is to estimate the value of θ by performing a measurement on ρ_θ.

The relationship between parameter estimation and the Fisher information is supplied by the well known quantum Cramér-Rao bound [33–35]. This states that the uncertainty of any unbiased estimator $\hat{\theta}$ must obey the inequality

$$\Delta^2\hat{\theta} \geq \frac{1}{nI_Q(\rho, G)},$$

where n refers to the number of independent measurements, and $I_Q(\rho, G)$ is the quantum Fisher information given by $I_Q(\rho, G) = 2\sum_{i,j} \frac{(p_i - p_j)^2}{p_i + p_j} | \langle i | G | j \rangle |^2$. It is known that for single parameter estimation problems, there always exists a quantum measurement that saturates this bound, so long as a sufficient number of independent measurements $n \gg 1$ is performed. The quantum Fisher information $I_Q(\rho, G)$ therefore sets fundamental limits on our ability to measure θ.

Consider now a system of N modes with creation and annihilation operators $a_i^\dagger$ and a_i where $i = 1, \ldots, N$. A N mode annihilation operator can be defined as $a_{\boldsymbol{\mu}} := \sum_{i=1}^N \mu_i a_i$ where $\boldsymbol{\mu} = [\mathrm{Re}(\mu_1), \mathrm{Im}(\mu_1) \ldots, \mathrm{Re}(\mu_N), \mathrm{Im}(\mu_N)]$ is a $2N$ dimensional real vector of unit length, i.e., $|\boldsymbol{\mu}|^2 = \sum_{i=1}^N |\mu_i|^2 = 1$. From this, we define the N mode field quadrature operator

$$X_{\boldsymbol{\mu}} := \frac{a_{\boldsymbol{\mu}} + a_{\boldsymbol{\mu}}^\dagger}{\sqrt{2}}.$$

$X_{\boldsymbol{\mu}}$ is the generator of the the unitary evolution

$$D(\theta, \boldsymbol{\mu}) := e^{-i\theta X_{\boldsymbol{\mu}}},$$

which performs an N mode displacement operation. Setting $G = X_{\boldsymbol{\mu}}$, we can consider the Fisher information:

$$I_Q(\rho, X_{\boldsymbol{\mu}}) = 2\sum_{i,j} \frac{(p_i - p_j)^2}{p_i + p_j} | \langle i | X_{\boldsymbol{\mu}} | j \rangle |^2.$$

Generally speaking, the larger the Fisher information, the more useful it is for estimating the parameter θ in $D(\theta, \boldsymbol{\mu})$, so we are interested to find the maximum Fisher information that we can extract from a quantum probe ρ over all possible quadrature directions given by $\boldsymbol{\mu}$, so we can consider for our figure of merit

$$M(\rho) := \max\{\frac{1}{2} \max_{\boldsymbol{\mu} \in S} I_Q(\rho, X_{\boldsymbol{\mu}}) - 1, 0\}, \tag{2}$$

where the second maximization ensures the non-negativity of the measure. In the first maximization, S is simply the set of all possible $\boldsymbol{\mu}$ satisfying $|\boldsymbol{\mu}|^2 = 1$. $M(\rho)$ is called the metrological power and can be shown to satisfy Properties 1, 2, and 3 under the resource theory of Tan et al. The interesting aspect of the metrological power is that it directly relates the nonclassicality of a state to the state's usefulness in a parameter estimation task. For pure states, the metrological power can identify every nonclassical state, thereby establishing that every nonclassical pure state is useful in this particular parameter estimation problem, and the usefulness of the state is directly related to the amount of nonclassicality. One limitation is that the metrological power does not identify every mixed quantum state, i.e., there may be some mixed, nonclassical states where the measure returns zero.

More recently, Ref. [36] considered the extension of negativity to cover the set of all s-parametrized quasiprobabilities $P_s(\alpha)$. The s-parametrized quasiprobabilities where $s \in [-1, 1]$ are a family of phase space distribution functions satisfying $\int \mathrm{d}^2\alpha \, P_s(\alpha) = 1$ but may possess negativities. Setting $s = 1$

we get the Glauber-Sudarshan P-function, setting $s = 0$ we retrieve the Wigner function, and setting $s = -1$ we get the Husimi Q-function. One may show that the negativity of all such distributions

$$N_s(\rho) := \frac{1}{2}\left[\int d^2\alpha |P_s(\alpha)| - 1\right] \tag{3}$$

are nonclassicality measures belonging to the resource theory of Tan et al. [23]. In particular, as s decreases, $N_s(\rho)$ becomes increasingly weaker as a nonclassicality measure in the sense that the negativity decreases and fewer nonclassical states are identified by the measure.

In Ref. [31], Yadin et al. also considered a resource theory where $\mathcal{O}$ is expanded to include the set of linear optical operations, plus operations allowing for the feed forward of measurement outcomes. We note that, by definition, linear optical operations belong to this expanded set of operations. As such, any measure of nonclassicality under the resource theory of Yadin et al. [31] will monotonically decrease under linear optical operations and thus also belongs to the resource theory of Tan et al. [23]. Similar arguments can also be made for the resource theory of Gehrke et al. [14,15].

6. Convex Resource Theories of Non-Gaussianity

A Gaussian state is a special class of optical quantum states whose Wigner function is a Gaussian function [37]. According to this definition, every Gaussian state ρ_G has a Wigner function that can be written in the form

$$W(\mathbf{r}) = \frac{1}{2\pi\sqrt{\det V}} \exp\left[-\frac{1}{2}(\mathbf{r}-\bar{\mathbf{r}})^T V^{-1}(\mathbf{r}-\bar{\mathbf{r}})\right], \tag{4}$$

where $\bar{\mathbf{r}} := (\langle x\rangle, \langle p\rangle)$ and V is the covariance matrix, which is given by

$$\begin{bmatrix} \Delta^2 x & \left\langle \frac{1}{2}\{x-\langle x\rangle, p-\langle p\rangle\}\right\rangle \\ \frac{1}{2}\langle\{x-\langle x\rangle, p-\langle p\rangle\}\rangle & \Delta^2 p \end{bmatrix}.$$

Quantum states such as coherent states, thermal states, and squeezed vacuum states are Gaussian. We observe that the coherent and thermal states are both classical states, while squeezed vacuum is an example of a nonclassical Gaussian state.

Based on the above definition of Gaussian states, we can also define the set of Gaussian operations. A Gaussian unitary U_G is any combination of displacement operations, phase shifters, beam splitters and squeezing operations [38,39]. More generally, Gaussian operations Φ_G can be written in the form

$$\Phi_G(\rho_1) = \mathrm{Tr}_2[U_G \rho_1 \otimes |0\rangle\langle 0|_2 U_G^\dagger].$$

We see that such maps always map a Gaussian state to another Gaussian state. Many Gaussian states can also be produced under laboratory settings [8]. As a result, the properties of Gaussian states are particularly well understood. At the same time, Gaussian states comprise only a small subset of the possible quantum states, and many quantum protocols are not possible if one stays strictly within the Gaussian regime [40–49]. This has motivated the study of non-Gaussian states as quantum resources, so there is considerable interest to quantify and characterize non-Gaussian effects.

In the strictest sense, every quantum state whose Wigner function is not a Gaussian distribution is non-Gaussian by definition. One may try to formulate non-Gaussianity measures based on this definition. Examples of such measures are the geometric based measures similar to the nonclassicality distance [50–55] which quantifies the distance of a state ρ to the closest Gaussian state ρ_G. However, employing this strict definition necessarily means that many non-Gaussian states are classical. An example of this is the mixed coherent state $\rho = (|\alpha_1\rangle\langle\alpha_1| + |\alpha_2\rangle\langle\alpha_2|)/2$, which is clearly classical but has two peaks and cannot be written in the form of Equation (4). Such states can be created using only classical states and classical statistical processes, and should not lead to any genuine quantum effects.

In order to circumvent this, there are recent proposals to formulate a convex quantum resource theory of non-Gaussianity [56,57] where genuine nonclassical effects are captured. In such proposals, the definition of non-Gaussianity is now stricter, and only includes any quantum state that is not within the convex hull of Gaussian states. Such states are said to display genuine non-Gaussianity, which means that non-Gaussianity that emerges from classical statistical mixing processes is now excluded. Such genuinely non-Gaussian states cannot be written in the form

$$\rho = \sum_i p_i \rho_G^i,$$

where p_i is a probability distribution and ρ_G^i is some Gaussian state. Since coherent states are Gaussian, every state with a classical P-function will lie within the convex hull of Gaussian states. As such, genuinely non-Gaussian resource states must also have nonclassical P-functions.

In such resource theories, the set of classical states $\mathcal{C}$ is therefore the set of states that can be written in the form $\rho = \sum_i p_i \rho_G^i$. As for the set of free operations $\mathcal{O}$, one possibility is to define it as any quantum operation Φ that is composed of the following elementary operations:

a. Gaussian unitary operations U_G.
b. Composition with a free state $\rho \otimes \sigma$, where $\sigma = \sum_i p_i \sigma_G^i$, p_i is some probability distribution, and σ_G^i is some Gaussian state.
c. Partial trace of a subsystem.

In addition to the above elementary operations, the resource theories of Ref. [56] and Ref. [57] also considered Gaussian operations conditioned on measurement outcomes. In this respect, their definition of a free operation differs slightly in that the former considered projective measurements onto general Gaussian pure states $|\psi_G\rangle\langle\psi_G|$ while the latter considered homodyne measurements.

One example of a genuine non-Gaussianity measure is the negativity of the Wigner function [56,57]. Given the Wigner function $W(\mathbf{r})$ of some given state ρ, the logarithmic negativity of ρ is defined as

$$L_w(\rho) := \log \int \mathrm{d}^2\mathbf{r} |W(\mathbf{r})|.$$

Recall from Equation (3) that the family of negativity measures N_s are all nonclassicality measures. Setting $s = 0$, N_0 is nothing more than the negativity of the Wigner function, so we have $\log[2N_0(\rho) + 1] = L_w(\rho)$. We therefore see that the negativity of the Wigner function is both a measure of genuine non-Gaussianity and nonclassicality.

In fact, every convex measure of genuine non-Gaussianity must necessarily be a nonclassicality measure. This is because coherent states are Gaussian states, and every linear optical unitary operation is also a Gaussian preserving unitary operation. From Properties a, b and c, we see that any linear optical operation of the type in Equation (1) must also be a free operation in the convex resource theory of non-Gaussianity. This implies that measures of genuine non-Gaussianity must also be linear optical monotones in the resource theory of Tan et al. [23].

We note that in such convex resource theories of non-Gaussianity, the convex hull of Gaussian states $\mathcal{C}$ necessarily contains many states that have nonclassical P-functions, with the most prominent being the squeezed coherent states. Any genuine non-Gaussianity measure will therefore exclude such states. Furthermore, since the squeezing operation is a Gaussian operation, it makes no attempt to capture any increase in nonclassicality due to squeezing. This is a feature that stems from the definition of the resource theory. As such, the starting point of the resource theory of non-Gaussianity is necessarily qualitatively different from resource theory of nonclassicality discussed in the preceding sections. For this reason, it is perhaps more appropriate to consider the concept of genuine non-Gaussianity as something that overlaps significantly with the notion of nonclassicality, rather than an independent approach of quantifying nonclassicality in light.

7. Conclusions

In this article we summarized recent trends in the characterization and quantification of nonclassical states of light. Along these lines, there has been considerable interest in developing a resource theory of nonclassicality for quantum light. Several prominent proposals are discussed. First, there is the resource theory of Gehrke et al. [14,15], which is based on an abstract definition of the set of free operations $\mathcal{O}$. We then discuss the resource theories of Tan et al. [23] and Yadin et al. [31], which are based on linear optical operations. Finally, we discuss proposals by Albarelli et al. [56], and Takagi and Zhuang [57] to develop convex resource theories of non-Gaussianity. Non-Gaussianity approaches are not technically resource theories of nonclassicality as the set of free states are not the set of states with classical P-functions, but there exists significant overlap.

It is interesting to note that all the measures discussed across these various schemes naturally fall under the resource theory of Tan et al. [23]. This is because the free operations under this resource theory, the set of linear optical operations, is the simplest among the resource theories that are discussed thus far.

Most of the nonclassicality measures that belong to these resource theories have geometric or other fundamental interpretations, with the exception of the metrological power [32], $M(\rho)$, which has an interpretation in terms of the operational usefulness of the state. This measure is sufficient to identify every nonclassical pure state as useful for a parameter estimation task, but does not identify every nonclassical mixed state in general. It is an open problem whether there is some parameter estimation problem that can identify every nonclassical state as useful. More generally, a promising future direction is to find nonclassicality measures that directly quantify the state's usefulness in some meaningful operational task.

Author Contributions: Writing—original draft, K.C.T.; Writing—review & editing, K.C.T. and H.J.

Funding: This work was supported by the National Research Foundation of Korea (NRF) through a grant funded by the the Ministry of Science and ICT (Grant No. NRF-2019R1H1A3079890). K.C. Tan was supported by Korea Research Fellowship Program through the National Research Foundation of Korea (NRF) funded by the Ministry of Science and ICT (Grant No. 2016H1D3A1938100).

Conflicts of Interest: The authors declare no conflict of interest.

References

1. Schrodinger, E. Der stetige Übergang von der Mikro- zur Makromechanik. *Naturwissenschaften* **1926**, *14*, 664–666. [CrossRef]
2. Glauber, R.J. Coherent and Incoherent States of the Radiation Field. *Phys. Rev.* **1963**, *131*, 2766. [CrossRef]
3. Sudarshan, E.C.G. Equivalence of Semiclassical and Quantum Mechanical Descriptions of Statistical Light Beams. *Phys. Rev. Lett.* **1963**, *10*, 277. [CrossRef]
4. Tan, K.C.; Jeong, H. Nonclassical Light and Metrological Power: An Introductory Review. *arXiv*. Available online: https://arxiv.org/abs/1909.00942 (accessed on 26 September 2019).
5. Heisenberg, Z.W. Über den anschaulichen Inhalt der quantentheoretischen Kinematik und Mechanik. *Physic* **1927**, *43*, 172.
6. Robertson, H.P. The Uncertainty Principle. *Phys. Rev.* **1929**, *34*, 163. [CrossRef]
7. Grosshans, F.; Assche, G.V.; Wenger, J.; Brouri, R.; Cerf, N.J.; Grangier, P. Quantum key distribution using gaussian-modulated coherent states. *Nature* **2003**, *421*, 238. [CrossRef] [PubMed]
8. Braunstein, S.L.; van Loock, P. Quantum information with continuous variables. *Rev. Mod. Phys.* **2005**, *77*, 513. [CrossRef]
9. Taylor, M.A.; Bowen, W.P. Quantum metrology and its application in biology. *Phys. Rep.* **2005**, *615*, 1. [CrossRef]
10. Berchera, I.R.; Degiovanni, I.P. Quantum imaging with sub-Poissonian light: challenges and perspectives in optical metrology. *Metrologia* **2019**, *56*, 024001. [CrossRef]
11. Chitambar, E.; Gour, G. Quantum resource theories. *Rev. Mod. Phys.* **2019**, *91*, 025001. [CrossRef]

12. Horodecki, R.; Horodecki, P.; Horodecki, M.; Horodecki, K. Quantum entanglement. *Rev. Mod. Phys.* **2009**, *81*, 865. [CrossRef]
13. Streltsov, A.; Adesso, G.; Plenio, M.B. Colloquium: Quantum coherence as a resource. *Rev. Mod. Phys.* **2017**, *89*, 041003. [CrossRef]
14. Gehrke, C.; Sperling, J.; Vogel, W. Quantification of nonclassicality. *Phys. Rev. A* **2012**, *86*, 052118. [CrossRef]
15. Sperling, J.; Vogel, W. Convex ordering and quantification of quantumness. *Phys. Scr.* **2015**, *90*, 074024. [CrossRef]
16. Schmidt, E. Zur Theorie der linearen und nichtlinearen Integralgleichungen. *Math. Ann.* **1906**, *63*, 433. [CrossRef]
17. Terhal, B.M.; Horodecki, P. Schmidt number for density matrices. *Phys. Rev. A* **1999**, *61*, 040301. [CrossRef]
18. Bennett, C.H.; Di Vincenzo, D.P.; Smolin, J.A.; Wootters, W.K. Mixed-state entanglement and quantum error correction. *Phys. Rev. A* **1996**, *54*, 3824. [CrossRef]
19. Uhlmann, A. Optimizing entropy relative to a channel or a subalgebra. *Open Syst. Inf. Dyn.* **1998**, *5*, 209. [CrossRef]
20. Hillery, M. Nonclassical distance in quantum optics. *Phys. Rev. A* **1987**, *35*, 725. [CrossRef]
21. Marian, P.; Marian, T.A.; Scutaru, H. Quantifying Nonclassicality of One-Mode Gaussian States of the Radiation Field. *Phys. Rev. Lett.* **2002**, *88*, 153601. [CrossRef]
22. Ozawa, M. Entanglement measures and the Hilbert-Schmidt distance. *Phys. Lett. A* **2000**, *268*, 158. [CrossRef]
23. Tan, K.C.; Volkoff, T.; Kwon, H.; Jeong, H. Quantifying the Coherence between Coherent States. *Phys. Rev. Lett.* **2017**, *119*, 190405. [CrossRef] [PubMed]
24. Kim, M.S.; Son, W.; Bužek, V.; Knight, P.L. Entanglement by a beam splitter: Nonclassicality as a prerequisite for entanglement. *Phys. Rev. A* **2002**, *65*, 032323. [CrossRef]
25. Wang, X.-B. Theorem for the beam-splitter entangler. *Phys. Rev. A* **2002**, *66*, 024303.
26. Asbóth, J.K.; Calsamiglia, J.; Ritsch, H. Computable Measure of Nonclassicality for Light. *Phys. Rev. Lett.* **2005**, *94*, 173602. [CrossRef]
27. Streltsov, A.; Singh, U.; Dhar, H.S.; Bera, M.N.; Adesso, G. Measuring Quantum Coherence with Entanglement. *Phys. Rev. Lett.* **2015**, *115*, 020403. [CrossRef] [PubMed]
28. Tan, K.C.; Choi, S.; Kwon, H.; Jeong, H. Emergence of synchronization and regularity in firing patterns in time-varying neural hypernetworks. *Phys. Rev. A* **2018**, *97*, 052304. [CrossRef]
29. Tan, K.C.; Kwon, H.; Park, C.-Y.; Jeong, H. Unified view of quantum correlations and quantum coherence. *Phys. Rev. A* **2016**, *94*, 022329. [CrossRef]
30. Tan, K.C.; Jeong, H. Entanglement as the Symmetric Portion of Correlated Coherence. *Phys. Rev. A* **2018**, *121*, 220401. [CrossRef]
31. Yadin, B.; Binder, F.C.; Thompson, J.; Narasimhachar, V.; Gu, M.; Kim, M.S. Operational Resource Theory of Continuous-Variable Nonclassicality. *Phys. Rev. X* **2018**, *8*, 041038. [CrossRef]
32. Kwon, H.; Tan, K.C.; Volkoff, T.; Jeong, H. Nonclassicality as a Quantifiable Resource for Quantum Metrology. *Phys. Rev. Lett.* **2019**, *122*, 040503. [CrossRef] [PubMed]
33. Helstrom, C.W. *Quantum Detection and Estimation Theory*; Academic Press: Cambridge, MA, USA, 1976.
34. Holevo, A.S. *Probabilistic and Statistical Aspects of Quantum Theory*; Springer: Berlin, Germany, 1982.
35. Braunstein, S.L.; Caves, C.M. Statistical distance and the geometry of quantum states. *Phys. Rev. Lett.* **1994**, *72*, 3439. [CrossRef] [PubMed]
36. Tan, K.C.; Choi, S.; Jeong, H. Negativity of quasiprobability distributions as a measure of nonclassicality. *arXiv*. Available online: https://arxiv.org/abs/1906.05579 (accessed on 26 September 2019).
37. Adesso, G.; Ragy, S.; Lee, A.R. Continuous variable quantum information: Gaussian states and beyond. *Open Syst. Inf. Dyn.* **2014**, *21*, 1440001. [CrossRef]
38. Ma, X.; Rhodes, W. Multimode squeeze operators and squeezed states. *Phys. Rev. A* **1990**, *41*, 4625. [CrossRef] [PubMed]
39. Cariolaro, G.; Pierobon, G. Bloch-Messiah reduction of Gaussian unitaries by Takagi factorization. *Phys. Rev. A* **2016**, *94*, 062109. [CrossRef]
40. Lloyd, S.; Braunstein, S.L. Quantum Computation over Continuous Variables. *Phys. Rev. Lett.* **1999**, *82*, 1784. [CrossRef]
41. Eisert, J.; Scheel, S.; Plenio, M.B. Distilling Gaussian States with Gaussian Operations is Impossible. *Phys. Rev. Lett.* **2002**, *89*, 137903. [CrossRef]

42. Giedke, G.; Cirac, J.I. Characterization of Gaussian operations and distillation of Gaussian states. *Phys. Rev. A* **2002**, *66*, 032316. [CrossRef]
43. Fiurášek, J. Gaussian Transformations and Distillation of Entangled Gaussian States. *Phys. Rev. Lett.* **2002**, *89*, 137904. [CrossRef]
44. Bartlett, S.D.; Sanders, B.C. Universal continuous-variable quantum computation: Requirement of optical nonlinearity for photon counting. *Phys. Rev. A* **2002**, *65*, 042304. [CrossRef]
45. Cerf, N.J.; Krüger, O.; Navez, P.; Werner, R.F.; Wolf, M.M. Non-Gaussian Cloning of Quantum Coherent States is Optimal. *Phys. Rev. Lett.* **2005**, *95*, 070501. [CrossRef] [PubMed]
46. Menicucci, N.C.; van Loock, P.; Gu, M.; Weedbrook, C.; Ralph, T.C.; Nielsen, M.A. Universal Quantum Computation with Continuous-Variable Cluster States. *Phys. Rev. Lett.* **2006**, *97*, 110501. [CrossRef] [PubMed]
47. Niset, J.; Fiurášek, J.; Cerf, N.J. No-Go Theorem for Gaussian Quantum Error Correction. *Phys. Rev. Lett.* **2009**, *102*, 120501. [CrossRef] [PubMed]
48. Zhang, S.L.; van Loock, P. Distillation of mixed-state continuous-variable entanglement by photon subtraction. *Phys. Rev. A* **2010**, *82*, 062316. [CrossRef]
49. Ohliger, M.; Kieling, K.; Eisert, J. Limitations of quantum computing with Gaussian cluster states. *Phys. Rev. A* **2010**, *82*, 042336. [CrossRef]
50. Genoni, M.G.; Paris, M.G.A.; Banaszek, K. Measure of the non-Gaussian character of a quantum state. *Phys. Rev. A* **2007**, *76*, 042327. [CrossRef]
51. Genoni, M.G.; Paris, M.G.A.; Banaszek, K. Quantifying the non-Gaussian character of a quantum state by quantum relative entropy. *Phys. Rev. A* **2008**, *78*, 060303R. [CrossRef]
52. Genoni, M.G.; Paris, M.G.A. Quantifying non-Gaussianity for quantum information. *Phys. Rev. A* **2010**, *82*, 052341. [CrossRef]
53. Ivan, J.S.; Kumar, M.S.; Simon, R. A measure of non-Gaussianity for quantum states. *Quantum Inf. Process.* **2012**, *11*, 853. [CrossRef]
54. Ghiu, I.; Marian, P.; Marian, T.A. Measures of non-Gaussianity for one-mode field states. *Phys. Scr.* **2013**, *T153*, 014028. [CrossRef]
55. Park, J.; Lee, J.; Ji, S.-W.; Nha, H. Quantifying non-Gaussianity of quantum-state correlation. *Phys. Rev. A* **2017**, *96*, 052324. [CrossRef]
56. Albarelli, F.; Genoni, M.G.; Matteo, M.G.A.; Ferraro, A. Resource theory of quantum non-Gaussianity and Wigner negativity. *Phys. Rev. A* **2018**, *98*, 052350. [CrossRef]
57. Takagi, R.; Zhuang, Q. Convex resource theory of non-Gaussianity. *Phys. Rev. A* **2018**, *97*, 062337. [CrossRef]

MDPI
St. Alban-Anlage 66
4052 Basel
Switzerland
Tel. +41 61 683 77 34
Fax +41 61 302 89 18
www.mdpi.com

Quantum Reports Editorial Office
E-mail: quantumrep@mdpi.com
www.mdpi.com/journal/quantumrep

CPSIA information can be obtained
at www.ICGtesting.com
Printed in the USA
LVHW070845111220
672560LV00045B/272